Emily Gowers is a lecturer in Latin at University College London.

The Loaded Table

The Loaded Table

Representations of Food in Roman Literature

EMILY GOWERS

CLARENDON PRESS · OXFORD
1993

Oxford University Press, Walton Street, Oxford OX2 6DP
Oxford New York Toronto
Delhi Bombay Calcutta Madras Karachi
Petaling Jaya Singapore Hong Kong Tokyo
Nairobi Dar es Salaam Cape Town
Melbourne Auckland
and associated companies in
Berlin Ibadan

Oxford is a trade mark of Oxford University Press

Published in the United States
by Oxford University Press Inc., New York

British Library Cataloguing in Publication Data
Data available

Library of Congress Cataloging in Publication Data
The loaded table: representations of food in Roman literature
Emily Gowers.
Includes bibliographical references and index.
1. Latin literature—History and criticism. 2. Food in literature.
3. Dinners and dining in literature. 4. Literature and society—Rome.
5. Dinners and dining—Rome. 6. Food habits—Rome. I. Title.
PA6029.F66G69 1992 870.9'355—dc20 92–4420
ISBN 0–19–814695–7

1 3 5 7 9 10 8 6 4 2

Typeset by Latimer Trend & Co. Ltd, Plymouth, Devon
Printed in Great Britain by
Bookcraft (Bath) Ltd,
Midsomer Norton, Avon

almis parentibus

Preface

THE urge to peer into the sinister entrails of Roman civilization seized me first as an undergraduate, when my stomach was turned by Petronius' account of the boar that had to be gutted at Trimalchio's table: it has led me since then into some of the stranger and more neglected areas of Latin literature. Those years of investigation were spent at Trinity College, Cambridge, where a Research Fellowship awarded in 1988 provided me with real meals more stylish than any I can conjure up on paper. By indulging my researches, the college was at least continuing a long tradition, since even its two most famous classicists were not above an unhealthy interest in food. One was renowned for his tactless extravagance, the other for his acrid criticisms, some of which still linger in an old copy of the kitchen suggestions book: 'The rhubarb is a vegetable not fit for human consumption.'

Among the many people I would like to thank for helping to stir this broth are: John Henderson (my research supervisor and most inspiring teacher), Elaine Fantham, Don Fowler, Rebecca Gowers, Neil Hopkinson, Richard Hunter, Ted Kenney, Geoffrey Lloyd, Jamie Masters, Michael McGann, Edmund Thomas, Malcolm Willcock, Gareth Williams, Hilary O'Shea (my OUP editor), and my parents. My husband, Timothy Gowers, sprinkled it at every stage with a generous pinch of salt, and tried in vain to instil in me his family's taste for plain words. As for what remains after their scrutiny, in Terence's words: 'You made the mess: you've got to eat it.'

There is a curious Observation concerning the diversity of *Roman* and *British* dishes, the first delighting in Hodge-podge, Gallimaufreys, Forced Meats, Jussels, and Salmagundies; the latter in Spare-ribs, Sirloins, Chines and Barons; and thence our Terms of Art, both as to Dressing and Carving become very different; for they lying upon a sort of Couch could not have carved those Dishes which our Ancestors, when they sat upon Forms used to do. But since the Use of Cushions and Elbow chairs, and the Editions of good Books and Authors, it may be hoped in time we may come up to them.

(William King, *Letters to Dr Lister and Others*, 1712)

Contents

Abbreviations

Acta Fratrum Arvalium	W. Henzen (ed.), *Acta Fratrum Arvalium* (Berlin, 1874).
Anth. Lat.	D. R. Shackleton Bailey (ed.), *Anthologia Latina*, i. *Carmina in Codicibus Scripta* (Stuttgart, 1982).
Buecheler	F. Buecheler (ed.), *Petronii Saturae* (Berlin, 1922).
Burman	P. Burman (ed.), *Petronius Arbiter Satyricon* (Amsterdam, 1743).
CIL	*Corpus Inscriptionum Latinarum* (Berlin, 1862–).
Dindorf	G. Dindorf (ed.), *Poetarum Scaenicorum Graecorum . . . Fragmenta* (Oxford, 1851).
DK	H. Diels and W. Kranz (eds.), *Die Fragmente der Vorsokratiker* (Dublin/Zürich, 1951–2).
Ernesti	J. C. G. Ernesti (ed.), *Lexicon Technologiae Latinorum Rhetoricae* (Leipzig, 1795).
Ernout–Meillet	A. Ernout and A. Meillet (eds.), *Dictionnaire étymologique de la langue latine* (Paris, 1959).
GLK	G. T. H. Keil (ed.), *Grammatici Latini* (Leipzig, 1857–80).
GP	A. F. S. Gow and D. L. Page (eds.), *The Greek Anthology: Hellenistic Epigrams* (Cambridge, 1965).

K	T. Kock (ed.), *Comicorum Atticorum Fragmenta* (Leipzig, 1880–8).
Keller	O. Keller (ed.), *Pseudacronis Scholia in Horatium Vetustiora* 2 vols. (Leipzig, 1902–4).
Leo	F. Leo (ed.), *Plauti Comoediae* (Berlin, 1895–6).
Leutsch–Schneidewin	E. L. Leutsch and F. G. Schneidewin (eds.), *Corpus Paroemiographorum Graecorum* (Göttingen, 1839–51).
Lindsay	W. M. Lindsay (ed.), *T. Macci Plauti Comoediae* (Oxford, 1904–5).
LP	E. Lobel and D. L. Page (eds.), *Poetarum Lesbiorum Fragmenta* (Oxford, 1955).
M	F. Marx (ed.), *C. Lucili Carminum Reliquiae* (Leipzig, 1904).
Malcovati	H. Malcovati (ed.), *Oratorum Romanorum Fragmenta* (Turin, 1967).
OLD	*Oxford Latin Dictionary*.
Pearson	A. C. Pearson (ed.), *The Fragments of Sophocles* (Cambridge, 1917).
Pf.	R. Pfeiffer (ed.), *Callimachus* (Oxford, 1965).
PLG	T. Bergk (ed.), *Poetae Lyrici Graeci* (Leipzig, 1878–82).
RE	A. Pauly, G. Wissowa, and W. Kroll (eds.), *Real-Encyclopädie der classischen Altertumswissenschaft* (Stuttgart, 1894–1980).
Ribbeck	O. Ribbeck (ed.), *Scaenicorum Romanorum Poesis Fragmenta*: i. *Tragici;* ii. *Comici* (Leipzig, 1897–8).
TLL	*Thesaurus Linguae Latinae*.
Usener	H. Usener (ed.), *Epicurea* (Leipzig, 1887).

Ussing	J. L. Ussing (ed.), *Plauti Comoediae* (Hanover, 1875–6).
W	E. H. Warmington (ed.), *Remains of Old Latin* (London/Cambridge, Mass., 1935–40).

Periodicals are abbreviated as in *L'Année philologique*. For other editions, commentaries, etc. see References.

All quotations from the main authors discussed are from Oxford Classical Texts, unless otherwise stated.

I
An Approach to Eating

1 Introduction

One of the more bizarre episodes in the travels of Peregrine Pickle involves a Roman banquet staged by a learned doctor and attended by an absurd collection of foreign guests. The dishes re-created for their diversion include boiled goose in anchovy sauce, salacacabia, dormouse and poppy pie, stuffed sow's stomach, and fricassee of snails. The dinner is a fiasco. Fits of choking, weeping, spluttering, and eventually vomiting ensue, and one guest speaks for the whole party when he exclaims: 'Christ in heaven! what beastly fellows these Romans were!'[1]

As a modern reaction to Roman food, this is fairly typical. The curious tastes of that civilization tend to inspire in us a mixture of fascination and disgust. Anyone who admits to an interest in the subject provokes hushed and disbelieving questions. Did they really eat larks' tongues? Did they really vomit at their orgies? However, the Romans who wrote about food apparently found it difficult to stomach too; they are themselves responsible for our mixed feelings. Smollett's banquet is based on scenes from Petronius and Horace, and we only think of Roman food as baroque and nauseating because that is the impression that these writers chose to leave to posterity.

This book will not be about what the Romans ate. It will provide no answers to the questions usually asked about Roman food, particularly when a great deal of recent work

[1] Smollett, *Peregrine Pickle*, ch. 48.

on the subject has diluted the usual hyperboles. Most Romans, we now know, lived at subsistence level, and would never have seen the kind of banquet Smollett describes.[2] The literary sources represent a tiny fraction of the élite, which ought to make us suspicious of their historical value. Even so, there is still room for interpreting them in a new way. The literary medium need not be seen as an obstruction; indeed, the kinds of evasions and prejudices that seem to cloud it can be illuminating in themselves. This book offers a number of new interpretations of food in Roman literature, food lodged irrevocably in a written form; it takes a more positive attitude to the elements of bias and selection involved, which in themselves show us how the Roman writer picked his way through the material things around him. It aims to reinstate a number of works, from Plautus to Pliny the Younger, which have been overlooked or under-read for the very reason that they are dominated by food. Even though these are confined to the kitchen, cookshop, or dining-room, and hedged around with apologies and disclaimers, they still preserve something of the savour of Roman civilization.[3]

This is the first of many contradictions that cluster around the subject: food, for the Roman writer who chose to discuss it, was simultaneously important and trivial.[4] Indeed, the idea that food is a negligible subject is a prejudice which the classical world has passed on to us; putting the pleasures of the mind above those of the body, for those who could afford it, was an essential tenet of classical culture. The following dictum of Epictetus, for example, is characteristic of most ancient philosophy: 'It is a sign of a stupid man to spend a great deal of time on the concerns of his body—exercise, eating, drinking, evacuating his bowels, and copulating. These things should be done in passing; you should devote

[2] See Garnsey (1988). There will always be a market, it seems, for more sensational accounts of Roman eating.

[3] See e.g. Pliny's ironic apology for writing about cabbages (*NH* 19. 59).

[4] Aug. *Civ. Dei* 6. 9 records that Varro in his catalogue of gods dealt first with gods who represent man himself, and then with the gods of things that concern man—food, clothing, and so on.

your whole attention to the mind.'[5] This principle was also part of the more nebulous codes of social behaviour. Theophrastus, in his catalogue of unbalanced types, makes his garrulous man itemize everything he had for dinner the previous night; and Petronius' boorish monumental mason Habinnas, although he claims to have an appalling memory, can remember every detail of a funeral feast.[6]

It is easy to see the effect of these views on the literary record. Plutarch, for example, adds his voice to the general disdain for people who can remember past meals too clearly, people who are 'over-enthusiastic about small and trivial pleasures'. In another context, he says that to reminisce about eating and drinking, the sorts of pleasures that are 'as fleeting as yesterday's perfume or the lingering smell of cooking', is not the mark of a free-born man: only the delights of philosophical discussion will remain perennially fresh, feasts that can be enjoyed again and again. If pleasure were only a physical thing, he goes on, Xenophon and Plato would have left a record in their *Symposia* not of the conversation, but of the relishes, cakes, and sweets served at Callias and Agathon's houses.[7] Plutarch's assumption was that food is material, trivial, perishable, and therefore not worth preserving, and in some ways his prognosis was correct. Other ancient works have been salvaged by posterity, but many of the cookery books and treatises on entertaining that we know existed have been jettisoned: even the transmission of classical texts is affected by this cultural bias.[8]

However, despite these losses, the smell of cooking is still one of the lingering souvenirs of Roman civilization; the banquet is as much part of its popular mythology as philosophy, laws, or monuments. And it is the literary record, not the wizened loaves and prunes of Pompeii, that keeps Roman food fresh in the imagination. Food, after all, comes to hand as the most immediate example of what man absorbs from

[5] *Ench.* 41.

[6] Theoph. *Char.* 3; Petr. *Sat.* 66. Cf. Athenaeus 4. 147c (= Philoxenus, *Banquet*): No one could recount all the dishes that came before us.

[7] Plut. *Mor.* 1094c, 686c–d.

[8] *RE* s.v. '*Kochbücher*'; Varro ap. *Gell* 13. 11; Col. 12. 4. 2, 12. 46. 1.

the world and how he divides it with other people. There is something about eating, despite all the inhibitions, that gives us a uniquely graspable sense of the relationship between the Romans and their universe. The fact that what a man ate appears so often in the Roman sources shows what great potential food had for projecting an individual's moral and cultural values. But this embracing of food as a literary subject always went hand in hand with squeamish contempt for the substance itself.

Nowhere is this clearer than in ancient biography, where there seems to be a running dispute about the value of eating habits as worthwhile testimony. One biographer of Saturninus, for example, passes over the physique and diet of his subject on the grounds that they are trivial. It would be tedious, he says, to record such frivolous information, quite beneath him to describe Saturninus' height, his physique, his appearance, what he ate and drank: that sort of thing, which has no use as a moral example, can be left to others.[9] In other accounts, however, the sheer amount of detail we are given about dietary preferences makes it plain that food had plenty of moral value.[10] Suetonius, in his life of Augustus, includes Augustus' diet, but only as an afterthought: 'In his food—for I would not even omit that—he was extremely frugal.' And we are told that Augustus liked to munch small figs and second-class bread while travelling in his litter, and that he revived state dinners for his more important subjects, at which he usually turned up late.[11]

This brings us to another of the contradictions associated with food: it can be mentioned both for its own sake and as a symbol of something else. An apple on a table is graspable and obvious. The fact that we can reach out and touch it, smell and taste it, makes it seem like the essence of uncomplicated matter. Indeed, a piece of fruit on a table figures again and again in ancient anecdotes as a test of the nature of reality. As a character in Macrobius' *Saturnalia* says, something that looks like an apple may be just be a craftsman's

[9] *SHA Firmus, Saturninus, Proculus et Bonosus* 11.

[10] *SHA Claudius* 32–3; *Vitellius* 13; *Elagabalus* 19–30; *Gallienus* 16; *Geta* 5; Plin. *Paneg*. 49. 5–8 (Trajan and Hadrian); Philostr. *Apollonius of Tyana* 1. 8.

[11] Suet. *Aug*. 74; 76–8.

fake, made of wax or stone; only our sense of smell can tell us whether or not it is genuine. A story is told by Diogenes Laertius about the philosopher Sphaerus, who was taken in by a wax pomegranate in the middle of a discussion about knowledge.[12] The apple seems, then, to epitomize matter for its own sake. But is it only that? It might suggest original sin, or the judgement of Paris. A pile of apples might suggest harvest-time abundance. An apple with a worm in it might be an intimation of mortality. The significance of food in its literary representations lies both in its simple existence *and* in a bundle of metaphorical associations, a capacity to evoke a whole world of wider experience.

In recent years, a great deal has been done to reverse the prejudice against food that we have inherited from the ancient world. The simple-seeming procedures of everyday life have begun to be taken seriously, both for their own sake and for what they can tell us about the larger structures of social organization.[13] There have been many different ways of interpreting food: semiological, socio-historical, cultural.[14] But the most influential work in this direction has been in the field of social anthropology, which has taught us that the classification of food, the rituals of cooking, and the arrangement of meals hold clues to notions of hierarchy, social grouping, purity and pollution, myths of creation and cosmogony, and the position of man in relation to the world. Lévi-Strauss's notorious 'culinary triangle' of raw, roasted, and boiled food (1965) was devised as a universal model for the mythological organization of culture.[15] For Mary Douglas (1970; 1975), the dietary prohibitions of Leviticus enshrine the same principles of cleanliness and abhorrence of boundary-crossing that mark out Jewish society as a whole;

[12] Macrob. 7. 14. 22–3; Diog. Laert. 7. 177. See Pliny *NH* 35. 155 on the wax and stone models of fruit made by Posis.

[13] Stoller (1989) pleads for a still more sensuous approach to ethnography, conveying the 'savour' of a culture by immersing oneself in its cuisine.

[14] See esp. Barthes (1979); the *Annales* collections by Hémardinquer (1970); and Forster and Ranum (1989); Henisch (1976); Goody (1982); Clark (1975); Revel (1979); Mennell (1985).

[15] Lehrer (1972) expands the triangle into a tetrahedron, but even this does not account for all the differentiation in cooking styles.

closer to home, however, the English working-class meal can only be analysed as a self-contained system of rigid patterns and contrasts. Vidal-Naquet, Vernant, and Detienne have revealed the uses of food as a conceptual tool in the ancient Greek world, showing, for example, that Greek sacrifice re-enacted the primeval division of the world between men and gods, and that dietary labels in the *Odyssey* marked out the cultural differences between outlandish tribes and the civilized, agricultural Greeks.[16]

Structural analysis tends to over-simplify for the sake of clarity or impact: it has been criticized for being anachronistic, for tending to offer a particular case as a general model, and for assuming that one city, or even one tribe or class, has a single mentality.[17] Models for studying food nowadays tend to be much more complex, accommodating the mode of production and the class structure, as well as cultural factors such as changes in taste.[18] Inspiring though it is, structural analysis of the original kind, and even its corrected versions, would not be the right approach for the texts that will be discussed here. There are two main reasons for this.

First of all, the fact that food is being presented in a literary form has special consequences. There may of course be a number of simple oppositions involved—nature and culture, raw and cooked, native and foreign—but the meaning of food is rarely so elemental; there are a host of complicating factors superimposed on any universal codes. Mary Douglas, in her pioneering essay 'Deciphering a Meal', says at one point that a meal is like a poem, because it is a carefully created artefact offered by one person to another.[19] That takes no account, however, of what happens when a meal *is* a poem. For a start, there is the civilized inhibition about choosing food as a literary subject, which is bound to affect the scope and nature of its representations.

[16] Lévi-Strauss (1965; 1970); Douglas (1970; 1975); Vidal-Naquet (1971; 1981a), Detienne (1977); Detienne and Vernant (1979).

[17] See Gordon in *Comp. Crit.* 1 (1979), 279–310, esp. 302–4.

[18] Goody (1982: 10–39) follows a socio-economic approach; Mennell (1985: 1–19) presents a developmental history of taste following Elias (1978). This is also D'Arms's approach in his outline for studying Roman meals (1984).

[19] Douglas (1975: 261).

For all the vast stores of information that have survived,[20] we have no straightforward, detailed description of a normal Roman meal. Instead, there is usually a split between disgusting or extravagant meals described from a hostile point of view,[21] and innocent meals where food is at its least gross, or is omitted altogether.[22] The tantalizing glimpse we are given of Horace's private supper in one of his satires, leek and chick-pea lasagne and half a litre of wine, is meant to be just that.[23] Martial's urban menus, are, as we shall see, heavily ironic: we sense the embarrassment of the author and his concern to avoid 'bare' representation.[24]

It is perhaps a mistake, then, to use literary sources simply as evidence for what the Romans ate.[25] The uneasy stance of the writer and the imbalanced distribution of food across the literary genres can tell us just as much about the Romans' attitudes to the subject as any catalogue of dishes. Texts that contain food are not just repositories of information: they are often evasive and compromising stabs at a tricky subject. Different forms of evasion—the disclaiming menu, the dinner party without food, the festival blow-out set in another

[20] Especially Pliny's *Natural History*; the medical writings of Celsus and Galen; the table-talk of Plutarch, Macrobius, and Athenaeus. See the list of sources in André (1981: 9).

[21] e.g. Lucilius (see Shero 1923: 127–34); Hor. *Sat.* 2. 8; Petr. *Sat.* 30–78; Juv. *Sat.* 5; 11. 1–20; Macrob. *Sat.* 3. 13. 12; *Sen. de Ben.* 1. 10. 2; *de Prov.* 3. 6; *Cons. ad Helv.* 10. 2–11; *Ep.* 95. 13–42; Lucian, *Nigr.* 31. 72; Amm. Marc. 28. 4. 13, 34; *SHA Elag.* 19–30.

[22] e.g. Hor. *Epod.* 2. 53–60; *Sat.* 2. 2. 116–25; Virg. *Ecl.* 1. 80–1 2. 11; F. Bibaculus (= Suet. *Gramm.* 11); Ov. *Fast.* 4. 545–8, 5. 505–22; Ov. *Met.* 8. 637–78 ('Baucis and Philemon'); Juv. 11. 64–76; Sil. 7. 162–211 (a bloodless meal in the middle of military epic); Mart. 1. 43; 3. 58. 33–44; *Moretum; Copa* 17–22; Manil. 4. 387 ff.

[23] Hor. *Sat.* 1. 6. 115–18.

[24] See Ch. 3 below. The best discussion of embarrassment in talking about food is Ricks (1974: 115–42), 'Taste and Distaste'.

[25] This has obviously been the assumption behind works which do try to establish what the Romans ate: e.g. the lexical works of Thompson (1936, 1957) and André (1981); Veyne's history of largesse (1976); Garnsey's economic study of food and famine (1988); the Brothwells' biological study (1969); Deonna and Renard's survey of table superstitions (1961); and the general surveys of Dosi and Schnell (1984), Ricotti (1983), and Vehling (1977). D'Arms (1984) has provided an outline for a sociological analysis of Roman meals; Edmunds for a semiotic one (1980). More recently, it has been realized that a cultural approach is needed, which integrates literary and figurative representations: see Murray (1990).

country—are all shaped by the cultural restraints that lie behind them.

Turning food into language has its own complications too. However much a Roman writer might want to disdain contact with the body, the fact remained that the Greek and Latin vocabulary of taste, appetite, consumption, satisfaction, pleasure, and disgust rested squarely on physical metaphors, as it still does in English. The Greek word for 'pleasure', for example—ἡδονή—comes from ἡδύς, 'sweet', and originally meant a taste or flavour. The Latin word for 'wise'—*sapiens*—originally meant 'juicy' or 'flavoursome'.[26] This meant that the act of eating managed to steal its way even into the more abstract or sublimated writing that tried to avoid it. Eating was linguistically ubiquitous, which is an indication of how useful it was as a conceptual parallel. It is not enough, then, simply to extract the 'real' food from a description, as this ignores the engagement between material and metaphorical food that is often vitally significant. In fact, similarities and differences between eating and the less material pleasures of a meal, or between eating and experience in general, are a central part of the meaning of food in Roman literature.

The uneasiness of food descriptions was a symptom of wider cultural inhibitions, but prejudices closer to the author's heart must also have been responsible. Literary works are, after all, designed as a perennial alternative to less enduring forms of enjoyment. In Roman society, where gift-exchange was a central ingredient of social relations,[27] a poem and a meal were often interchangeable currency. Literary patronage, boiled down to its lowest terms, consisted in just such an exchange: Persius, in the most damning exposé of the system, lets slip that the stomach is the real muse of poetry.[28] At the same time, the transaction between poet and patron was a paradoxical one. A piece of paper was in many ways the antithesis of the material support that paid for it. If the paper contained food, the food was illusory, like

[26] Gosling and Taylor (1981: 18); Onians (1954): 61 ff.

[27] See Veyne (1990: 5–8).

[28] *Sat.* prol. 10–11: 'tutor for arts and sugar-daddy for talent—the stomach' (*magister artis ingenique largitor/venter*).

the wax fruits that deceived the philosophers. A grateful description of a present or an invitation to a meal answered the challenge laid down by real food; it excited a different kind of appetite and gave a different sort of pleasure.

The distinction between these sorts of pleasure has been pondered by Roland Barthes, one of the most openly voluptuous of writers,[29] in *The Pleasure of the Text*. At one point (1976: 45), he lights on a passage quoted by Stendhal:[30]

> In an old text I have just read . . . occurs a naming of foods: milk, buttered bread, cream cheese, preserves, Maltese oranges, sugared strawberries. Is this another pleasure of pure representation (experienced therefore solely by the greedy reader)? But I have no fondness for milk or so many sweets and I do not project much of myself into the detail of these dishes. Something else occurs, doubtless having to do with another meaning of the word 'representation'.

Barthes does not have a sweet tooth, but he still finds this a 'sweet' description. At the opposite extreme, Peregrine Pickle arranges the disgusting Roman banquet for the pleasure of seeing the guests squirm; Smollett's readers get the same sort of warped delight from a menu which is exactly the opposite of what they would enjoy in real life.

These are a few of the extra dimensions, then, which make it impossible to analyse literary food rigidly: tensions between its trivial and significant, material and metaphorical, real and illusory aspects contribute as much to its meaning as any universal cultural code. I shall have more to say later about specific problems of interpretation.

The other obstacle to rigid analysis is Rome itself. If it is difficult to generalize about a small state or a tribe, it is impossible to comprehend a vast cosmopolis, which for many Roman writers was not just a city but a globe in itself.[31] Athenaeus, in his rambling collection of table-talk, the *Deipnosophistae*, describes Rome as the 'epitome' of the

[29] See Bauer in Bevan (1988: 39–48).

[30] *Episodes de la vie d'Athanase Auger publiés par sa nièce*. Stendhal *Mémoires d'un touriste*, in *Œuvres complètes* (Paris, 1891), i. 238–45.

[31] e.g. Ov. *Fast*. 2. 137–8.

civilized world: all the cities of the world are settled there, Alexandria, Antioch, and Athens (to take only the letter A; it would take a whole year to list each one).[32] It is not surprising, then, that Plutarch says of Rome in the time of Cato the Elder (the starting-point of this book) that it was a city which had already lost its purity (*τὸ καθαρόν*) and had no single moral system: 'Its conquests provided it with a great mixture of customs and ways of life of every kind.'[33] *Romanitas*, the essence of Roman virtue, had always, it seems, been threatened by more cosmopolitan ideals.[34]

Rome was the archetype of the eclectic and absorbent civilization, in every way the opposite of Mary Douglas's Israelites, who were only one of many separate tribes massed together under the empire. Exogenous marriage, for example, was enshrined in the myth of the Sabine women, citizenship was relatively easy to come by (even to former slaves), and Roman religion was an accretion of different exotic cults. A Roman writer might be a Spaniard, an African, or a Greek. At the same time, racial distinctions were not forgotten or blurred: xenophobia, or at least a sense of society as a melting-pot, sharpens many writers' outbursts of resentment. The same capacity for boundless abundance and variety holds for Roman food too. *Imperium* had turned Rome into the world's emporium: its alimentary choices are presented as almost infinite, from the turnips of Romulus to the larks' tongues of Elagabalus.

To complicate matters further, any notion of the 'mentality' of a single individual has to be replaced with a confusion of moral systems. Cicero, for example, preaches the old-fashioned virtues of food as fuel for the body in his philosophical writings.[35] In his more urbane and intimate role as a letter-writer, he dabbles in the gastronomic knowledge proper for a cultivated man.[36] In a defamatory speech, he

[32] Ath. 1. 20b. This may be an allusion to convivial alphabet games: cf. *SHA Antoninus Geta* 5. 8, for meals based on different letters of the alphabet.

[33] *Cato Maior* 4. 2.

[34] Just as Aristotle had modified the ascetic principles of Plato: Newman (1887) speaks of his 'not unfriendly reference to the art of cooking' at *Pol.* 1255b. 25.

[35] *Tusc. Disp.* 5. 34.

[36] *Fam.* 9. 18. 3; *Att.* 13. 52.

sneers at Piso's squalid dinner parties, with their piles of half-rancid meat; here he is forced to say: 'All luxury is vicious and degrading, but there is still a kind which is not unworthy of a free-born man.'[37] In other words, he is uncomfortably aware that the increase in material wealth had changed standards of entertaining and honouring people for ever, and compromised the frugal ideals of the past: one man's simplicity was another man's meanness. The same meal could be represented in very different ways: the simple and friendly Epicurean lunch proposed by Philodemus to Piso (*Anth. Pal.* 11. 44) was presumably offered in return for the boorish dinners at Piso's house that Cicero ridiculed. Augustus is said to have left a dinner which sounds like the perfect example of unaffected frugality (*cena satis parca et quasi cotidiana*) murmuring to the host, 'I had no idea I was such a good friend of yours.'[38] And Lucilius found material for satire even in the kind of rustic menus which were traditionally exempt from abuse.[39]

Clearly, then, there was a good deal of disputing about tastes in Rome. Any attempt to do justice to the variety of Roman consumption and the variety of responses to it begins to look impossibly bewildering. The sheer distance of this civilization and the gaps in its written legacy are obvious impediments. And of course the Proustian element, the particular redolence of food for the individual who wrote about it, is lost or altered in any reading. It is surprising, even so, what close reading can still capture: ruminating over a few neglected Latin texts can tell us more about the Romans' approach to eating than a skimming survey. But the very nature of the subject, the shifting preferences of a myriad people, demands the most flexible kind of approach. We have to recognize how much these works manipulate, superimpose, and often parody universal codes, and how much each author's presentation of food reflects his own concerns as a writer. It may be helpful to begin with a very general framework for looking at these texts.

[37] Cic. *Pis.* 67.

[38] Macr. *Sat.* 2. 4. 13: *non putabam me tibi tam familiarem.*

[39] See Shero (1929).

2 *The Consuming Empire*

All the texts discussed here will be from fiction. Of course, the kind of works we loosely call non-fiction, Suetonius' *Lives of the Emperors* or Pliny's *Natural History*, for example, are just as biased, selective, and loaded, but in fiction the author has most control over his choice of material. The sifting process involved gives us a better idea of the applications of food or its use as a focus for other ideas than, say, a book on dietetics, a farming manual, or a cookery book. For a start, all ancient rhetoric is founded firmly on antithesis, which provides us with some essential polar distinctions, between simple and luxurious, raw and cooked, foreign and native food. Without any more objective source, we have to assume that these sharp polarities do represent conceptual divisions, and give us a broad outline for considering Roman culture, however varied the shades of meaning in each case. Another advantage of fictional evidence is the writer's use of metaphor, which picks out correspondences across wide fields of experience. Fiction may have its own private concerns and idiosyncrasies, and these deserve full attention, but that is no reason to separate it from the environment in which it was produced. In fact, the treatment of food in the texts discussed here is always a response, whether evasive or aggressive, to the monstrous entity of Rome.

The literary metaphor that most clearly links individual consumption, the Roman empire, and, I shall argue later, the literary text itself, is that of the human body. This had been an image for the political state at least since Aristotle, and provided an analogy for any combination of whole and parts, together with a model of scale, growth, excess, and limits.[40] Over the three centuries I deal with here, from Plautus to Pliny the Younger, Rome was completing its expansion from a city into a cosmopolis of global proportions. Writers often described this cosmopolis as an over-consuming body. Once self-sufficient in its consumption ('adequately prosperous and adequately powerful', in Sallust's words[41]), Rome had

[40] e.g. Tac. *Ann.* 1. 12: 'that the body politic was a single whole' (*unum esse rei publicae corpus*).

[41] Sall. *Cat.* 6. 3: *satis prospera satisque pollens.*

expanded with self-fuelled greed, or so the social myths insisted. *Luxuria*, the black name given to the trappings of Roman expansion, was originally a metaphor from organic growth. Pliny, for example, uses it of cabbage-plants that have grown too tall.[42] So when, in another context, he uses the phrase *luxuriantis iam rei publicae* ('now that the city was in the thick of luxury'[43]), he is implying a link between physical growth and moral abandonment, and transferring the image of a sprouting plant to the state itself.

Conversely, the individual body could be seen as the small-scale incarnation of national *luxuria*.[44] Indeed, a passage from Livy shows how easily the distinction between the growth of the state and the growth of its citizens *en masse* was blurred: 'We only grow in the directions we strive towards, that is, in wealth and luxury (*adeo in quae laboramus sola crevimus, divitias luxuriasque*).'[45] City and citizen were invertible figures. Gellius tells us that Cato made an example of a knight deprived of his horse because he was too fat: 'His body had swollen and spilled over in such an excessive manner' (*cuius corpus in tam inmodicum modum luxuriasset exuberassetque*).[46] Plutarch gives us the reasons for his disgust: 'How can such a body be useful to the state, when everything between the gullet and the groin is given over to the stomach?'[47] Like the state that had exceeded its proper boundaries, the body that had gone to seed was not thought to be effective as a political or military machine: Caesar, we remember, preferred men that were fat.[48] An over-indulged stomach was thought to disturb the equilibrium of a body where desires ought to be ruled by the head: this explains Menenius Agrippa's sophistical apology for the Senate as a stomach which seems to be idle but in fact supplies the rest

[42] *NH* 19. 139.
[43] *NH* 18. 17.
[44] Polhemus (1978: 9) traces the sociological theory of the physical body as analogue of the social body back to Hertz and Mauss; see also Turner (1984).
[45] 7. 25. 9.
[46] Gell. 6. 22. 4.
[47] *Cato Maior* 9. 5.
[48] Plut. *Caesar* 62. 5.

of the body,[49] and Cato's image of Rome as a 'belly without ears'.[50]

The individual could at least control his own consumption, even if he had no power over that of the state: Cato was clearly using the example of his personal habits as a model for Rome in his speech *de Sumptu suo*.[51] And Seneca pictures himself as an island of integrity in the swelling flood of luxury.[52] Prescriptions for diet and regimen seem to have fulfilled the same sort of function as they do today, treating the body as a potentially perfect microcosm over which the occupier had complete control, particularly when his political power had been eroded.[53] The emphasis in Rome, however, was more on the ideal balance of humours in an unbalanced world than on protecting the body and its food from the chemical pollution destroying the planet.

Writers also turned the city of Rome itself into a kind of bodily map, where the different regions fitted into the same hierarchy of head and stomach. For Livy, the Capitol was not only the uppermost part of the city but also its public face, its exemplary moral figurehead. According to legend, a human head was found on the site during its construction, which was taken as a prophecy that Rome would be the *caput rerum*, 'the head or capital of the world'.[54] It is significant, too, that the only record of a Roman eating taboo comparable to those of the Jews is the diet of the Flamen Dialis, the high priest of Jupiter, whose temple was on the Capitol: he could not touch flour, yeast, or raw meat, which Plutarch, anticipating Mary Douglas, identifies as boundary-crossing foods.[55] Livy's account of the building of Rome, which begins with the Capitol, ends with the Cloaca Maxima, the giant sewer.[56] The word *cloaca* is now the technical term for

[49] Liv. 2. 22. 8–12; Plut. *Coriolanus* 6. 2–4.

[50] Plut. *Mor.* 198d.

[51] See Malcovati fr. 173–5 for the remains of this speech.

[52] *Tranq. An.* 1. 4. 10: *circumfudit me ex largo frugalitatis situ venientem multo splendore luxuria et undique circumsonuit.*

[53] See Foucault (1990). Rousselle (1983) and P. Brown (1989) discuss the body as a focus for personal autonomy in antiquity.

[54] Livy 1. 55. 5.

[55] *Mor.* 289e.

[56] 1. 56. 2.

the excretory duct of birds, and a fragment of Varro suggests that it was linked in antiquity with intestines; in other words, it might be conceived of as the metaphorical excretory duct of the city.[57] Cicero speaks of sewers 'stuffed' with citizens, which is perhaps not an accidental metaphor (*corporibus civium . . . cloacas refarciri*); his use elsewhere of the proverb 'to make an arch out of a sewer' shows monuments and drains conceptually poles apart.[58] The story Dio tells about Agrippa riding through the cleaned-out Cloaca in a boat shows how it could be associated either with dirt or the purging of dirt. This main duct, according to Juvenal, was fed by a drain in the middle of the Subura, a bustling Soho-like district famous for its cookshops and low life.[59] Could we perhaps make the connection and see this area as the 'stomach' of Rome, by analogy with Zola's portrait of Les Halles?[60] The proximity of the restaurant areas to the sewer must have been a constant reminder of the transience of food, provoking a gut reaction similar to the one Agathias Scholasticus records, when he peers into a latrine in Byzantine Izmir and sees 'Pheasant, fishes, mixtures pounded in the mortar, and all the deceit of eating turned into dung.'[61]

Rome also had a moral topography to match the bodily one. Seneca maps out the two distinct regions haunted by Virtue and Pleasure in a deliciously vivid personification:

> Virtue is something high, exalted, and regal, unconquerable and indefatigable; pleasure is something lowly, servile, weak, and perishable, which haunts the brothels and cookshops. Virtue you will find in the temples, in the forum, in the senate-house, standing in front of the walls, dusty, tanned, with calloused hands; Pleasure you will find lurking more often in the shadows around the bath-houses and the sweat-rooms and the places which fear the aedile, delicate, effeminate, dripping with scent and wine, pale and sultry, and smeared with ointment. (Sen. *de Vit. beat.* 7. 3)

57 Var. *Men.* 290: *cloaca intestini*.
58 Cic. *Sest.* 77; *Planc.* 95.
59 Dio 49. 43; Juv. 5. 105.
60 On Zola's *Ventre de Paris*, see J. Brown (1984).
61 *Anth. Pal.* 9. 642.

In other words, Rome presented two different images simultaneously: an image of ritual purity—the frugal sacrifices, the hallowed agricultural surnames of its leading men, the primeval festivals; and an image of adulterated or contaminated confusion—the smoky cookshops, the profusion of foreign delicacies, the debauched dining-clubs. Cicero claims that the founders of Rome were wise to build the city five miles from the sea, as maritime cities were hopelessly polluted by an influx of foreign languages and morals.[62] Here he is clearly on shaky ground: Rome had its own corrupt and seedy face. Sallust compared it to a cesspool which absorbed all the debris of the countryside;[63] and images of the sewer and its effluence also clouded snobbish categorizations of the lower classes: 'the cesspool of the city' (*sentina urbis*, Cic. *Att.* 1. 19. 4); 'the dregs of the populace' (*faex populi*, Cic. *QF* 2. 5. 3). The extension of citizen rights to foreign 'scum' was similarly perceived as a kind of pollution. According to Suetonius, Augustus preserved the citizenship 'pure and intact from all foreign and servile filth' (*sincerum atque ab omni conluvione peregrini ac servilis incorruptum*).[64]

The notion of the Roman meal, too, was informed by ideas of bodily expansion and adulteration. The 'history' of the meal began with a simple unit, which was progressively amplified and adorned. Tacitus actually gives dates for the peak period of alimentary luxury, and Livy dates its onset to 187 BC, but the moral contrast between antique simplicity and present decadence was timeless.[65] The meal, it seems, had expanded in two directions, horizontally and vertically. Servius notes, for example, that the old Romans dined on two courses alone, while Juvenal protests at seven courses in his own day. Pliny bewails the fact that two or three boars, once the substance of a whole dinner, are now little more than hors d'œuvres. The dessert course, the *secundae mensae*, once exceptional, was now a standard embellishment; so

[62] *De Rep.* 2. 4. 7.
[63] Sall. *Cat.* 37. 5.
[64] Suet. *Aug.* 40. 3.
[65] Tac. *Ann.* 3. 55 (31 BC to AD 68); Livy 39. 6. 7–9.

were perfumes, garlands, sauces, and sideboards with removable shelves.[66] Macrobius actually describes one Republican pontifical dinner as though it were a body expanded by luxury. After two courses of roast and stewed meat, fish and poultry, the dinner, like the feasters, is distended with stuffing: *tot rebus farta*.[67] One Greek comic cook's 'Lévi-Straussian' history of civilization in cooking terms is typical of these myths:[68] first, man was a wild cannibal; then he discovered roast meat and instituted cities and sacrifices; finally he invented stuffed paunches and fish disguised with sauce, and became fully civilized.

The structure of the Roman meal as a unit can only be reassembled from fragments, but it also begins to look like a commemorative history of eating habits, a way of classifying food so that it becomes steadily more 'civilized': it begins with the *gustatio*, the initial tasting of pure roots, vegetables, fish and eggs (the containers of early life); culminates with the *cena* proper, with its 'sacrificial' meat; then either atones by returning to nature in the form of fruit and nuts (fruition), or declines further into the superfluities of the *secundae mensae*.[69] Pliny sees pastry-making as a luxury reserved for peaceful nations; and Athenaeus suggests that the *secundae mensae* were especially associated with the Saturnalia, when an extended version of the normal meal was needed. At the other extreme, *puls* (pottage), Rome's aboriginal food, was apparently used to celebrate both birthdays and primitive rituals, which suggests that parallels were made between events in the citizens' lives and the structure of Roman history.[70]

[66] Serv. ad *Aen.* 1. 726; Juv. 1. 94; Plin. *NH* 8. 210; Val. Max. 2. 6. 1: unguents, garlands, and *secundae mensae*; Plin. *NH* 19. 19. 57: sauces; Plin. *NH* 33. 140: sideboards; Sen. *Ep.* 114. 9 on meals where the order of courses was reversed in the cause of novelty.

[67] *Sat.* 3. 13. 13. The dinner is a magnified version of the individual stuffed dishes it contains: e.g. *gallinam altilem, altilia ex farina involuta, altilia assa.*

[68] Athenion, *Samothracians* (Ath. 14. 660e–661d = 3. 369 K).

[69] See Lovejoy and Boas (1965) on the variety of myths about 'primitive' eating. Schama (1987: 599) mentions a Dutch 18th-c. burgomaster who staged a banquet on the theme of Dutch history: red herring and cheese (the nation's infancy), puddings and roasts (its prime), French wines and delicacies (its decadent decline).

[70] Plin. *NH* 18. 105 (pastry); Ath. 14. 639b; Plin. *NH* 19. 83–4 (pottage).

The standard extremes of Roman eating, simple and luxurious food, were used, then, to mark out two different stages in the mythology of Roman civilization: the pure, rustic nature of the Romans' past imposed on over-sophisticated urban culture. The contrast could be geographical as well as historical. The diet of other races often threw Roman decadence into relief: when Tacitus says that the Germans eat fresh wild meat, wild fruit, and cheese, and satisfy their hunger without trappings and seasonings, that is really a reflection on his own society.[71] In some ways, though, these historical and geographical gaps were artificial: as Seneca's 'map' of the city suggests, Rome was large enough to allow images of simple and luxurious eating to coexist.[72] In the Roman forum, yards away from the pontifical banquets and steaming cookshops, a table was laid with frugal offerings commemorating the old way of life;[73] sacrifices to Vesta, Cybele, and Carna were also reminders of the Romans' primitive diet.[74] Some grand Roman houses had their own shrines to frugality, the so-called poor man's cell (*pauperis cellae*), in which the owners could follow an artificially meagre diet, the 'dinners of Timon' (*Timoneae cenae*).[75] Country life was another form of symbolic reversion to nature. Martial satirizes people who buy fruit in the centre of Rome to display on their rural estates. Picnics in caves, by waterfalls, in tree-houses, in storerooms full of apples: all these testify to the Romans' nostalgia for their origins.[76]

Writing against luxurious food and the superfluous desires of the body can now be explained as the most immediate and universally intelligible image of Rome's expansion. Like the

[71] Tac. *Germania* 23.

[72] Cf. Ov. *Fast.* 1. 222–6: the past and the present had equal value in shaping the meaning of Romanness.

[73] Dion. Hal. *AR* 2. 23–5. See Le Bonniec (1958: 160 ff.). Cf. Ath. 4. 137e: the Athenians offer the Dioscuri cheese, a barley-cake, olives, and leeks 'to remind them of their earlier way of life'; see Gill (1974).

[74] Ov. *Fast.* 6. 309–10: bread for Vesta ('a remnant of the old way of life': *aliquid de more vetusto*); 4. 367–72: *moretum* for Cybele ('ancient food': *priscos cibos*); 6. 169–70: bacon, beans, and spelt for Carna.

[75] Sen. *Ep.* 18. 7; cf. 100. 6; Mart. 3. 48.

[76] Martial 10. 94. 6; 7. 31. 12. Tac. *Ann.* 4. 59; Plin. *Ep.* 4. 30. 3; Plin. *NH* 12. 5. 9–10; Varro *RR* 1. 59. 2.

spoils heaped up haphazardly in a Roman triumphal procession,[77] lists of food, verbal 'heaps' which challenged the reader's or listener's own bodily capacity, graphically reproduced the amassing of goods in Rome,[78] whether on the tables of the rich or in the city's cookshops, where the wealth of conquered nations was translated into ingestible matter. These hostile descriptions often present food in the form of actual heaps: a monstrous boar surrounded by a mound (*strues*) of apples; cookshops propped up (*instruitur*) by foreign taxes.[79] Gluttony was an image of the Romans' uncontrolled appetite for power, their unlicensed absorption of the world. Antony, for example, is transformed into one giant maw engulfing the property of Pompey: *per fauces bona Cn. Pompei transierunt*.[80] Seneca the Younger follows the progress of fish and birds from the furthest reaches of the world into the all-consuming stomach: *si in ventrem suum longinqui litoris pisces et peregrina aucupia congereret*;[81] Sidonius describes the unsettled stomach of a parasite as a 'cesspool of regurgitated dinners'.[82] Vomiting and emetics, which were actually standard medical treatments in the ancient world, became distorted and abused in attacks on luxury because they were ways of extending the body's capacity indefinitely.[83] Quintilian gives Cicero's description of Antony vomiting in public and bespattering the whole tribunal with bits of food as a perfect example of hyperbole: *vomens frustis esculentis gremium suum et totum tribunal implevit*.[84]

Gastronomy was another kind of aberration from primeval simplicity. Here fussiness, rather than sheer quantity, was the deviant factor. Gourmets' tastes for far-fetched

[77] Josephus, *BJ* 7. 148 uses the adverb χυδήν—indiscriminately, wholesale—to describe the heaps of triumphal spoils.

[78] e.g. Varro, *Peri Edesmaton* (ap. Gell. 6. 16. 5; see Hense (1906: 1–18)); Ennius, *Hedyphagetica* (lines 1–11 ap. Apul. *Apol.* 39).

[79] Sen. *de Prov.* 3. 6: *si ingenti pomorum strue cingeret primae formae feras*; Sen. *Suas.* 6. 7: *popina a tributo gentium instruitur.*

[80] Sen. *Suas.* 6. 3.

[81] *De Prov.* 3. 6.

[82] Sid. 3. 13: *redundantium sentina cenarum.*

[83] Sen. *Cons. ad Helv.* 10. 3; Cels. 1. 3. 17.

[84] 8. 6. 68 = *Phil.* 2. 25. 63.

flavours reflected the range of choice, the multiplication of gastronomic possibilities. The baroque dishes that Pliny the Elder describes, flavours from India, Egypt, Crete, and Cyrene rolled into one, are miniature versions of Athenaeus' Rome: a conglomeration of the world's cities.[85] In Pliny's descriptions of plants and animals bred to mammoth size, the all-enveloping word *luxuria* covers physical size and the proliferation of types on the one hand, and gastronomic fussiness on the other.[86] Hothouse vegetables are like fattened animals: 'crammed cabbage' (*caule saginato*); 'gorged asparagus' (*altiles asparagi*);[87] differentiation stretches even to common or garden herbs.[88] Where Greece had scientific discoveries, Rome had culinary ones: Maecenas was the first to serve wild asses, Hortensius to serve peacocks; Lucullus and Hortensius were the first to keep reserves of wild boars; and so on.[89] Pliny is at his most satirical when he awards a special prize to a ridiculous dish made up of cocks' crests and goose feet.[90] Together, gluttony and gastronomy presented two deformed bodily images of decadent Rome, a stuffed and multiplied perversion of its original self.

The symbolic value of individual consumption found a focus under the empire when supreme power was removed from the citizens and concentrated in the hands of one man. For the Romans, this was an uncomfortable or grotesque anomaly, as Suetonius' description of Vitellius' favourite dish, the 'Shield of Minerva', suggests. This hybrid monstrosity was a mixture of pike livers, pheasants' and peacocks' brains, flamingos' tongues, and lampreys' milt,

[85] *NH* 15. 105.

[86] e.g. 17. 27. 220 (of maggots from worm-eaten trees): 'even these have succumbed to the influence of luxury . . . stuffed with flour, they too are fattened' (*et hoc in luxuria esse coepit . . . farina saginati hi quoque altiles fiunt*); 31. 95 (of *allec*, the sediment of fish-sauce): 'then it became a luxury, and the types expanded *ad infinitum*' (*transiit deinde in luxuriam creveruntque genera in infinitum*).

[87] 19. 54. Pliny refers to them as 'gastronomic prodigies' (*prodigia ventris*). *Prodigium*, a monster or monstrosity, can be (dubiously) connected with *prodigus*, extravagant or wasteful (cf. Lucan 4. 373–4: *o prodiga rerum / luxuries*).

[88] 9. 19. 54.

[89] Plin. *NH* 8. 170; 10. 45; 8. 211.

[90] Ibid. 10. 52: he puns on *palma*, a prize and the sole of the foot.

'brought by his captains and triremes from the whole empire, from Parthia to the Spanish Strait'.[91] All-embracing imperialist power (the parodic title conjures up glorious military exploits) is made miniature, confined in a dish, to fit the expanded limits of one man's consumption. What the emperor ate symbolized his demands on the state's resources. This is the force, for example, behind Augustus' claim that he ate only figs and second-class bread; by maintaining state dinners, he showed his calculated regard for republican protocol.

Roman writers had different responses to the 'embarrassment of riches' which they saw around them.[92] Some eliminated matter entirely from their work, others exposed it fully or with disclaimers. Some rose to the insurmountable challenge of summing up Roman food, like Ennius and Varro in their gastronomic Odysseys, for example, or Pliny in his *Natural History*, or Athenaeus in his feast of quotations. Others rejected materialism and offered their own frugal alternatives. Literary cataloguing of the products of luxury is always an ambivalent phenomenon. The sumptuary laws of 181 to 22 BC, which provided a functional model for this sort of writing, aspired to antique virtue with their prescriptions for a simple diet ('the fruits of earth, vine, and tree'),[93] but they also listed all the delicacies which were being prohibited, running to pages in their detail and stretching to all kinds of unheard-of delights in the process.[94] In other words, the laws conspicuously displayed the potential for consumption in the very act of renouncing it.[95]

[91] Suet. *Vit.* 13. 2. Cf. emperors' taste for *tetrapharmaca* (sows' udders, pheasant, peacock, ham, and wild boar): *SHA Aelius Verus* 5. 4; *Hadrian* 21. 4.

[92] The title of Schama's book (1987) sums up a very similar tension between opulence and puritanism in 17th-c. Dutch culture; cf. an earlier article on the same subject (1979).

[93] Macr. 3. 17. 9: *quod ex terra vite arbore sit natum.*

[94] Plin. *NH* 8. 27. 209: *hinc censoriarum legum paginae*; 36. 2. 4: dormice and other things too trifling to mention; Macr. 3. 17. 11: *exquisitis et paene incognitis generibus deliciarum.*

[95] Cf. L. Brown (1985: 8–43) on Pope's ambivalent treatment of mercantile expansion: he simultaneously parodies and glorifies the profusion of foreign imports. Vegetti (1981) likens Pliny's panorama of the works of nature (including *mirabilia* and the products of luxury) to the emperors' display of freaks and exotic

Fictional representations of food also tend to list exotic or expensive food only to deny it to the reader, a kind of negative gloating.[96]

The hierarchy of the head and stomach which can be seen in Roman culture as a whole also affected the ranking of literary genres. There is a striking correlation between a work's place in the generic hierarchy—where epic, tragedy, and history take precedence over satire, comedy, table-talk, epigram, epistle, and the picaresque novel—and the degree to which it represents or suppresses ignoble functions of the body. One frivolous emperor was given away by his bedside reading: the recipes of Apicius, Ovid's *Amores*, and Martial's epigrams, instead of more highminded Virgil.[97] The implication here that trivial literature was always a substitute for something grander is a clue to the place of food-writing in Roman culture. Food appears, as a rule, in texts which are mixed and miscellaneous, and set themselves up as trivial or parodic; it tends to be absent, except in its most solemn, sacred, and undefined terms, from the higher genres. Perhaps the most awkward exception to the rule is Seneca's *Thyestes*, where the moralist's obsessions spill over into tragedy, and the disguised limbs of the hero's infant sons take on the flavour of some contemporary evil stew.[98] The hero's intestinal eruptions and thunderous belches are magnified on a cosmic scale, as though the disarray of the Roman tragic universe could best be brought to the surface by an explosion of unnatural food.

Common food terms were often enough in themselves to debase a literary text. The orator Albucius, according to the Elder Seneca, was a polished enough speaker, but he was not above naming some extremely sordid things—vinegar, flea-mint, lanterns, and sponges. Seneca's verdict is that the habit polluted rather than reinforced his arguments: *his admixtis sordibus non defendi sed inquinari.*[99] So the same

beasts in the circus, a symbol of their power. Cf. Barthes (1964) on Dutch still life as an auditing of mercantile wealth.

[96] e.g. Hor. *Epod.* 2. 49–54; Stat. *Silv.* 4. 6. 8–10.

[97] *SHA Aelius Verus* 5. 9.

[98] 760–72, 1057–65.

[99] Sen. *Cont.* 7 prol. 3.

ideas of filth and corruption that were used to justify or criticize social and moral categories surface in the literary sphere as well. It is as though food in its most specific and ordinary terms made a stain on literary decorum. Quintilian goes straight to food when he needs an example of a euphemism, quoting the phrase one orator used as a periphrasis for salt fish: *pisces muria duratos*. Whatever the unmentionable word was, it has been lost for ever.[100]

The Romans' reluctance to mention food in their more serious writings is not just dictated by literary decorum. Decorum itself is shaped by larger constraints, and the greatest of these is perhaps a sense of cultural inferiority. Despite its great wealth and power, Rome was a cultural parasite, particularly, of course, on Greece. The contrast between the mind and the body that was normally used to mark differences within Roman culture also provided Sallust with the basic distinction between Greece and Rome: 'None of the Romans exercised their minds without involving their bodies. The best of them preferred action to speech.'[101] Without efficient bodies, or bodies that performed the duties of the state, the Romans' material superiority was worthless. This helps to explain the divide between literature that evades physical issues completely and literature that confronts them from a position of hostility. Many of the Romans' inhibitions about bodily pleasure were in fact inherited from Greece, but it would be hard to imagine a Roman making a speech as straightforwardly rejoicing in the products of empire as Pericles' funeral oration.[102]

As we shall see in the next chapter, it was in Greece that cooking was first represented as a mess-making and adulterating process,[103] but the theme had a special significance for Roman writers, when they saw their own culture as a mixture or amalgam. Quintilian tries in vain to rouse the Romans to the more expansive kind of rhetoric that their

[100] Quint. 8. 2. 3, who thinks the substitution absurd.

[101] Sall. *Cat.* 8. 5: *ingenium nemo sine corpore exercebat. optumus quisque facere quam dicere malebat.*

[102] Thuc. 2. 38; contrast Tiberius' letter to the Senate about Roman luxury (Tac. *Ann.* 3. 53–4).

[103] See Ch. 2, sect. 5, below.

civilization seemed to demand: if they could not aspire to the grace, subtlety, and precision of the Greeks, they did at least have strength, weight, and copiousness on their side.[104] But few Romans who wrote about food chose to advertise the weight and copiousness of their culture wholeheartedly. The historian Ammianus Marcellinus, for example, poured scorn on those who did: hosts who called for scales to weigh fish, birds, and dormice at the table, and marvelled at their unprecedented size, while a troop of secretaries stood by to record every detail.[105] Elagabalus, we are told, loved to hear the price of the food at his banquets overestimated, and claimed that this was the appetizer for his dinner.[106] And Roman cookshops were thought deceitful because they displayed eggs, liver, and onions floating in bowls of water so that they looked larger than their real size.[107] It also seems to be the case that the sorts of people depicted relishing food in Roman literature are children, slaves, parasites, cooks, gluttons, and gourmets; in other words, uncontrolled people who cannot be identified with the author or his accomplice the reader.[108] Without a special licence—for moral criticism or festive indulgence—the writer could not wallow in superfluous consumption.

3 *Making a Meal*

This special licence was transferred to the literary text by two Roman institutions: the dinner party and the festival. These are not only the settings for much 'bodily' writing: they also provide the protection of a playful or trivial context. The *convivium* is very often the implied frame of many rowdy or humorous works, even if no specific feast is

[104] Quint. 12. 10. 36: *non possumus esse tam graciles: simus fortiores. subtilitate vincamur: valeamus pondere. proprietas penes illos est verborum gratia certior: copia vincamus.*

[105] Amm. Marc. 28. 4. 13.

[106] *SHA Elagabalus* 29. 9.

[107] Macr. 7. 14. 1: *quippe videmus in doliolis vitreis aquae plenis et ova globis maioribus et iecuscula fibris tumidioribus et bulbos spiris ingentibus.*

[108] Children: Fronto, 2. 172 (181N) (a cloying description of a child eating grapes); slaves: e.g. Plaut. *Most.* 64, *Stich.* 689 ff.; parasites: e.g. *Capt.* 845–52; *Men.* 77–106.

mentioned.[109] This was the time when Roman behaviour ceased to be grave and austere, and taboos on eating, and talking about eating, were relaxed.[110] The *cena* was the climax of daily eating and a respite from daily business; festivals, and the Saturnalia is the consummate example, licensed overeating on a gargantuan scale. Both institutions were of ambivalent importance, marginal compared with 'normal' Roman activity—military, forensic, political—but still symbolically central.[111]

Cicero sees great significance in the literal meaning of the Roman word for a dinner party: *convivium*, or living together. In his eyes, the meal was the heart of Roman communal life: 'it is then that we truly live together' (*tum maxime simul vivitur*).[112] Compared with other activities, though, it was obviously peripheral. Plutarch, for example, divides his life of Lucullus into two clear-cut scenes of a play, one military and political, the other festive and convivial: 'As in an ancient comedy, you will read in the first part of political measures and military commands, and in the second part of drinking-bouts, banquets . . . and all manner of frivolity.'[113] This meant that any Roman who chose to represent or sum up Roman culture by describing a meal was using a conscious alternative to the usual channels, epic or historical narrative. Horace in *Ode* 1. 6 contrasts the proper material of the epic-writer, the stomach or wrath of Achilles (*gravem Pelidae stomachum*), with the feasts and amorous battles of his lyric poetry: *nos convivia, nos proelia virginum*. In his satires, he deals with real stomachs (for example the saga of the *iratum ventrem* at Nasidienus' dinner).[114]

The meal is often described loosely as a microcosm of the society in which it takes place. This is not strictly accurate.

[109] Koenen (1977) argues that Horace uses sympotic imagery in *Epod.* 10 (the shipwreck, an image of drunkenness), even when no symposium is explicitly named.

[110] Seneca (*Tranq. An.* 17. 7) points out that even in the old days great men would set aside time for a holiday.

[111] See Babcock (1978: 32): 'What is socially peripheral is often symbolically central.'

[112] *Fam.* 9. 24. 3; cf. *de Sen.* 13. 45.

[113] Plut. *Lucullus* 39.

[114] Hor. *Od.* 1. 6. 6; *Sat.* 2. 8. 5.

Public sacrifices in Rome may well have mirrored the social hierarchy, but at the private meal the principle was theoretically overruled.[115] This kind of meal was more of a licensed reorganization, the host's choice of his own world, and this cherished right was summed up in a well-known Pompeian graffito: 'The man with whom I do not dine is a barbarian to me' (*at quem non ceno, barbarus ille mihi est*).[116] Even the Greek *symposium* was not so much a cross-section of democratic society as an élite.[117] Plato's *Symposium* is an oligarchy or meritocracy in miniature, progressively disrupted by the *demos*. In Roman Republican society, where the principle of an equal élite was essential for political coherence, the domineering host (or *rex*) became a sinister reminder of monarchy. Under the empire, the *convivium* acquired a new significance as an isolated and precarious revival of *libertas*. Horace describes the guests at one of his rustic dinners as 'released from all unhealthy laws' (*solutus inanis legibus*). In the context, this refers to drinking laws, but it hints at wider restrictions too.[118] Divisiveness in the private meal is a perversion of the principle of *noblesse oblige* that ought to govern such meals, rather than a norm. To this extent, the meal *can* be used in literature as a symptom of social hierarchy, though it is usually an inadvertent or transgressive one.[119] By exposing feasts in this way, Roman writers were like censors enforcing the sumptuary laws, which decreed that citizens were to dine with their front doors open, and subject their own luxury to the public gaze.[120]

As for festivals, the Saturnalia is symbolically the most important. It was the Romans' equivalent of the Carnival in later Europe, the 'example par excellence of a festival as a

[115] See D'Arms (1984), Lincoln (1985); Veyne (1990: 220–21) on hierarchy at public banquets.

[116] Diehl (1910: 641).

[117] See Murray (1982). D'Arms (1984; 1990) discusses whether the theory of *hospitium* was put into practice in Rome.

[118] Hor. *Sat.* 2. 6. 68–9.

[119] See Bek (1983); Lucian, *Sat*: 17; Cic. *Att.* 13. 52; Plin. *Ep.* 2. 6, Juv. *Sat.* 5, Mart. 1. 20 on socially divisive meals.

[120] Macr. *Sat.* 3. 13. 17. 1: *et imperari coepit ut patentibus ianuis pransitaretur et cenitaretur*; Val. Max. 2. 1. 5 tells us that the old Romans always used to eat in public: *nam maximis viris prandere et cenare in propatulo verecundiae non erat.*

context for images and texts'.[121] A licensed restoration of the Golden Age, which temporarily toppled the social hierarchy (slaves were traditionally waited on by their masters at table[122]), the Saturnalia also sanctioned variations on the usual pattern of eating: the Romans would either return to primeval simplicity or stuff out the normal meal with superfluous food.[123] Consistent with the joking spirit of the festival were dishes of fake food, made of mud and other inedible substances.[124] Saturnalian lucky dips of unequal presents with punning titles made havoc of normal categories, while the rubbishy token presents parodied the normal structures of gift-giving.[125]

It was this capacity of the *convivium* and the Saturnalia to comment on society and rehearse alternative arrangements that gave them their power in imaginative literature.[126] The unruly celebrations of the Saturnalia became an image of what was abnormal at other times: it was as though the festival had exceeded its bounds and overflowed into everyday life. Seneca looks out of his window at the Roman people revelling below, and remarks that there is no point any more in celebrating what used to be an annual festival, as the season never ends.[127] According to a character at Petronius' *Cena Trimalchionis*, the 'big jaws' celebrate the festival all

[121] Burke (1978: 182) on Carnival. On the pervasiveness of carnival images in European culture, see Bakhtin (1968); Leach (1961); Babcock (1978); Stallybrass and White (1986).

[122] Lucian, *Sat.* 18; Sen. *Ep.* 47 14; Mart. 14. 79.

[123] The festival coincided with the winter slaughter of pigs (Mart. 14. 70, Lucian, *Sat.* 28).

[124] See Petr. *Sat.* 69. 9 ('fake dinners': *cenarum imagines*); and compare the capricious tricks of Elagabalus: *SHA Elag.* 25. 9, 27. 4. Perhaps we can see Commodus' fondness for mixing excrement into expensive food in the same context. Here the joke certainly seems to have been on him (*SHA Commodus* 10. 1).

[125] Suet. *Aug.* 75: *titulis obscuris et ambiguis . . . inaequalissimum rerum sortes*; Petr. *Sat.* 56. 8–10; Mart. 14. 1. 5; *SHA Elag.* 22. See Rankin (1962); and Tanner (1979: pp. 52–7) on puns at the Cena Trimalchionis.

[126] *Saturnalia* is the title of a mime by Laberius, a dialogue of Lucian, and a miscellany by Macrobius. It is also the explicit setting for Horace *Satires* 2 (2. 3. 5, 2. 7. 4), Mart. 11, 13, and 14, the *Testamentum Porcelli* (as I argue in Ch. 2; sect. 2), and the implicit setting for many other comic works, like Petronius' *Satyrica* and Seneca's *Apocolocyntosis*.

[127] *Ep.* 18. 1.

the year round.[128] In other words, writers used the setting of a dinner party or a festival to uncover underlying social trends, though these occasions were always meant, ironically, to be a relief from normal society. The *Cena Trimalchionis*, for example, uses the Saturnalian slave-dinner as an image of a latent threat to the social hierarchy: the freedman's travesty of an elegant dinner usurps everything but the cultural *savoir-faire* of the élite.[129]

The dinner party and the festival were also exceptional in that they relaxed the tight constraints on language that applied to other areas of activity. This atmosphere of relaxation again overflowed into literary depictions. The *convivium* and the festival, according to Quintilian, were the proper occasions for jokes, though only the lower classes (*humiles*) could make rude ones.[130] Cicero, similarly, defines obscenity as something 'unworthy not only of the forum, but even for a dinner party of free-born men'.[131] Jokes and puns were the linguistic equivalents of festival experiments with social categories: here again was a form of expression with scope for imaginative writers. Even so, Cicero's and Quintilian's distinctions between *humiles* and *liberi* are revealing. Clearly, a free-born man stopped short of the licensed behaviour acceptable in a man of lower birth. Once again, we need to recognize that there were many different levels of writing about dining: literature with a convivial setting spans the extremes of control and anarchy, civilization and bestiality, levelling and hierarchy, intellectual and bodily concerns embraced by the institution itself, and often makes an issue of the differences between them. Saturnalian mock 'laws' for dining, like the *Lex Tappula* and the *leges convivales*, exploit

[128] See Sen. *Ep.* 18. 1; Nauta (1987); Petr. *Sat.* 44. 4. Seneca in the *Apocolocyntosis* (esp. 8. 2, 12. 2) turns Claudius' reign into a perpetual Saturnalia with a nonsense king, though he himself claims the right of Saturnalian free speech to unclothe the previous emperor.

[129] See Veyne (1961).

[130] 6. 3. 16: *ipsae quas certis diebus festis licentiae dicere solebamus*; 6. 3. 8: *nam in convictibus et cotidiano sermone lasciva humilibus, hilaria omnibus convenient*; Gleason (1986) explains Julian's bizarre *Misopogon* as a retaliation against the popular abuse licensed during the New Year festival.

[131] *De Or.* 2. 62. 252: *non solum non foro digna, sed vix convivio liberorum.*

the double scope of the *convivium*, which was at once a formal ritual and a respite from normal rules.[132]

The literary *cena* was a loaded form for another reason: it provided a striking image of the difference between Greek and Roman civilization. The *cena* was to Roman culture what the *symposium* was to Greece. The main difference was that the *symposium* was a drinking party padded out with sweetmeats and relishes, while the Roman *cena* was inescapably weighed down by food.[133] That helps to explain why so many Roman writers use dinner parties to reveal the innate materialism of their culture; by inserting a large amount of food into the basic framework of Plato's *Symposium*, they are able to suggest the distance between Greece and Rome.[134] Other meal descriptions do their best to erase food: lyric poetry takes sympotic wine as its inspiration, and Virgil uses the untainted roast meat of Homeric heroes, instead of modern stews. Table-talk miscellanies are a rather ambiguous case. Athenaeus' *Deipnosophistae*, for example, is a rare example of a positive attempt to do justice to the size and abundance of Rome against the background of a dinner party. A Roman host of astonishing erudition in both languages opens his doors to an empire of scholars, and parades in front of them an infinite variety of dishes. But there are two levels on which we can understand the dinner. Either it is a Roman version of the *Symposium*, in which philosophy descends to the level of food; or else it is a sublimation of real eating. The reader's attention is diverted away from the food on the table towards airier philological subjects: words, not dishes, are the staple fodder of the meal.[135]

[132] Buecheler 266–7; cf. Lucian, *Sat.* 13–18; *CIL* 4, 7698; see Meiggs (1973: 429) for the fresco of Seven Sages uttering maxims on defecation.

[133] Murray (1985) emphasizes the differences between Greek and Roman versions of the *symposium*, though many of his examples appear to come from descriptions of *cenae*, not *symposia*. I do not wish to underrate the cultural importance of the hybrid Roman *symposium*, only to stress the very basic difference in content that made the two institutions of Greek *symposium* and Roman *cena* such useful symbols.

[134] Hor. *Sat.* 2. 8; Juv. *Sat.* 5; Lucian, *Symposium*; Lucian's Lexiphanes declares ἀντισυμποσιάζω (1), reducing Plato's dialogue to a list of comic food. See Cameron (1969); Dupont (1977) on Petronius and Plato.

[135] Athenaeus constantly plays with the idea that words, not food, provide the 'satisfaction' of the meal: e.g. 1. 4c, 1. 5a, 6. 270c, 7. 330c, 8. 354d, 9. 402d. See also

This emphasis reflects a wider cultural bias. Self-conscious references to singers and readers at Roman dinners, brought in to delight the mind rather than the stomach, suggest that civilized Romans were anxious to underplay the material pleasures that surrounded them.[136] In this suppression of bodily functions we can see the origins of the 'civilizing process' observed by Norbert Elias in Renaissance culture.[137] At the other extreme of convivial literature, however, particularly in satire, table manners are swept aside to expose the precarious or euphemistic aspects of civilized eating. We see the ancestors of the 'grotesque bodies'—spewing, spitting, stuffing, expanding—that Mikhail Bakhtin (1968) identified as a feature of European festivals and of the literature that adopted their licence.

A good example of the grosser extreme is the monumental mason's description of a slave's funeral feast at the end of the *Cena Trimalchionis*. This follows the classic outlines of a Bakhtinian bodily cycle, with its black confusion of human and animal, edible and inedible, slavery and freedom, festivity and death.[138] The first course of entrails, for example (pork crowned with sausage and surrounded by blood-pudding and nicely done giblets[139]) sets up an ambiguity between human and animal intestines later on, when Habinnas' wife tries bear's meat for the first time: 'She nearly threw up her tripes.'[140] For Habinnas, the eating of bear's meat is a logical inversion of the natural order: 'If a bear can

Lukinovich (1990). Cf. Gell. 7. 13. 2: 'we brought as our contribution not delicacies, but topics for discussion' (*coniectabamus ad cenulam non cuppedias ciborum, sed argutias quaestionum*); Varro, *Men.* 144; Stat. *Silv.* 4. 6. 4–16 says that the real dinner (i.e. the conversation) remains 'unconsumed' (*inconsumpta*) in his memory; Macrob. 1. 2. 15; Plut. *Mor.* 716e: 'For a group of men to say nothing at all while stuffing themselves with food would be positively swinish.'

[136] Nepos *Atticus* 14 (on readers at meals): 'to please the mind as well as the stomach' (*non minus animo quam ventre convivae delectarentur*); Plin. *Ep.* 9. 34, 9. 17. 3; *SHA Alexander Severus* 34. 6–8; Gell. 3. 19. 1.

[137] Elias (1978) does not in fact make enough of the classical precedents for this phenomenon.

[138] Petr. *Sat.* 66.

[139] *Porcum botulo coronatum et circa sanguinculum et gizeria optime facta*, 66. 2; cf. pieces of tripe (*chordae frusta*) at 66. 7.

[140] *Paene intestina sua vomuit*, 66. 5.

eat a man, surely a man should be able to eat a bear?'[141] But this also gives rise to the unpleasant suspicion that, somewhere along the cycle, human flesh has been consumed. Another dish, a culinary imitation of excrement (*catillum concacatum*, literally 'crap stew'[142]) follows Habinnas' ungentlemanly comments on his reasons for preferring wholemeal bread: it is better for the bowels.[143] A meal description that is not very polite even at first sight exposes on further scrutiny many of the unpleasant functions and grey areas of eating that the civilized meal normally glosses over. Bakhtin's 'festive' model is a helpful one for a great deal of Roman literature about food, because it successfully combines cultural history and literary criticism; it recognizes that literary texts are the products of their cultural background, but also allows for the heightened concentration of meaning and idiosyncratic experiments licensed by their extraordinary context. But to see a literary work as a manifesto for popular culture is too rosy: Petronius' dinner party is the smug wallowing of an aristocrat, who wipes unmentionable bodily filth and the botched products of a parasitic culture off on to the freedmen beneath him.[144]

To sum up: the marginal literary position of convivial or festive works reflects the status of these institutions in real life. To some extent, this position is perpetuated by the authors involved, who tend to put on an ironic and self-deprecating manner and dismiss their work as jokes or ephemeral trifles.[145] Even the table-talk or quiz-games that

[141] *Si . . . ursus homuncionem comest, quanto magis homuncio debet ursum comesse?* 66. 6.

[142] Burman's conjecture for *concagatum*, by analogy with ὀνθυλεύω —to make forcemeat (from ὄνθος, dung). Defecation and mess are connected at Sen. *Apoc*. 4: 'he made a shitty mess of everything' (*omnia certe concacavit*).

[143] 66. 2. Like other dishes at the *cena* itself, some of the dishes here are jokes on the theme of manumission, a reminder that death, for this slave, has at least brought posthumous freedom: eggs with freedmen's caps (*ova pilleata*) and a ham that was sent away, or given its freedom (*pernae missionem dedimus*, 66. 7).

[144] See Stallybrass and White (1986) on 'low' culture as an object of disgust and fascination for the European bourgeois class.

[145] Auson. *Cento* pref. links his patchwork with the *sigillaria*, cheap Saturnalian gifts; Mart. 4. 14. 12: 'books steeped in naughty jokes' (*lascivis madidos iocis libellos*); Cat. 50. 2 (at a drinking-party): 'we toyed on my writing-tablets' (*lusimus in meis tabellis*); Ov. *Tr*. 2. 471 on trivial Saturnalian poems on the themes of entertaining

were substituted for food description had the same marginal or miscellaneous quality: superfluous material which had no proper place in more serious literature. The background of a meal or festival is made to excuse the appearance of this material as a literary subject. Aulus Gellius, for example, marks off as suitable dinner party topics 'those commonplace subjects for which we do not have the leisure in the forum and in business'.[146] However, although the institutions of dinner party and festival, and the literature that depicts them, labelled themselves as supplementary, it would be more accurate to say that they *completed* Roman culture. Convivial works are worth taking seriously because they are among the most experimental forms of Roman writing: in them we see the Romans at ease, sharing off-guard confidences, playing at new ways of organizing society or language, or storing up extra repositories of material, all of which challenge the accepted view of what was normally considered worth recording.

4 The Loaded Table

Within this very general framework, the possibilities for manipulating individual foodstuffs were infinite. The chapters that follow may seem to take it for granted that food in Roman literature is always loaded with extra meaning, even when it looks as though it is meant to resist deeper interpretation. The contradiction is often difficult to resolve. On the one hand, the meaning of literary food in general may lie simply in the fact that it is evoking material pleasures, or taking the place of some more obviously 'meaningful' subject; on the other hand, the individual details have always been chosen for a purpose. This purpose may not, however, be apparent on the surface. The literary *convivium*, for example, has the sort of all-inclusive, open-ended quality that makes it look like a representation of things as they are, in an unsifted state. Quintilian, not surprisingly, chooses a

or rules for dinner parties: 'these are the sorts of skits written in the smoky month of December' (*talia luduntur fumoso mense Decembri*).

[146] Gell. 13. 11.

sketch of a *convivium luxuriosum* by Cicero to illustrate the rhetorical technique of *enargeia*, vivid description:

Videbar videre alios intrantis, alios autem exeuntis, quosdam ex vino vacillantis, quosdam hesterna ex potatione oscitantis. Humus erat immunda, lutulenta vino, coronis languidulis et spinis cooperta piscium. (Quint. 8. 3. 66)

I thought I could see people coming in and going out, some wobbling from too much wine, some still unsteady from yesterday's drinking. The floor was filthy, muddy with spilt wine, covered with drooping garlands and fishbones.

His conclusion: 'No one who had been there could have seen more.' Even this description, though, if we look at it more carefully, is ordered and contrived. Cicero's perspective has lurched from the groups at the table to the peripheries of the room, the unsettling mêlée of people coming in and out, the floor littered with puddles and wilting debris. The effect is dizzying and centrifugal; it recreates the drunken squalor and teetering movements of a long feast. In other words, the same text is both realistic and highly selective.

The same ambiguity occurs with individual foodstuffs. As we have seen, these are also quintessentially real and disposable, which is why they have been read literally for so long. Material things that appealed to the senses traditionally posed the greatest challenge to the writer who tried to reproduce them on paper, as the anecdotes about philosophers and wax fruit suggest. Ecphrasis, vivid literary description, had a figurative equivalent in still-life painting. Both presented a challenge to the supremacy of narrative, by presenting matter for art's sake alone.[147] In Philostratus' art-gallery of descriptions, the *Imagines*, where the writing is two stages removed from the original object, a *trompe l'œil* description of a painting of figs begins with the words *καλὸν καὶ συκάσαι*: either 'it is pleasant to pick figs', or 'it is clever to be able to deceive like a sophist'.[148]

[147] On still life, see Blanchard (1981); Bryson (1990); Sterling (1981); Gombrich (1963).

[148] *Imag.* 1. 31. *συκάζειν* has associations with *συκοφαντεῖν* (cf. Ar. *Av.* 1699), and therefore suggests deceiving.

If the art of realism or illusionism stands on the convincing powers of its superfluous details,[149] still life is nothing but superfluous detail, a sentence without a verb, or representation without meaning.[150] We are torn between recognizing its narrative possibilities and accepting the sheer absence of narrative as a statement with its own worth.[151] Philostratus claims that his ecphrases of still life give the pictures a voice, but that voice only puts into words the chronological shifts from nature to culture, raw to cooked, ripe to rotten, that the pictures capture anyway. On a more general level, however, the meagre and rustic *xenia* of Pompeii—a bowl of eggs, a brace of birds—take their meaning from what they exclude. They are not grand or heroic art, and they blot out the corrupt food that the moralists shun. In more senses than one, they are pictures to be lived with; they sustain an illusion of innocent hospitality.

Remarks about the status of still-life painting in antiquity have something to tell us about the status of food-description too. It was certainly accepted that there were figurative alternatives to heroic or narrative painting. Rhyparography (the painting of sordid subjects) had its own paradoxical heights. Pliny tells us that Piraeicus' paintings of *obsonia* and other trivial subjects achieved *summa gloria*, afforded *consummata voluptas*, and fetched higher prices than more prestigious works. This suggests that simple illusions were appreciated for their own sake, and gave an alternative kind of pleasure.[152] Against this, however, we have to weigh the story of Hadrian, who interrupted a discussion of a building scheme for Rome, and had to be told scornfully; 'Go back to your gourd-painting.'[153] Catalogues of food, like still life, are a teasing alternative to the monuments: a two-line epigram

[149] On the importance of detail in realist literature, see Booth (1961); Watt (1976: 9–37, 104–5); on illusionist art, see Bryson (1981: 1–28).

[150] Bryson (1981: 23): 'a sentence without a verb'.

[151] See Blanchard (1981).

[152] *NH* 35. 112.

[153] Dio 69. 4. 2. Even if, as F. E. Brown (1964) suggests, the speaker was in fact referring to Hadrian's plans for domed buildings, to describe architectural drawing in terms of vegetable-sketching is still contemptuous.

by Martial posing as a gift-tag tied to a bunch of sprouts is provocatively mundane and ephemeral.[154]

The pleasure of represented food, however, lies in its connotations as well as in the mimesis, and the aim of this book is to recover some of these connotations. The danger in the past has been to try to treat Roman food as a possible lexicon of symbols. It has been argued, for example, that the significance of Roman food, unlike that of modern American food, is clear from its 'raw' state.[155] That is, a writer only has to mention the bare name of a foodstuff—oysters, say, or cabbage—to make its meaning immediately obvious; each item has its place on a scale as a social marker. As we shall see, however, the meaning of food in Roman literature is rarely just social or economic. Roman writers, like the gourmets they despise, choose from a vast repertoire: the significance of the food they use is determined not just by wider cultural codes but also by personal manipulation; the particular connotations evoked are specific to each context. As we saw in the extract from Petronius, the food there was chosen to match the peculiar phenomenon of a slave's funeral feast or to flirt with eating taboos, rather than to follow any social prescription. So it is hardly surprising that, in the rare cases where classical writers put their own interpretations on the same foodstuffs, these are completely different. Athenaeus, we find, claims that figs in Axionicus' *Lover of Euripides* mean sycophancy, and salt fish mean lewd sexual acts, while Artemidorus (*On Dreams*) interprets figs as deceit or good and bad weather, and salt meat and fish as delay and time-wasting.[156]

The approach to Roman food in its 'raw' state might be adequate for a relatively unadorned representation, like Apicius' cookery-book, but it is not good enough for morally loaded works, where many other important oppositions (between simple and composite, raw and cooked, native and foreign food) are in play, and still less appropriate for the

[154] Mart. 13. 17.

[155] Edmunds (1980), inspired by Barthes (1979), who speaks of food as a 'system of communication' (169). Halliday (1961: 277) sees the meal as a patterned activity by analogy with linguistics, but this is a sterile approach for literary texts.

[156] Ath. 8. 342c; Artem. 73, 71.

special tricks of imaginative fiction. Denatured stews, for example, have a special redolence in Roman literature because of their potential to suggest the hybrid excesses of Rome itself. Seneca makes a dish of expensive seafood the focus of a tirade against the impatience, greed, and confused morals of the Roman people,[157] and Vitellius' favourite dish, the 'Shield of Minerva', is an imperial conquest of the world in miniature.[158] In the culinary system of the eighteenth-century French *Encyclopédie*, stew has two apparently contradictory functions. In the moral sections, it is treated as a negative symptom of an over-civilized society; in the recipe sections, it becomes a completely harmless dish.[159] The two roles are not hard to reconcile, if we realize that only the most introspective person who cooks or eats stew will think twice about its moral connotations. That does not mean, though, that stew is not extremely prone to moral or aesthetic 'loading' in literary representations, given that it is the supreme culinary example of mixture, spicing, mystery, and so on. After all, the word 'stew' comes naturally to our lips when we want a word for any messy situation. Simon Schama wants to see the Dutch national dish, *hutsepot*, as the perfect embodiment of national self-sufficiency, the old oppositions of fast and feast held precariously in balance, the judicious ruminations of Northern humanism. But elsewhere he says that a cheese is sometimes just a cheese.[160] These are the sorts of conflicting impulses that affect any decision to interpret or to leave well alone.

Roman writers always mould food, then, into their own impressions. However, particular images still tend to recur. Certain foodstuffs seem to have a wide appeal because of their organic qualities, especially their anthropomorphic potential. Bloated paunches or guts, for example, are prominent in satire not just because they are the kind of festival food that has now become commonplace, but because they suggest a Saturnalian confusion between grossly over-fed

[157] *Ep*. 95. 26–9. Compare J. Brown (1984: 96 ff.) on the stew as metonym of the 19th c. Parisian ghetto.

[158] Suet. *Vitell*. 13.

[159] See Bonnet (1979).

[160] Schama (1987: 176–7, 162).

food and the gross bodies of its human consumers.[161] Gourds or pumpkins tend to crop up in contexts of lush abundance, or inflated hollowness and inanity.[162] Often, specifically 'convivial' jokes lie behind the choice of food. A list of paltry, mangy foodstuffs, for example Statius' Saturnalian gift-list (*Silv.* 4. 9), is not just perversely banal and unliterary[163]: it also contains all the fun of the festival. For example, it exploits the traditional confusion between man in his most unflattering state and his food: a weak ham, or an onion stripped of its ragged tunic. In other words, the style of the meal itself is often what ought to determine our interpretation of its ingredients: we have to bear in mind the author's genre, tone, the degree of licence he allows his work (when, for example, sensing sexual innuendo), adjectives, irony, parody, puns, the composition of the guests, Saturnalian confusions, even the letters of the alphabet, as these are all aspects of the convivial spirit that informs the whole.

The matter of genre is particularly intriguing. We have already seen that there was a striking correlation between the weight given to food in a literary text and the status of that text. Many of the works discussed here are comic or at least light-hearted, and that is a function of their convivial or festive contexts. Even so, descriptions of food tend to steal language from the spheres of activity furthest removed from the *convivium*—war, triumphs, monumental building, religion, the law, philosophy, astrology, cosmology, and so on—though of course the degree to which these spheres are really distant from the *convivium* is ambiguous and variable. The related pursuit of sexual pleasure is often in the air as well.[164]

161 See below, chs. 2 and 3; cf. Bakhtin (1968: 162 ff.) on tripe.

162 e.g. Hadrian's silly drawings: Dio 69. 5. 2; Columella's pregnant garden vegetables: 10. 379–99; Martial's empty culinary fakes: 11. 31; or Seneca's pumpkinification of a fatuous emperor: *Apocolocyntosis*. Cf. one man's description of the 'sordid' orator Albucius: 'He'll give you a declamation on why thrushes fly, but not pumpkins' (Sen. *Cont.* 7 pref. 8). Coffey (1989: 168) says that the Romans found pumpkins as funny as we find sausages (though, as we shall see, the Romans seem to have found sausages funny too). For a full treatment of the subject of literary gourds, see Normann and Haarberg (1980).

163 Enumeration itself lends a mocking dignity: cf. Lucilius' parody of a rustic dinner, a catalogue of herbs (*enumeratis multis herbis*).

164 See Adams (1982: 138f.); J. Henderson (1975: 47–8, 52, 60–1, 129–30, 142–4, 174).

Like many other 'minor' forms of representation, the literary *convivium* scrutinized other spheres from an alternative perspective; it was compendious in its own way.[165] By nature peaceable, private, and cheerful, the written *convivium* could become a battlefield,[166] a circus,[167] a temple,[168] a stage,[169] a lawcourt,[170] an anatomy theatre,[171] or a morgue.[172]

The 'iconography' of the Roman meal was, in any case, a parody or miniature of larger celebrations. Like the variegated tarts and monumental 'subtleties' of medieval feasts, which exploited heraldic emblems, Roman food was shaped in the images of theatre and military triumph.[173] One host 'had himself wreathed and took part in extravagant banquets, at which he gave toasts in his *toga triumphalis*, while

[165] See Cèbe (1966) on Roman parody in general; see Shero (1929: 60) on mock-heroic food catalogues in Greek comedy and Lucilius.

[166] The paradoxical links between combat and feasting date back at least to Matron's Homeric parody of a meal: Ath. 4. 134d–137c.

[167] On the meal as a circus, see Dupont (1977); Vegetti (1981); Rosati (1983. *Exhibere* is a standard term for serving food (*OLD* s.v.), and the distinction between wild beast and freak shows and exhibitionist feasts—Nero's banquet of marine animals from the ends of the earth (Tac. *Ann.* 15. 37), the tragic actor Aesopus' pie of songbirds that imitated the human voice (Plin. *NH* 10. 141–2), or Commodus' brace of hunchbacks smeared with mustard (*SHA Commodus* 10. 2)—was only one of scale. See Varro, *RR* 1. 12. 11 for a pun on *ova*, the egg-shaped bollards which marked the boundary-posts in the circus, and Mart. 1. 43. 7 for *meta* used of a cone-shaped cheese. Severus Alexander (*SHA* 34. 8) refused to eat in public because it was like eating in a theatre or a circus.

[168] E.g. Macr. 3. 16. 8: a fish served up like a god, with troops of attendants and a fanfare of flutes.

[169] See Rosati (1983).

[170] See e.g. Vespa, *Anth. Lat.* 1. 1. 190 (*Iudicium coci et pistoris*) and Milazzo (1982). See below Ch. 3 on Hor. *Sat.* 2. 8.

[171] Amm. Marc. 28. 34 describes Romans who rush to the kitchen screaming like a flock of peacocks, standing on tiptoe, gnawing their fingers in suspense, while others crowd over the nauseous piles of meat as intently as students in an anatomy lesson.

[172] Sen. *Cont.* 9, a rhetorical exercise on the scandal of Flaminius, who executed a criminal at a dinner party, contains many aphorisms based on the differences between the forum and the dining-room, death and the feast. See Herzog (1989) on death at Petronius' feast; Plut. *Mor.* 994e–995d on similarities between spicing meat and embalming. Cf. Dio 67. 9 on Domitian's macabre funeral-feast cum mystery-tour: the senators were summoned at dead of night to a room hung with black, with tombstones and funeral food at each place, and the emperor holding forth on ghoulish themes. This meal seems to have been restaged in 18th-c. France by the gourmet Grimot de la Reynière, a complicated act of revenge on his parents: see Wheaton (1983: 227).

[173] Medieval food: see Henisch (1976: 103, 163).

cleverly contrived Victories hovered above and extended golden trophies and garlands to him.'[174] The word *pompa* is often used satirically of the processions of dishes to more humble tables, a metaphor helped by the double meaning of the word *ferculum*, which was both a dish and, originally, a platform to carry spoils in a triumphal parade.[175] Other words used to describe food—*coronatus* (garlanded), *miniatus* (reddened), *laureatus* (laurel-wreathed), *ornatus* (adorned)—all invest it with triumphal imagery.[176]

Conversely, food was seen as a contaminating influence on the grand institutions of Rome. The favourite puns of convivial literature, on *sapiens* (wise, tasty, or tasteful) and *ius* (sauce, gravy; or the law, what is right), will come up again and again in the course of this book. They are often no more than superficial jokes. But they also suggest, in a comic form, the same sense of indignation at the central place cooking had acquired in Roman culture (with gastronomy absurdly codified and material values intruding into the lawcourts, temples, and philosophical schools) that moralists condemn more seriously. The invertibility of cooking and these more dignified spheres may be paradoxical, but it is still a vital part of many representations of food.[177]

174 Plut. *Sert.* 22. 2.

175 *Pompa*: e.g. Varro, *RR* 1. 2. 11; and see Rosati (1983); *ferculum* as a platform: Cic. *Off.* 1. 6. 131. Cf. Plut. *Mor.* 200e for an anecdote about a hubristic citizen who served up a honey-cake in the shape of Carthage and invited his guests to plunder it (ἁρπάζειν, 'to plunder', or, 'to reach out for food'), which exploits the fact that models of cities were frequently among the *fercula* in triumphal processions. Silius Italicus specifically names Carthage among these models (17. 635). Cf. also Gallienus, who built castles out of fruit (*SHA Gallienus* 16: *de pomis castella composuit*); and the parodic tomb of the baker Eurysaces, which has a frieze of bread-making scenes parodying military exploits, with circular loaves in place of rosettes: see Bianchi Bandinelli (1970: pl. 73, 164). Cf. also Polyb. 6. 25. 7: embossed Roman shields look like sacrificial cakes.

176 Petr. *Sat.* 66. 2: *porcum botulo coronatum* (cf. Mart. 10. 48. 11); Cic. *Fam.* 9. 16. 8: *polypum miniati Iovis similem*, of an octopus smeared with red like a triumphal Jupiter (see Versnel (1970: 57, 59, 78–82) on the associations of *minium* with triumphs); *Acta Fratrum Arvalium: panes laureatos*; Ath. 14.647e: κάτιλλος ὀρνᾶτος, a well-known Roman dish.

177 Sen. *Ep.* 95. 23: the philosophical schools have been deserted for the restaurants; C. Titius ap. Macr. *Sat.* 3. 16. 15–16: drunken gourmets witness in court; Suet. *Aug.* 70: Augustus' guests dress up as gods when there is a national famine.

Why was food so often used to parody the vocabularies of these other spheres? And why was it funny? The obvious reason is the discrepancy between lumpish matter and the dignified language of important institutions. But another explanation may lie in the same concepts of dirt and pollution that were used to uphold or criticize many of the other categories of Roman culture. Our word 'mess' comes to us from food: it originally meant a serving of food or a course of a meal; then, via homogeneous dishes of mashed food (like Esau's mess of pottage), came to mean any form of confusion.[178] The classic definition of dirt or mess is 'matter out of place'.[179] Just as food becomes mess when it is dropped on the floor or thrown away, it is also incongruously material or disgusting when it is smeared physically on to the pages of a book, or finds its way verbally into writing. Martial uses an old cliché about using poems to wrap take-away food before perversely filling his own paper with mottoes for edible gifts.[180] A famous mosaic of antiquity depicts an unswept floor spattered with cherries, nutshells, crusts, and chicken-bones, the same sort of debris Cicero records in his description of the squalid feast.[181] Food that had touched the ground was taboo for the Romans: like the food in the mosaic, it could not be picked up.[182] This is the kind of context in which we ought to look at food in Roman literature: as an artful and often comic illusion, which is also 'matter out of place'.

5 *A Diet of Words*

There is one extra area of analogy that is of special concern to writers and will be given corresponding weight in this book:

[178] Like English 'stew' or Italian 'pasticcio'.

[179] See Douglas (1970: 12, 48, 189). The definition goes back at least to Jeremy Bentham.

[180] Mart. 13. 1. Occasionally, edible substitutes for paper were used: see Isid. *Orig*. 6. 12 for large-format books, histories, etc., written on elephants' tripes and mallow or palm leaves (Isidorus quotes from Cinna's poem written on mallow, *levis in aridulo malvae descripta libello*, a suitably delicate medium for an Alexandrian work); Plin. *NH* 7. 20. 85 for the *Iliad* written in a nutshell.

[181] See Plin. *NH* 36. 60. 184; a version by Heraclitus after Sosus of Pergamum is in the Vatican Museum.

[182] Deonna and Renard (1961).

the analogy between food and literature itself. Literary production, as we have seen, presented itself to varying degrees as an ever-fresh substitute for perishable matter. Yet the vocabulary of literary style, like that of mental sensation, took many of its metaphors from the bodily experience it rivalled.[183] This is a dimension of literary food that most interpretations of food in literature ignore,[184] even though the structural similarities between meals and poems have been recognized.

As an ordered form created out of selected, transformed ingredients and offered to others, the meal has a great deal in common with literary composition. The Greeks and Romans could describe the whole process of creating, presenting, and consuming a literary text in alimentary terms. Writers often characterized themselves as cooks or caterers serving feasts of words.[185] The style of a literary text could be described in the language of tastes—sweet, bitter, or salty. Attic purity, for example, was the sharp scent of thyme; Cato's bluntness was like pine-nuts; Seneca's pungent aphorisms were like febrile plums.[186] The principles of cooking, too, were transferred to literature, above all proportion and variety: for example, Quintilian recommends the minor poets for those who are sated with the classics.[187] Body and ornament within a text could be seen as parallels for staples and seasonings in cooking.[188] Literature could be tasted, sampled, or

[183] Bramble (1974: 45–59) has the most comprehensive survey; I give only a few representative examples here. See also Van Hook (1905: 28 ff.), Assfahl (1932: 26 ff.).

[184] Though see now the introductions to Tobin (1985); J. Brown (1987); and Bevan (1988).

[185] Bramble (1974: 50): e.g. Ar. *Eq.* 538–9; Astydamas 4N2 (= *Com. Adesp.* 1330K); Teleclides 39K; Metagenes 14K; Plaut. *Poen.* 6–10; Mart. 9. 81; Agathias, *Anth. Pal.* 4. 3; Hor. *Ep.* 2. 2. 58–64. On the philosophical 'feast of words' see Bramble (1974: 51): e.g. Plato, *Rep.* 9. 571d; Cic. *Top.* 5. 25; *de Div.* 1. 29. 61; and add Plato, *Phdr.* 227b; Varro, *Men.* 144 (Buecheler). Cf. Plato's comparison between cookery and rhetoric (*Gorg.* 465d).

[186] See Bramble (1974: 50); Van Hook (1900: 28 ff.). Salt: Cic. *Orat.* 26; *de Or.* 1. 34. 159; Quint. 6. 3. 18–19. Honey: see Rocca (1979). Attic thyme: Quint. 12. 10. 25; pine-nuts and plums: Fronto 2. 102 (155N).

[187] Quint. 10. 1. 58: *quod in cenis grandibus saepe facimus ut, cum optimis satiati sumus, varietas tamen nobis ex vilioribus grata sit.*

[188] e.g. Ar. *Av.* 462 ff.; *Eq.* 215 ff.; *Thesm.* 162; Socrates ap. Stob. 34. 18; Diog. Laert. 4. 18–19; Dio Chryst. 18. 13; Mart. 7. 25; Julian, *Or.* 7. 207b.

devoured.[189] Education was a process of nourishment and assimilation: Seneca forgets his prejudice against mixed dishes when he speaks of the need to digest and absorb a whole library of books.[190] This analogy between food and literature became a European tradition. To give just three examples: Fielding prefaces *Tom Jones* with a 'bill of fare'; Proust compares *A la recherche du temps perdu* to a favourite childhood dish, *bœuf à la mode*, rich pieces of meat compacted in jelly;[191] and *Finnegans Wake* may be a pun on 'Hooligan's cake', a mythical Irish confection so crammed with ingredients that it was inedible.[192]

These sorts of comparisons are made explicit so often in Latin literature that we can reasonably assume that, even if a Roman poet offers a meal to his readers *without* making the analogy directly, this is still a metaphor or programme for the literary composition that contains it. After all, imaginative writing was still the main forum in Rome for aesthetic debate. And we can see how, in reverse, the 'typical' meals of tragedy and comedy were somehow absorbed into the biographies of ancient writers and actors: the tragic actor Claudius Aesopus was said to have staged a quasi-cannibalistic banquet out of birds that imitated the human voice, as though he were dining on the tongues of men; while the mime-writer Publius, we are told, pigged himself on sow's belly at every meal.[193]

While other disciplines have been more flexible in seeing the connections between bodily and literary consumption,[194] this kind of interpretation is still controversial among classicists. However, pioneers of this approach have included Mette (1961), who first saw the connection between Horace's

[189] Bramble (1974: 50): e.g. Plaut. *Most.* 1063; Quint. 12. 2. 4; 10. 1. 104; 4. 1. 14; Sen. *Ep.* 46. 1; Gell. 5. 16. 5.

[190] e.g. Quint. 10. 5. 14; 10. 1. 19; 10. 1. 31. Sen. *Ep.* 84.

[191] Proust, *A la recherche du temps perdu* (Paris, 1961), iii. 1035; quoted by Clarke (1975: 40).

[192] See Hodgart and Worthington (1959: 85).

[193] Plin. *NH* 10. 141–2; 8. 209.

[194] Trimpi (1962: 187) reads the invitation poems of Ben Jonson and his classical predecessors as statements of literary resources; Tucker (1984) sees alimentary functions as a parallel for the creative process in James Joyce; and Jeanneret (1988) considers Renaissance convivial works as lavish feasts of words in the classical tradition.

slender means, the *mensa tenuis*, and his slender style, the *genus tenue*, in the *Odes*; Bramble (1974), whose catalogue of stylistic metaphors (which includes food and eating) laid the foundations for a complete reassessment of the relations between style and content in Latin literature; and Race (1978), who recognized that classical invitation poems were expressions of the writer's limited but congenial poetic resources.

This is the kind of approach that will be used to shed new light on many of the texts in this book, but with some important reservations. I shall take it for granted that, when food appears in Roman literature, it always has some connection with the style of the work to which it belongs. In some cases, it is simply that the dignity of the work dictates the kind of language or kind of food that can be used: highly flavoured or messy food only appears in works that are willing to leave that kind of taste in the reader's mouth. In other cases, there are more specific connections: the cooking of a dish or the laying of a table is often a concealed description of the work itself. There is certainly enough evidence from the rhetorical tradition and ancient literary theory to suggest that this kind of interpretation is not subjective or fanciful: there was as much disputing about tastes in Roman literature as there was in gastronomy.

There are reasons, however, for thinking that the analogy between food and literature is not a simple one. For a start, this analogy was not just a self-contained literary trope among writers who were indifferent to the outside world. The texts that deal with food need to be seen as contributions to the wider issue of Roman consumption, an issue whose difficulties put their own significant constraints on the analogy between food and writing.

In particular, the metaphor of bodily consumption I discussed earlier also stretched to literary aesthetics.[195] When ancient poets deign to include details about their slender material resources or their meagre physique and diet, these are metaphors for their own slender consumption

[195] See Van Hook (1905: 18 f); Bramble (1974: 35–8); Assfahl (1932: 4 f.).

of literary space or material, as well as polemical complaints about their lack of wealth or substance.[196] Food could be used as a device for measuring scale, proportion, and taste; books themselves could be seen as consumers of words, fat or slender. Most of the writers I deal with here, Horace, Catullus, Persius, and Martial, subscribe in some degree to 'Callimachean' principles (though Juvenal consciously pits himself against them). They use the stylistic metaphors *pinguis* and *tenuis* to revive the Hellenistic antagonism between Callimachus' slender Muse (*Μοῦσαν λεπταλέην*) and his rival Antimachus' *Lydian Woman*, which Callimachus called a 'fat book' (*παχὺ γράμμα*).[197] When such writers apply these adjectives to bodies or to food, they are using them to refer self-consciously to the quality of the text itself.[198] This aesthetic of slimness was not simply a revival of a rarefied system of literary allusions. In Rome, it also came to represent the stand of the discriminating individual against the tide of materialism and conspicuous consumption. Alternatively, one could say that these poets justified their aesthetic tastes by appealing to the physical images which informed broader moral issues: above all, the unfair distribution of resources, and the solidarity between individuals whose bodily or textual 'consumption' did not mirror that of the body politic. While aesthetics was largely a private or exclusive concern, it took its rhetoric from wider cultural disputes.[199]

Critics who do allow that the literary meal is an aesthetic figure have tended to regard the comparison as uncomplicated. Bramble is one of the most confident: 'It emerges quite clearly that literature was frequently conceived as some kind

[196] Hor. *Ep.* 1. 20. 5: *[liber] nutritus* ('a book nourished'); Mart. 10. 59. 1: *consumpta lemmate* ('pages eaten up'); see Ch. 3 below on Juv. 1. 4–6 and 86 (*farrago libelli*: 'the fodder of my book').

[197] Call. *Aet.* prol. 24; fr. 398 Pf.

[198] See Clausen (1964: 194–5) on *tenuis* as a programmatic image at Virg. *Ecl.* 1. 2, 6. 8. See also Bramble (1974: 56–8) on fat (*pinguis, turgidus, opimus, adipes*) and thin (*gracilis, tenuis*) as literary terms, 156–64 on physical images for insubstantial poetry versus its bulky enemies. Parker (1987: 8–35) discusses fat women as images for rambling or expansiveness in later European literature.

[199] Bramble (1974: 54–5): Persius (5. 1–6) fuses literary and moral polemic in his attack on the gluttonous swallower of turgid epic.

of foodstuff, to be seasoned before consumption.'[200] However, as we saw earlier, literature was also the antithesis of material 'stuff'. Strictly speaking, a writer had more in common with the singer or reader at a meal than with the cook. A poem on the subject of a meal might be given in exchange for a real meal from a patron, but the gifts are contrasting rather than equivalent, like 'feasts' of table-talk which distract attention from real food. And the act of 'consuming' a work was never a perfect parallel for real eating, especially, in fact, if the work was about food. Indeed, writers are very often concerned that material display was a rival for rhetoric. Martial says of one pretentious host, 'It's your dinner that's eloquent, not you'; while Cicero jokes about exchanging rhetorical lessons for cookery lessons (punning on *ius*, 'sauce' or 'the law') with his pupils Hirtius and Dolabella.[201] A writer could invite his readers to a generous banquet while being fully aware that this was an insubstantial substitute for real largesse. And even the least complicated occurrence of food in literature, a simple list of foodstuffs, may actually be ironic, intended to show us that the essence of a meal or friendship cannot be summed up in a recipe.[202]

The comparison has its limits anyway.[203] Plato's Gorgias compares cooking and rhetoric on the grounds that they are both arts of deceit.[204] For the Romans to compare their writing, a route to immortality which bypassed triumphs or monuments, to food, which embodied their materialist society, was deliberately to underestimate it. Athenaeus, who openly compares his vast miscellany of learning to the lavish feast on the table, is perhaps the only food-writer to take such a positive attitude to the material abundance of Rome.[205] As we saw, though, he is more interested in words than in food. Plautus is also generous in his comparisons, but he makes his own excuses, as the next chapter will show.

[200] Bramble (1974: 54).

[201] Mart. 6. 48. 2; Cic. *Fam*. 9. 18. 3.

[202] See below Ch. 4, Introduction.

[203] See Douglas (1975: 261).

[204] Plato, *Gorg*. 465d.

[205] Ath. 1. 1.

Once again, the hierarchy of genres seems to be decisive: the lower down the literary scale we go, the less inappropriate the comparison seems to be, or the less it seems to matter. In the *Odes*, for example, Horace's slender style can only be represented by ethereal wine and light rustic vegetables;[206] in the *Satires*, he portrays himself as a much more sophisticated and dubious cook.[207] Satire, which takes its name from a fat woman, a mixed dish, or a sausage, will be given the most ample space in this book because it is a special bearer of meaning in Roman culture: it is uniquely unrestrained in exposing the real links between food, the mixed literary text, and the mixed, expanded city of Rome. The works that are most fragmented or parodic are the ones that tend to make the parallel between text and food most prominent. The grotesquely hybrid dishes of Petronius' *Cena Trimalchionis*, for example, remind us that the book itself is a bogus pastiche, as well as the society it depicts. And Smollett's Roman banquet, with its Roman 'pickles and confections' and assembly of picaresque guests, has more than an incidental link with the title *Peregrine Pickle*.

6 *Salad Dressing*

Before turning to detailed discussions, let us look briefly at one small text where food is used as both a real and a metaphorical subject, and also as a marker of the literary status of the text which contains it. The *Moretum* is an anonymous poem of the first century AD, which describes nothing more than a peasant baking bread and flavouring a cheese with herbs. It is usually read simply as a realistic picture of the hardships of rural life.[208] However, we could regard the poem's emphasis on poverty and tininess[209] as a Callimachean metaphor which suggests the exiguous scale

[206] Griffin (1985: 82): wine, not food, is Horace's sustenance in the *Odes*. There is in fact food, but only of a light kind, unlikely to weigh down the poetic spirit: *Od.* 1. 31. 15–16: olives, chicory, and light mallows. See Mette (1961).

[207] See Ch. 2, sec. 3, on Hor. *Sat.* 2. 4.

[208] See Kenney (1984: xix).

[209] *Ibid.* 14.

of the poem and its creation out of limited resources. The central line of the poem,[210] which describes a garden, epitomizes the general impression of variety within a small compass: *exiguus spatio, variis sed fertilis herbis* (62, 'small in space, but flourishing with a variety of plants'). The flavoured cheese, the *moretum* of the title, is another internal image of variety: *tot variatur ab herbis* (104, 'it was variegated with all those herbs'). The most frequently used words in the poem, *opus* (work), *manus* (hand) and *orbis* (round shape), reinforce parallels between culinary and poetic creation: the poem ends when the food is made.[211] The round cheese (*orbis*) is also a universe in miniature: Virgilian and Lucretian cosmogonic imagery (42–7) inform its creation.[212]

Microcosm and macrocosm, cooking and poetry, are all contained within the small compass of the poem. The phrase *e pluribus unum* (102), which describes the *moretum*, suggests the eccentric mixture of common and poetic diction and different literary scraps which the poet stirs together.[213] It is a phrase that reappears on US banknotes, implying that a

[210] Allowing 35a; other lacunae have been suggested.

[211] See Johnson (1985) on Ronsard's *Salade tourangelle*, a description of a salad which he made while on a country picnic (the poem may well have been an imitation of the *Moretum*, though Johnson does not discuss this). He argues that the salad is an internal metaphor for the poem itself, a fresh and varied creation from diverse literary reminiscences: 'Il devient clair . . . que le poète, sans jamais rendre explicite sa comparaison, investit d'un pouvoir métaphorique considérable la salade.' Sydney Smith's well-known recipe for a salad has less claim to be such a self-conscious production, as we know from his letters that it was simply a dressed-up version of a recipe he was passing on to a friend; though the way this dressing jars with the trifling theme is typical of the mock-important air of many ancient food epigrams.

[212] The ancestor of the *moretum* is clearly, in concept if not genetically, the κυκεών offered to Demeter, a mixed drink of grain, wine, milk, honey or oil, and herbs. See Richardson (1979: 344–8). Cf. Hom. *Il.* 11. 624–41; *Od.* 10. 234 ff., 316 ff. It is cited by Tert. *Adv. Val.* 12 (*Nestoris Coccetum*), as a miscellany, and is used as an image by Lucian, *Vit. Auct.* 14 (the contradictions of life), *Icar.* 17 (a panorama); Diog. Laert. 10. 8: Epicurus calls Heraclitus κυκητής, 'muddler'. Heraclitus (DK fr. 125; cf. Plato, *Tim.* 41d on the Creator's mixing bowl) uses the drink as a simile for the cosmos, which stays together only because it is perpetually stirred. Plutarch (*Mor.* 511c) describes him drinking a κυκεών to illustrate the virtues of concord and simplicity (presumably because it was a 'self-sufficient' dish, a complete meal of different foodstuffs, grains, herbs, and milk, rather than, as Richardson (1979: 344) suggests, because it was a symbol of frugality). On the self-sufficient dish, see Douglas (1975: 258).

[213] Cf. Montaigne, *Essais* 1. 46 (tr. Florio): 'What diversitie soever there be in herbs, all are stuffed together under the name of a salad.'

united America is made of diverse states,[214] and it could equally describe Rome.[215] It would be hard in this case, of course, to make a direct link between the mixed poem, the mixed cheese, and the heterogeneous city, which the author in fact puts at a distance from his rural farm (at 81 Simulus visits the urban market). But in other ways the poem is a typically motley and trivialized product of Roman culture.[216] Cooking, not poetry, is the obvious activity of the poem; the poet is disguised as a shaggy peasant; the cheese is a ridiculous parody of the cosmos.[217] The poem's material surface limits it to the status of a miniature.

Any interpretation of food in Roman writing must always take account of its ambiguous cultural status, as this has a great deal to do with its significance as a literary subject. Even so, this grotesque and often trivial-seeming perspective can be one of the most illuminating from which to approach the Romans' attitudes to their bodily functions, their material wealth, and their mixed and messy civilization.

In the chapters that follow, I have chosen four literary genres where food is prominent: comedy, satire, epigram, and iambics. Chapter 1 considers Plautus' comedies as public celebrations of Roman culture at its most heterogeneous, a mixture of native frugality and foreign excess. Chapter 2 concentrates on satire: the culinary or bodily etymology of *satura* shows a clear link between mixed food and mixed literature, which must make us think again about the role of food within this verse-form. Horace tries to create satire that is paradoxically slim and neat, Persius boils its flavour into a concentrate, while Juvenal fulfils its true capacity. Since Trimalchio has been amply exposed in recent

[214] See Kenney (1984) *ad loc.* A similar phrase occurs in a convivial riddle to which the answer is *conditum*, a mixed drink of three different flavours: *Anth. Lat.* 1. 1. 281. 82: 'one is made of three, and three are mixed in one' (*ex tribus est unus, et tres miscentur in uno*: irreverent in a Christian era?).

[215] The French word for fruit salad, *macédoine*, was coined by analogy with Philip's Macedonian empire, a harmonious collection of potentially warring states.

[216] The name of the hero, Simulus, 'snub-nosed', suggests a link with the satyric figures who parodied serious culture in Greek satyr-plays and in Roman triumphal processions. The name of his African servant, Scybale, 'dung', 'scrap', 'refuse', unloads Roman cultural inferiority on a subordinate nation.

[217] The reverse of Heraclitus' famous comparison.

years, he will be mentioned only in passing, even if a book on Roman food without him may seem like *Hamlet* without the prince.[218] Chapter 3 considers three very different authors' versions of the dinner invitation, and argues that the food they contain is a good indication of the ratio of ethereal and material things in each author's work as a whole. In Chapter 4, a single foodstuff, garlic, the food of Horace's iambic rage in *Epode* 3, is used to illustrate the complexity of literary food metaphors.

[218] Arrowsmith (1966) and Dupont (1977) have brilliantly demonstrated the part played by food in complementing decayed rhetoric and culture at the *Cena Trimalchionis*.

2

Barbarian Spinach and Roman Bacon

THE COMEDIES OF PLAUTUS

1 *Introduction*

Latin literature begins in bulk with Plautus, and many later writers look back to his world as a Land of Cockaigne, where cooks, parasites, and slaves wallowed among mounds of tripes, hams, and delectable sweets and shellfish. Food, above all else, seems to preserve the savour of Plautus for later generations. Fronto, for example, conjures up a menu out of Plautus' 'ancient crammings' (*veterum saginarum*), capons, seafood, cakes, and wine, to tempt the puritan Marcus Aurelius. Apuleius' own opulent style is made flesh in cloying lists of sausages and calamitously sticky sweetmeats straight out of comic recipes. And Gellius thinks some verses on a parasite's appetite are genuine because they have the authentic 'smack' of Plautus (*propterea resipiant Plautinum stilum*).[1]

Plautus' carnival world may seem comfortably cut off from reality, but his *floreat* coincides with a period which, for later historians at least, was decisive for Rome's development into an imperialist power. In the middle of it is that hallowed date 187 BC, which Livy marked as the beginning of all Roman sin, including the arrival of cooks in Rome and a marked change in attitude to cuisine:

[1] Fronto 2. 224N; Apul. *Met.* e.g. 1. 4. 1: *polentae caseatae* (cheesecake); 2. 7. 2: *viscum fartim concisum et pulpam frustatim consectam* (mincemeat); 10. 13. 6: *ille porcorum, pullorum, piscium et cuiusce modi pulmentorum largissimas reliquias, hic panes, crustula, lucunculos, hamos, laterculos et plura scitamenta mellita* ('One brought me generous leftovers of pork, chicken, fish, and so on, the other bread, pastries, and other honeyed sweetmeats'); Gell. 3. 3. 13.

Epulae quoque ipsae et cura et sumptu maiore apparari coeptae. tum coquus, vilissimum antiquis mancipium et aestimatione et usu, in pretio esse, et quod ministerium fuerat, ars haberi coepta. (Livy 39. 6. 7–9)

Banquets became more elaborate and extravagant. And it was then that the cook, who had formerly had the status of the lowest kind of slave, first acquired prestige, and what had once been a service-industry came to be thought of as an art.

Whether or not this date was really significant, the period was certainly one of increased self-consciousness for Rome. One curious consequence of Hellenization had been to force the Romans to construct a sense of their own national identity independent of the Greeks, whose influence was seen as alternately civilizing or corrupting. And it is this 'identity crisis' that lies behind many of Plautus' jokes. Food in his plays is not only part of the riotous world of comedy, but is also a focus for many of the Romans' anxieties about their whole culture: how to separate it from the Greeks', how to be tough and civilized at the same time. A basic example of the cultural transformation and labelling of natural products, food is used by Plautus as a boundary-marker, and it is a symptom of his times that he is constantly shifting those boundaries: the words 'barbarian' and 'Roman' in the title of this chapter refer to one and the same people.

The notion of boundaries is also helpful for explaining why food has always been an integral part of comedy. Overeating was not just an obvious feature of the festivals in which comedy originated. As I suggested earlier, food may be there because of its potential to become 'mess'.[2] Smeared onto the dignified pages of drama and inflated by the cooks and parasites who replace messengers and seers, it helps to transform tragedy into the comedy which parodies it. Comedy, in short, is filled with 'matter out of place'.

Mess, matter, and festivals all loom large in Plautus' work. His *palliatae*, comedies in Greek dress, are themselves dislodged boundary-crossers, expanded versions of lost Greek originals, whose 'pure' Greek setting is openly compromised by allusions to Roman life. This chapter explores

[2] See Ch. 1, sect. 4, above.

the ways in which Plautus' comedy, and in particular his presentation of food, exposes the unmarked boundaries of contemporary Rome and defines the author himself as a product of a mixed civilization. Food not only identifies the different cultures which Rome indeterminately contains and excludes, Greek and Roman, festival and everyday. It also reinforces the sheer messiness of Plautine comedy, with all its excesses and inconsistencies. The assorted flavours that waft from the plays—porridge, pepper, or pig's meat—give us tastes of a diverse and cosmopolitan culture which only the most hybrid form of representation, furthest removed from 'pure' Greek tragedy, can hope to describe.

As for the figure of the cook, it will be argued that he appears so often in Plautus, and in comedy in general, because he is a parallel for, or parody of, the comic author. Recent criticism has transformed Plautus from a bungling amateur into a sophisticated and self-conscious playwright, who is constantly pressing on us analogies between the internal actions of his plays—invention, imbroglio, muck-raking—and his own acts of creation.[3] If we recognize this self-conscious dimension, which has come to be known as 'meta-theatre', we can begin to understand the metaphorical potential of food and cooking in Plautus' plays. This is not to say that Plautus' use of the cook-figure is simply a private act of self-reference: it is contemporary Roman ambivalence towards the new cuisine, a civilizing but tainting phenomenon, that gives a sharper edge to his pose as the shoddy or dubious entertainer who contaminates pure culture.

Another modern influence on my interpretation will be the Bakhtinian concept of 'festive' literature: works seen as part of the riotous, ephemeral context of the festivals during which they were performed.[4] Plautus' work lends itelf naturally to this kind of perspective, as Segal (1987) has shown.[5] Several of Plautus' contemporaries named their plays after festivals,[6] and in his own *Cornicula* there is an explicit

[3] See Barchiesi (1970); Knapp (1979); Slater (1985).

[4] Bakhtin (1968: esp. 198–276) on the relations between popular festive culture and 'festive' literature.

[5] Using Barber (1959) on Shakespeare as a model.

[6] e.g. Laberius' *Saturnalia* and *Compitalia*, Pomponius' *Kalendae Martiae*.

parallel made between *ludi* on the stage and the wider context of the festival: *quid cessamus ludos facere? circus noster ecce adest* ('Why should we stop playing games: look, we have our theatre here'). But there are significant distortions involved in Plautus' 'Greek-style' Roman holidays, and these will need some attention later.[7]

2 *The Mess of Pottage*

Plautus' comic perspective is of course idiosyncratic as well as of its time, and later it will be necessary to bring on to the scene the perfect straight man, Cato the Elder. Chalk and cheese perhaps, but together they show the *range* of responses, however different, to the same cultural conditions, and in particular the difficult problem of Greek influence.

Plautus helps us by being very self-conscious about his role as the Roman adapter of Greek plays. He presents himself as a clod-hopping provincial mangling a Greek text into the barbarian tongue: *Plautus vortit barbare*, as he sometimes puts it.[8] *Barbarus* is of course the classic *Greek* term for defining all foreign races, and Plautus is exploiting an irony offered by his Greek context: to portray the Romans as others see them, others in this case being specifically the Greeks, from their vantage point of cultural superiority. Again and again, 'barbarian' in Plautus means primitive, incompetent, and cack-handed. This might be just an ironic joke from a position of indifference or even conscious superiority, but it has the potential as well for exposing the Romans' underlying lack of confidence in their own mastery of the world. They may have had military and material dominance, but they were still culturally wanting.[9]

The jollity of Plautus' plays is often disturbed by spiteful side-references to barbarian incompetence. For example, in the *Mostellaria* (The Haunted House), a well-built pair of pillars is contrasted with the jerry-building of barbarian

[7] Plaut. fr. I.

[8] e.g. *Trin.* 19: *Philemo scripsit, Plautus vortit barbare*; *As* 11: *Demophilus scripsit, Maccus vortit barbare.*

[9] *Vortere*, I will argue later, also has negative associations.

workmen: *non enim haec pultiphagus opifex opera fecit barbarus* (828, 'No barbarian porridge-eating craftsman did this job'). In the context, this looks like a parallel with Plautus' own craftsmanship,[10] especially since, at *Poenulus* 54, Plautus seems to be labelling himself specifically as *Plautus . . . pultiphagonides* ('Plautus of the race of porridge-eaters').[11] Indeed, it has been argued that the name Maccius Plautus, which may mean something like 'flat-footed son of a pottage-eating clown', is so improbably suitable for a comic writer that it is actually a pseudonym.[12] Maccus is a stock clown from Atellan farce, while *maccum* is found in a late glossary translating Greek κοκκολάχανον, a mixture of grain and vegetables similar to pottage; Plautus, literally 'flat-footed', is a synonym of *planipes*, a common expression for a mime-actor. T. Maccius Plautus, then, suggests some clod-hopping, mash-eating barbarian, a typically Saturnalian travesty of Roman nomenclature.[13]

This explanation seems quite plausible, in the light of the ludicrous gastronomic names given to the supporters of the fictional *Lex Tappula*, which parodies the form of a sumptuary law—Tappo, Multivorus, Properocibus, and Mero[14]—and the name of the pig who makes his will in another Saturnalian text, the *Testamentum Porcelli*—M. Grunnius Corocotta.[15] It also throws into doubt the authenticity of another comic writer's name, Fabius Dossennus, which also implies connections between protuberant bodies, humble food, and low comedy. Dossennus was a stock character of

[10] *Opus* can mean both 'architectural structure' and 'literary work', especially ambiguous here when the play is called 'Haunted House'.

[11] There is a textual crux here, which Leo resolves by printing a lacuna between 53 and 54, to read: *Carchedonius vocatur haec comedia* † . . . † *latine Plautus patruus pultiphagonides* (the gap would include an antithesis between *graece* and *latine* of the kind usual in Plautine prologues). Gratwick (1973) rejects Lindsay's *latine Plautus 'Patruos' Pultiphagonides* because *vocat* cannot be supplied from *vocatur*, and the hyperbaton is unnatural anyway.

[12] Gratwick (1973: 78–84).

[13] Ibid. 82: 'The absurd contrast between the portentous shape and the clownish associations of the *tria nomina* would have been in tune with the Saturnalian world of the *ludi* at which everyday values were stood on their heads.'

[14] Buecheler, 266–7; see also Cèbe (1966: 269).

[15] Buecheler, 268–9.

Atellan farce, possibly a hunchback;[16] the *faba mimus*, or bean farce, was a notorious mime (probably on the theme of the apotheosis of Romulus).[17] In any case, the words *patruus pultiphagonides*, 'the porridge-eating uncle', at *Poen*. 54, can be explained as a paraphrase of Maccius Plautus himself, rather than of his play, *The Little Carthaginian*.[18]

Plautus' name, then, whether by chance or design, seems to cement the connections between native farce and lumpish Roman food. Any objection that there is no real connection between *Maccus* and *maccum*, 'clown' and 'pap',[19] disappears if we only consider the age-old links between clownishness, physical protuberance, and gluttony. Atellan characters had hunched backs, large jaws or fat cheeks, few brains, and greedy appetites.[20] Plautus' protuberant slaves and parasites are the ancestors of the Renaissance grotesque bodies Bakhtin connects specifically with festive literature.[21] Their lumpishness traditionally betokens stupidity: food inside the body seems to remain food, bulging through the skin in the form of unthinking ballast.[22] Besides, there may also be linguistic links between the Atellan Maccus and Maison, the earliest Greek cook-figure, depicted with a large head and a fat stomach.[23] The chances are, then, that Maccus' name

[16] See Beare (1955: 129).

[17] Cic. *Att*. 1. 16. 13. Sen. *Apoc*. 9. 3. For an interesting later parallel, see Henisch (1984: 18) on the English 'Twelfth cake', which contained a bean that turned the person who found it into a Saturnalian king.

[18] Gratwick (1973: 78): *pultiphagonides* is actually a more appropriate name for a Roman than for a Carthaginian (Cato's recipe for *puls Punica* at *Ag*. 85, cited by Leo, suggests that *puls Punica* was a variation on what was normally a Roman dish). This makes Copley's suggestion (1970), that *pultiphagonides* is a makeshift pun on *Carch-edonius*, combining the Aramaic *qarh/karh*, 'pea, chick-pea', with *edere*, 'to eat', look far-fetched.

[19] See Beare (1955: 129).

[20] See Hor. *Ep*. 2. 1. 73 on Dossennus; Fest. s.v. 128 and Varro, *LL* 7. 95 on Manducus; Bucco is from *buccae* (Plaut. *Bacch*. 1088 uses the word to mean 'fool'). Compare 'buffoon' and '*opera buffa*', from Italian *buffare*, to puff out the cheeks.

[21] Bakhtin (1968: 302–5). E.g. *Curc*. 231: *conlativo ventre* ('pot-bellied'); *Merc*. 638–9: *ventriosum, bucculentum* ('pot-bellied, fat-cheeked'); cf. *Rud*. 317; *Pseud*. 1218. See Cèbe (1966: 119).

[22] *Crassus* and *acutus* describe mental as well as physical qualities: *OLD* s.vv.

[23] See Giannini (1960: 140), who connects *mattus* (= *fatuus*) and *maccum* (– It. *macco*, bean porridge), and compares It. *matto* (stupid) and the words *polentone* and *maccherone* (fool), which link stupidity with 'padding' food.

implied that he was stuffed with *maccum*, and lacked mental acumen, *sapor*.[24]

Puls, or pottage, was in any case part of the mythology of Roman beginnings, offered in rituals celebrating the origins of Rome and the birthdays of its citizens: in other words already a quaintly anachronistic way of labelling Plautus' contemporaries.[25] The term *pultiphagonides* smacks of tribes from the *Odyssey*, where Homer, as Vidal-Naquet (1981a) has taught us, uses diet to cement the differences between the Greeks and the rest of the world: cannibals, lotus-eaters, or Phaeacians, against the civilized, corn-growing, bread-eating Greeks. Of course the Romans, as Plautus portrays them, are not cannibals or living off trees. What humbles them in the eyes of the Greek observers is their humdrum agricultural diet of pottage: baby food or peasant food. Here, then, is a very different picture of 'barbarism' from the kind that held sway in Greece, barbarians who were cruel, tyrannical, and luxurious.[26] Though foreign, Plautus' Romans are not exotic or picturesque primitives. They are quintessentially bland, bovine, and flavourless. That is the picture that emerges from all other references to barbarians in Plautus. A sideways glance in *Poenulus*, for example, takes in Roman cattle-mash, as though that typified the natives' diet as well: *macerato hoc pingues fiunt auro in barbaria boves* (598, 'This is the mash they use to fatten the cattle in the land of the barbarians'). *Macerare*, which means 'to soak in water', sounds as though it has connections with Plautus' name Maccus (especially as *maccum* is recorded as the name of a mashed cereal dish).[27] Elsewhere in Plautus it is used as a

[24] Cf. *Pseud.* 737–42, where this double meaning of *sapere* is exploited; see Onians (1954: 61–5) on the belief that intelligence was related to bodily juice (*sapor*).

[25] Plin. *NH* 18. 19. 83–4: *pulte autem, non pane, vixisse longo tempore Romanos manifestum . . . et hodie sacra prisca atque natalium pulte fritilla conficiuntur; videturque tam puls ignota Graeciae quam Italiae polenta* ('The Romans clearly lived for a long time on porridge, not bread . . . and today ancient rites and birthdays are celebrated with gruel-porridge; it seems that porridge is as unknown in Greece as polenta is in Italy'); Varr. *LL.* 5. 105: *de victu antiquissima puls* ('porridge is the most ancient of foods'); cf. Juv. *Sat.* 14. 120–1; Val Max. 2. 5. 5.

[26] See Hall (1989: 99–100).

[27] Despite differences in vowel-length (cf. the pun at *Amph.* 723 on *malum*, 'apple' or 'evil'). See Ahl (1985: 54–60) on other puns which ignore vowel length.

metaphor for mental confusion, being marinated with doubts or inner turmoil, being 'in a stew'.[28] These verbal or metaphorical links between mental confusion, cookery (and rustic barbarian mashes in particular), and comic confusion or messiness in general will be explored in more detail later.

3 *Barbarian Spinach*

As for the barbarians' diet, Plautus conjures up harsh, unenjoyable food, *asper victus* (*Capt.* 79), pepped up perhaps with a few pungent and antisocial condiments. The commonest of these is garlic, the food of Roman galley-slaves and rustic peasants.[29] Another of the Roman staples, mentioned in *Casina*, is ***barbaricum bliteum***, 'barbarian trash', the 'barbarian spinach' of my title. The word is related to ***blitum***, 'spinach', a food which later grammarians consistently define as the essence of blandness, and etymologize from the Greek word *βλάξ*, meaning stupid.[30] The overriding impression, then, supposedly in the eyes of supercilious Greeks, is of an incompetent, insipid cuisine, typical of the under-civilized society of Rome. What Plautus is exposing is perhaps not so much Roman complacency of the kind that eighteenth-century English commentators showed when they compared their robust and unfussy English roasts with fancy French concoctions,[31] but rather the real cultural anxiety that underlay the Romans' lordly attitudes to their own provincial inferiors, their own 'barbarians'.

[28] e.g. at *Trin.* 225, where it is joined with *coquere*. See Fantham (1972: 59); and West (1977: 65–8) on Hor. *Od.* 1. 13. 8: *quam lentis . . . macerer ignibus* ('I am boiled alive on a slow fire').

[29] *Most.* 39, 48; *Poen.* 1314; *Pseud.* 814.

[30] Paul. Fest. 34: *blitum genus holeris a saporis stupore appellatum esse a Graeco putatur, quod ab his βλαξ dicatur stultus* ('spinach-beet as a kind of vegetable is thought to take its name from Greece, where they use the word *blax* to mean "stupid" '); here, as in the case of Maccus and *maccum*, there is a link being made between plants without flavour—*sapor*, lit. 'juice'—and human beings without intelligence—also *sapor*). Cf. Plin. *NH* 20. 252; Isid. 17. 10. 15; Varro, *Men.* 163; Non. 550.

[31] Mennell (1985: 126) quotes Addison (*Tatler* 148, 21 Mar. 1709) on seeing a sirloin moved to a side-table. 'I . . . could not see, without some indignation, that substantial *English* dish banished in so ignominious a manner, to make way for *French* kickshaws.'

Here, then, is a good example of what anthropologists call 'displaced abjection'. Plautus' joke at the Romans' expense is only tolerable because it is based on Roman stereotypes about people who are rougher, more inept, more tasteless than the Romans themselves. Diet once again plays its part as a marker of difference. In *Rudens*, the fishermen of Cyrene supply urban tables with fish and send silphium, or devil's dung, to Italy, but their own lives are harsh and desolate.[32] Nearer to Rome, in a fragment of Naevius, good cooking is thought to be pearls before swine for the municipal Praenestines and Lanuvians, who only appreciate sow's belly and nuts.[33] Similarly, Plautus' Periplectomenus boasts about his elegant table manners: he is an Ephesian, not a spitting, dribbling Apulian.[34] In later Roman depictions of barbarians, too, the barbarians are always differentiated by their gargantuan, uncontrolled appetites, the comparative simplicity of their diet (a positive, utopian characteristic in Tacitus' *Germania*), and finally by their lack of etiquette (for example the Burgundians who belch fumes of garlic at breakfast in one of Sidonius' poems).[35] This picture of modern barbarian life is also interchangeable with myths of the Romans' own rugged past. Varro, for example, in one of his satires, mentions the Romans' forefathers, whose breath stank of onion and garlic, but who were, even so, men of spirit (*bene animati*, an untranslatable pun on 'spirit' and 'breath').[36]

With the name Maccius and the epithet *pultiphagonides*, Plautus makes himself the representative of this backward and under-civilized race, which survives on primitive

[32] *Rud.* 630.

[33] Ribbeck 2. 9 f. = Macr. *Sat.* 3. 18. 6: *Quis heri / apud te? Praenestini et Lanuvini hospites. / Suopte utrosque decuit acceptos cibo, / altris inanem volvulam madidam dari, / altris nuces in proclivi profundier* ('Who was at your house yesterday?' 'Some guests from Praeneste and Lanuvium.' 'They should each have been welcomed with their own particular food: a dripping farrowed sow's womb for the Lanuvians, and an apronful of nuts for the Praenestines'). See Ramage (1973: 33) on this and the Praenestines' other weak points: their oafish boasting and their old-fashioned turn of speech.

[34] *Mil.* 653.

[35] See Montanari (1988: 19–21); Tac. *Germ.* 23; Sidon. *Carm.* 12. 14; Ramage (1973: 74–6, 116–17, 124–5).

[36] Varro *Men.* 63.

mashes scarcely distinguishable from cattle-food and peppered with noisome flavourings. Another area where food embodies worthlessness and lowers Plautus' own self-esteem in the process can be found in the range of words he uses for 'rubbish'. *Nugae, frit, naucus*, and *ciccum* are all linked by ancient and modern etymologists with humble foodstuffs: pumpkin-seed, corn-ear, bean-seed, and pomegranate husk respectively.[37] In *Stichus*, a banquet of slaves, which is in itself a Saturnalian theme (though here the point is that such behaviour is everyday in Greece), consists of nuts, beans, small figs, lupins, and pastry:

> SAG. hoc conviviumst
> pro opibus nostris sati' commodule nucibus, fabulis, ficulis,
> oleae †intripillo† lupillo, comminuto crustulo. (689 ff.)

SAG. This dinner is the neatest we can stretch to on our incomes: nuts, little beans, tiny figs, oil . . . lupins, crumbs of pastry.

This food is meant to be covering a Greek table, but it would have had a variety of conflicting associations for a Roman audience. Paltry, insubstantial, and low, it was also a hallowed symbol of the most ancient festivals of Rome, in particular the Floralia, the Saturnalia, and the Kalends of December.[38] The two sorts of associations, trifling and festive, can be reconciled if we recognize that these are both references to the simple diet of the Golden Age. Nuts, beans, and so on, were both the 'popcorn' of the circus and the commemorative token of Rome's origins: a double 'souvenir'. In *Bacchides*, a slave threatens to roast his master like a chick-pea, and puncture him with more holes than a shrew's

[37] *Nugae* (*OLD* s.v.): = nonsense, *Amph.* 604; *Aul.* 830; *Curc.* 452; = rubbish, *Most.* 1088; *Per.* 718; *Curc.* 199; = con-man's talk, *Pseud.* 1204; *Trin.* 856; possibly related to Italian *nogina*, pumpkin-seed: see Ernout-Meillet s.v. *Naucus: Bacch.* 1102; *Most.* 1041; *Truc.* 611. *Ciccum: Rud.* 580; as bean-seed: Festus, 166. *Frit: Most.* 595; *ne frit* (*nec erit* codd.) *quidem*; as the tip of a corn ear: Varro, *RR* 1. 48. 3. *Ciccum* as the inner husk of a pomegranate: Varro, *LL* 7. 91. Festus defines *hilum*, another rubbish word, as the skin of a bean (101).

[38] Lupins as farm fodder: Varro, *RR* 2. 1. 17. Lupins scattered at festivals: see schol. on Hor. *Sat.* 2. 3. 182 (*cicere . . . faba . . . lupinis*): the kind of food the aediles scattered to the crowd at the Floralia. Nuts and the Saturnalia: Mart. 5. 30. 8: *Saturnalicius nuces*. Figs and the Kalends of December: Statius *Silv.* 1. 6. 15. *Crustulum* as the standard currency of imperial largesse: e.g. *CIL* 11. 3613; 14. 3581.

gut, *soricina nenia*. This word *nenia*, later at least, comes to mean 'rubbish', a bit like our words 'tripe', 'garbage', and 'baloney'.[39]

Once again, Plautus' metaphors of worthlessness and the frugal celebrations where they become real are intimately connected with his own identity as an author: he aligns himself with all the trivial and worthless aspects of the festivals at which his plays were performed. This gesture is typical of so-called 'Saturnalian' or 'festive' writing, which tends to be infused with a sense, however ironic, of its own worthlessness. The preface to Ausonius' *Cento Nuptialis* (a patchwork of august Virgilian lines naughtily rearranged to suggest the events of a wedding night), shows the survival of the links between festive works, cheap Saturnalian puppets (*sigillaria*), and the old comedians' words for trifling things: *Pro quo [centone], si per sigillaria in auctione veniret, neque Afranius naucum daret, neque ciccum suum Plautus offerret* ('If this patchwork quilt came up for sale in an auction, Afranius would not give his pip for it, nor Plautus his husk'). These connections are made explicit in the prologue to *Poenulus*, where Plautus identifies the play itself with the worthless snacks of the festival theatre. He apologizes for offering no real nourishment to the hungry audience except for the comedy itself: *qui non edistis, saturi fite fabulis* (8, 'If you haven't eaten, you'll have to fill up on the play'). After a long stretch of puns on real and spiritual nourishment (*essurientes, saturi, impransum, sapientius*), we cannot quite tell whether *fabulis* means the play alone: it might also be the little beans that we saw on the slaves' table, which were also common as theatre snacks.[40]

4 The Spice of Life

So much for Romans as frugal barbarians and Plautus as their representative. Obviously the twisted perspective, the

[39] *Bacch.* 767, 889. See Ch. Appendix on *nenia*.

[40] Scullard (1981: 11): at the Floralia, vetches, beans, and lupins were scattered among the crowd. *Fabulus*, with a short *a*, is recorded as a metaphor for a pellet of dung in Lucilius: see *OLD* s.v. The pun survives more explicitly at Apul. *Met.* 1. 26, where the hero goes to bed hungry but filled with stories: *cenatus solis fabulis*. NB also the so-called *faba mimus*, or 'bean mime': see above, sect. 2.

ironic point of view, are the marks of a concealed sophistication, which is borne out by the very fact that Romans are demanding exotic plays with a Greek setting. There is also another joke to be got out of the foreign context. The Greek characters are often found describing their own licentious behaviour as 'Greek-style' morality: they use the verb *pergraecari* of their own nasty foreign habits, which include delicate, exquisite meals of pigeon, fish, and poultry, or seafood with Greek names. In *Casina*, the kind of food contrasted with barbarian spinach is dainty limpets and other types of shellfish; a parasite in *Captivi* dreams of a feast not to be found in harsh and barbaric Italian cities; two slaves in *Mostellaria* bicker about the virtues of exotic perfumes and farmyard dung, fish and poultry versus rustic garlic-flavoured dishes (*aliatum*).[41]

What is peculiar is that these scenes, so far as we can tell, are probably Plautus' own invention, and not part of the Greek source. Or at least their meaning changes dramatically when they are transferred to a Roman context. For example, the scene between the two bickering slaves may have originally been a town–country debate when and if it appeared in Plautus' Greek source, but under Roman eyes it is transformed into a cultural debate about degrees of acceptance or rejection of Greek-style manners. There has been a topographical shift westwards in the layout of the moral universe: the Greeks have themselves taken over the role of luxurious Orientals that they gave to the Persians.[42]

If that is not enough, the whole problem of distortion has further complications. Plautus' audience may have appreciated all the ironies of his perspective, but it was artificial, even so, to unload immorality and luxurious living on to Greece, and leave the Romans with only cattle food and stinking garlic. Relief from puritan living had always been incorporated into Roman life, whether in the structure of the year, in the form of festivals, or every other day, in the form of the relaxing dinner. In other words, the traditional

[41] *Cas.* 493–4: 'baby cuttlefish, limpets, little squid, barley-fish' (*sepiolas, lepadas, lolliguncułas, / hordeias*); *Capt.* 851: 'salt fish, mackerel, sting-ray, and porpoise' (*horaeum, scombrum et trygonum et cetum*); cf. *Most.* 23–4, 39–48, 64–5.

[42] See Hall (1989: 99–100).

dichotomy between simple and luxurious food is to some extent a contrived one: there was actually room for both in the Romans' diet, in the see-saw of feasting and fasting that shaped the calendar. So Plautus is making extraneous, banishing to a distant 'elsewhere', the areas of pleasure which already had their place in Rome, and which were already making Roman civilization complete. By the time he was writing, the most luxurious food was not necessarily Greek or even foreign, and food with Greek names could be the lowest of the low at Rome. Witness a string of insults in *Poenulus* (1314 ff.), a catalogue of stinking trash, where garlic-stinking Roman galley-slaves, Rome in its humblest form, are jumbled up with Greek-named broken olives and flayed sardines:

> deglupta maena, sarrapis sementium,
> manstruca, halagora, sampsa, tum autem plenior
> ali ulpicique quam Romani remiges.

A flayed sardine . . . [unknown foodstuffs] . . . a dish more stuffed with garlic than a whole galley of Roman oarsmen.

Another example, though admittedly a century later, comes from Cicero's letters to Paetus, where he contrasts a luxurious meal of mushrooms and octopus with a plebeian cheese and anchovy dish called, *à la grecque, tyrotarichum.*[43]

Conversely, any empire which adopts the food of its subjects has already made that food its own. Plautus is also famous for his mouth-watering catalogues of dinner-party food. For every sneer at barbarian churlishness, there is also a luscious description of a feast; the audience is either invited to participate or regretfully excluded. This is the other side of festival eating: the blow-out, a celebration of excessive consumption. It is an area where Plautus seems more at home. As Segal points out, his language, his 'wanton waste of words' and 'inexcusable verbal extravagance', is itself prodigal.[44] Plautus' expansion or amplification of the material can be especially sensed in the drooling speeches of his parasites, where he coins absurd names or resists the natural

[43] Cic. *Fam.* 9. 16. 4; cf. *Att.* 14. 16.

[44] Preface to 1st edn. of *Roman Laughter* (1968).

limits of his audience's appetites in a breathless and unstomachable imagined swallowing of food.[45] The word *scitamenta*, baubles or titbits, which Lucilius applies to verbal ornaments, such as assonance, alliteration, and rhyme,[46] is used by Plautus of edible dainties (*Men.* 209 ff.):

> . . . aliquid scitamentorum de foro opsonarier
> glandionidam suillam, laridum pernonidam,
> aut sincipitamenta porcina aut aliquid ad eum modum.

I'll go and buy something scrumptious from the market: Sir Pigling Sweetbread, Lord Hogg, Temple Swinehead, or something of the kind.

But the passage is full of Plautus' own embellishments. Crammed into it are the silly staccato of *scitamenta* and *sincipitamenta* (a monumentalized pig's head) and the nonsense rhyme of two mock-heroic patronymics, *glandionidam* and *pernonidam*.

Again, as far as we can tell, this highly 'spiced' style is Plautus' own rather than the Greek original's. In the one case where it is now possible to compare Plautus with his Greek source, *Bacchides* versus Menander's *Dis Exapaton*, it is clear that Plautus exchanged bland stereotypical names for more colourful and humorous ones, and added more jokes. In his study of the two, Handley concludes: 'We have seen, on a small scale but by direct observation, how Plautus likes his colours stronger, his staging more obvious, his comedy more comic.'[47] Plautus does not, in other words, Romanize his comedy in the direction of blandness. Another good example comes from *Curculio*, with its satire on Greek philosophers (*Graeci palliati*), those old windbags who creep out of snack-bars (*thermopolia*), stuffed not only with book-learning but also with picnic-baskets (*suffarcinati cum libris, cum sportulis*), letting off polenta-flavoured farts (*crepitum*

[45] Compare the Greek party-game in Athenaeus (14. 647d), where the guests had to contribute in turn to a verse-list on a chosen subject: if the subject in question was food (food and types of dessert are among the subjects listed), the conflict between the guests' desire not to exhaust the theme and their imagined fullness, as the list expanded *ad nauseam*, must have made the game doubly preposterous.

[46] Ap. Gell. 18. 8. 1.

[47] Handley (1968: 18).

polentarium) as they go.[48] Polenta, which *we* know as an Italian food, was then the Greek equivalent of *puls*, another ancient staple.[49] So the reference to humdrum windy food looks like a plausibly *Greek* joke, a way of demystifying philosophy for the common Greek philistine. And in any case, the word *thermopolium* seems to give the game away: this ought to be part of the Greek source. However, Fraenkel points out that *thermopolium* is not extant in classical Greek, and was probably originally a southern Italian institution. In other words, the whole passage seems to have been dreamed up by Plautus himself.[50]

It is clear, then, that the overall 'flavour' of Plautus' plays is much more peppered than all his own dull labelling would suggest. The presence of exotic spice-names alone suggests that the Romans' outlook was already cosmopolitan. In *Casina*, for example, the heroine, whose name means 'cinnamon', provides the elusive spice in a goatish old man's life, and we find more oppositions between oriental perfumes and farmyard oafishness.[51] The old man, Lysidamus, plasters himself with exotic scents (*exotica unguenta*), and his slave Olympio orders delicate shellfish for Casina as an alternative to barbarian spinach: both suggest the Romans' clumsy reachings towards cosmopolitan *savoir-faire*. Lysidamus' paean to love (217 f.) extols it in culinary terms as the quintessential spice (*condimentum*, 220) of life, the transforming element which turns bile to honey (223), makes things *salsum* or *suave* ('savoury' or 'sweet', 222). In his case, the pursuit of exotic tastes ends in their disastrous opposite: a bride in drag with a goatish smell and a vegetable-like phallus (*num radix? . . . num cucumis?*, 911).

Rudens, by contrast, takes us to Cyrene, where the stinking crop silphium, or devil's dung, destined for Roman tables,

[48] *Curc.* 288 f. See Petrone (1983: 170–2).

[49] Plin. *NH* 18. 19. 84.

[50] Fraenkel (1960: 149).

[51] Another character is named Myrrhina, after myrrh. Puelma (1988: 22–7) proposes *fragranti nomine* ('with its sweet-smelling name'), a reference to the title, instead of *latranti nomine* ('with its barking name') at *Cas.* 34: *Plautus [scripsit hanc comediam] cum fragranti nomine*. See also Detienne (1977) on the contrast in Greek culture between sex-evoking perfumes and spices and the sexual suppressants, lettuce and garlic.

complements the bitter life of the fishermen.[52] In *Pseudolus*, a slave's fertile brain is compared to a well-stocked *pantopolium*, ready with grape-syrup, honey, vinegar, or smelly goats, for any eventuality.[53] This word *pantopolium* (a grand bazaar) turns out to be a much better metaphor than a dish of porridge for describing Plautus' Rome. It is as though the need for external flavouring was already an essential part of Roman culture. But Plautus, as we saw, introduces *Poenulus* as nothing better than popcorn. And he ends the play by completing the metaphor of blandness, telling the audience to provide their own relish (*condimentum*): their applause.[54] One Roman cultural product, drama, is taking its metaphors from another, Rome's native cuisine, which needs external spicing to make it palatable. But this view of Rome and its playwright as insular and inadequate is of course given the lie by the play itself, which has a Greek plot and even includes a monologue in Punic.

One explanation of this blurring and distorting of national boundaries is that Plautus is devising his own dichotomy between crude 'barbarian' food and luxurious 'Greek' food, one which did not exist, or fully exist, in the original Greek texts, but which instead highlights problems of identity or 'split identity' within Roman culture itself. Neither of the two dietary poles alone represents Rome. In fact, both are traditionally contrasted with 'normal' Roman food: one belongs to the distant and primitive past (or the festivals which commemorated it), the countryside, or the peripheries of the empire; the other is part of the corrupting influence of over-civilized and effete peoples 'encroaching' on Rome from the East. On the other hand, however marginal they are, these are both, perversely, ways of representing the Romans' national identity. 'Greek' influence was by now so firmly entrenched in Roman culture, or indeed was already *equivalent* to Roman culture, that it was artificial to mark it off as foreign, or to ascribe the excesses of festivals to Greek luxury.

[52] *Rud.* 630.
[53] *Pseud.* 742.
[54] *Poen.* 1370.

Plautus' own Greek elements are, similarly, hard to separate from his Roman ones. For a start, his language, like all Roman colloquial speech from early times, is liberally sprinkled with Greek words. And the nowhere world inhabited by his self-aware nasty foreign Greeks and his far-away barbarian Romans has a shaky nonsensical topography: a 'map' of Epidaurus in *Curculio*, for example, sounds like a description of Rome, complete with localized satire on the tradesmen of a particular district.[55] Plautus' inconsistencies may look like a blatant example of barbarian incompetence. From another point of view, however, his representation of the world is less contrived and, even, more accurate than his successor Terence's perfectly manicured dramas, which only highlight the contradictions of 'pure' Roman drama that is also faithful to its source. Plautus shows Rome as a fantastical cosmopolis, hybrid and adulterated, with all its inconsistencies intact. He escapes by using an alias (the personae of slaves, cooks, and parasites), an alibi (Greece), and a special licence (carnival), but his other world is really Rome when it's at home. This double perspective on Rome, first the pure Rome contaminated by Greek culture, then Rome as a dirty outsider from the Greek point of view, adds up to a more appropriately 'contaminated' representation.

5 *Roman Bacon*

These boundary-marking notions of 'purity' and 'contamination' are not just my metaphors: they are also part of the rhetoric of Cato the Censor, the perfect foil for Plautus' comic voice. As we saw earlier, Plutarch praised Cato for his moral steadfastness at a time when Rome had lost its purity (*τὸ καθαρόν*).[56] In other words, Cato's integrity was already something of an anomaly in cosmopolitan Rome, and that was what elevated it to mythical status. Cato was prominent in the backlash against cooks and foreign cuisine, and even against obesity.[57] His account of his own military-style diet

[55] *Curc.* 470–83. Cf. the reference at *Capt.* 489 to oil-dealers in Rome.
[56] *Cato Maior* 4. 2.
[57] *Carm. Mor.* Jordan fr.2 = Gell. 11. 2. 5; Malcovati, 96.

(he says that he ate meat only in order to be fit to serve the state) set an example to the state as a whole, which was once a lean and fit military machine, now a greedy and questionably sprawling 'belly without ears'.[58]

A famous letter from Cato to his son gives us a different perspective on the problem of who contaminated whom. The last straw, he says, will be the corrupting influence of Greek literature; the Greeks already call us barbarians, and stain us even more filthily than they do other people by calling us philistines (Opici, or Oscans): *nos quoque dictitant barbaros et spurcius nos quam alios opicon appellatione foedant.* The Opici were, later at least, famous for their ignorance and oafishness.[59] Here we have another example, then, of displaced abjection, a Roman asserting his cultural supremacy or purity at the expense of a provincial Italian tribe. Rome is doubly stained, once by Greek culture itself, once by the Greeks' view of Roman culture.

How different Cato's idea of contamination is from that of another writer, who defines it, at least, in a strikingly similar way: *'contaminare' proprie est manibus luto plenis aliquid attingere . . . 'contaminari' tangi . . . polluta manu ac per hoc velut foedari aut maculari.* ('To contaminate' technically means 'to touch something with muddy hands' . . . 'To be contaminated' means 'to be touched with filthy hands and thus be stained or besmirched'). These are of course the scholiast Donatus' glosses on the word that Terence uses in several of his prologues, while strenuously rebutting the charge of having 'contaminated' Greek plays for the Roman stage.[60] In a literary context, 'contamination' seems to mean amalgamating two Greek originals to make one Roman play.[61] But there is no reason to suppose that the word was

[58] *De Sumptu suo* = Malcovati fr. 173–5; Plut. *Mor.* 198d.

[59] e.g. Juv. 3. 205.

[60] *And.* 15–16: *id isti vituperant factum atque in eo disputant / contaminari non decere fabulas* ('His critics attack him for doing this, and argue that two plays should not be mixed together into one'). See Beare (1955: 300–3). Cf. *Heaut.* 17–18: *Nam quod rumores distulerunt malevoli / multas contaminasse Graecas* ('As for the malicious rumours which have torn him apart, that he has mixed together several Greek plays . . .').

[61] Donatus further glosses *contaminare* at *Heaut.* 17: *id est ex multis unam facere* ('That is, to make one play out of many').

ever a technical term, and scholars are increasingly doubtful whether *contaminatio* was ever so tightly defined. As one has written: 'If *contaminatio* as a technical term is to be used in future, it should be redefined to denote all those ways in which a Roman playwright might "mess about" with his model.'[62] The concept of 'messing about' is a useful one, as it both implies that the original is being distorted, and makes the activities of the Roman playwright seem trivial or inept. Terence bends over backwards to refute this charge. Plautus, on the other hand, as we have seen, makes no attempt to defend either his literary actions or his national identity. In fact, he positively exploits ways in which Roman culture mixes Greek originals or makes a parodic mess of them. He allows his Roman elements to corrupt and be corrupted by the Greek setting; by calling the Romans *barbari*, Plautus presents his own civilization as 'matter out of place'.

Now of course the main paradox in all Cato's brave assertions is that Roman culture or civilization had only really come into being by first absorbing Greece. Cato's personal ideology was itself probably shaped by Greek concepts of purity. One of his books, *de Agricultura*, provides a good example of these contradictions. In many ways, it is clearly an ideologically loaded tract, carrying the mythology of *Romanitas* in embryo: the harsh lifestyle, the unliterary log-book form, the traditional rural calendar, and so on. But the book is just as much a child of its time as Plautus' plays, and it is often the most 'Roman' moments in the book that are most sophisticated. It is odd, for example, that Cato's ingredients for rural sacrifices, if anything the supremely Roman ritual, include recipes for cakes with Greek names (*placenta* and *spaerita*) and Carthaginian pottage (*puls Punica*).[63] It looks as though Cato the Greek-hater and Cato the author of the slogan *delenda est Carthago* is being compromised by his own list into admitting that Rome has absorbed the outside world. And what about Cato's hymn of praise to the cabbage, another famously wholesome Roman

[62] Gratwick (1982: 117).

[63] *De Agr.* 76, *placenta* (= Gk. πλακοῦς); 82, *spaerita* (= Gk. σφαιρίτης); 85, *puls Punica*.

passage?[64] It turns out to be largely lifted from Greek medical sources.[65] The end of the book, it seems, is where Cato most solidly identifies himself with salty Romanness. Despite the rough-and-ready lack of a proper finish, it can be no accident that the last, redolent image of the book is of a side of smoked bacon dangling from Roman country rafters. And who better than someone called Marcus *Porcius* Cato to shoulder the weight of Roman mythology?[66]

With Plautus, bacon is an area where, not surprisingly, he goes the whole hog. Tripes, sweetbreads, trotters, heads, hams, and so on, strings of pork names dangling out of reach of the parasites who dream of them: these are a constant feature of Plautus' feast menus. Pork was a prominent part of the Roman carnivore's diet and it was rarely eaten in Greece, so scholars of Plautus can put lines through these passages with some relief: here really is proof of Plautus' own invention, a sop for the Roman audience with no regard for local authenticity.[67]

In fact, the appearance of pork adds more complications to the contrasts Plautus draws between Greek luxury and barbarian frugality. To go back to *Captivi*, for example, we find a typical contrast being drawn between a parasite's ideal meal and the harsh fodder (*asper victus*) offered by barbarian (that is, Italian) cities. The imagined menu includes fish and other seafood, in other words typically Greek food.[68] But the parasite also has visions of pork: *laridum* (847), *porcinam* (849), *pern[ul]am* (850). From the point of view of a Greek, the pork is simply an anomaly. From the point of view of a Roman, however, the imagined menu makes perfect sense: a composite or eclectic mixture of native and foreign food. In

[64] *De Agr.* 56.

[65] See Astin (1978: 162).

[66] Romans had a special place in their affections for families with homely-sounding names: e.g. Fabii, 'beans', Cicero, 'chick-pea', Lactucae, 'lettuces' (see Plin. *NH* 19. 19. 59; and cf. Varro, *RR* 2. 4. 1 on the name Scrofa, 'Sow'). Or were they clinging on to them to clear their consciences?

[67] See Lowe (1985*a*: 76–9) on pork; and cf. Plut. *Cato Minor* 46. 3 for Cato's presents for theatre-goers: beets, lettuces, radishes, and pears for the Greeks, wine-jars, pork, figs, cucumbers, and bundles of wood for the Romans.

[68] *Capt.* 851: *horaeum, scombrum et trygonum et cetum* ('salt-fish, mackerel, sting-ray, and porpoise').

other words, pleasure for a Roman audience is already much more cosmopolitan than simple spinach and pottage, or even pork on its own. And it is interesting that Rome does not figure among the list of harsh Italian cities: it is already too much of a cosmopolis to be credible as a barbarian outpost.

What is the significance, then, of all these pork dishes? Segal's ingenious suggestion, which supports his view of Plautus' comedy as 'festive' literature, is that they are specifically flouting the current sumptuary laws—a naughty shared joke between excluded audience and guzzling actors.[69] There is, as he suggests, a striking similarity between any pork list in Plautus, for example *Curc.* 323 (*pernam, abdomen, sumen suis, glandium*: ham, belly, sow's udder, sweetbreads) or *Men.* 210–11 (*glandionidam suillam, laridum pernonidam, / aut sincipitamenta porcina*: sweetbreads, bacon, cheeks), and the lists of forbidden pork which, Pliny tells us, filled page after page of the sumptuary laws: *hinc censoriarum paginae, interdictaque cenis abdomina, glandia, testiculi, vulvae, sincipita verrina* (*NH* 8. 77. 209, 'Hence all those pages of sumptuary laws, and the ban on eating paunches, sweetbreads, testicles, wombs, and cheeks made of pork').

The trouble with this explanation is that, according to what we know of the sumptuary laws, it is anachronistic.[70] The only law valid in Plautus' time, the Lex Oppia of 215, dealt exclusively with luxury in clothing. The Lex Orchia, the first to deal with table luxury,[71] did not come into force until 182 BC, after Plautus' death. It is not known which sumptuary laws contained the restrictions on pig-meat mentioned by Pliny, but the Lex Fannia of 161, too late for Plautus, is the first for which there are any details:[72] no poultry could be served on ordinary days, save one unfattened hen, a restriction which became a repeated motif in Roman sumptuary legislation.[73] C. Titius spoke in favour of

[69] Segal (1987: 48).

[70] See Clemente (1981); *RE* s.v. *Sumptus*.

[71] According to Macr. *Sat.* 3. 17. 2: *prima autem omnium de cenis lex ad populum Orchia pervenit.*

[72] See Macrob. 3. 17. 3.

[73] Plin. *NH* 10. 71. 139: *ne quid volucrum poneretur praeter unam gallinam quae non esset altilis*; Tert. *Apol.* 6. 2: *nec amplius quam unam inferri gallinam et eam non saginatam.*

this law, and used as an example of a luxurious dish the so-called *porcus Troianus*, a pig stuffed with other animals like the Trojan Horse.[74] Stuffed dormice were banned in the Lex Aemilia and in the Claudian sumptuary laws.[75] With this sort of chronology, it looks more likely that Plautus influenced the sumptuary laws than vice versa. Or, if we look at it in another way, both were in fact exploiting the same bundle of associations.

What gave crammed meat its peculiar significance? The fact that Cato, a fanatical supporter of the sumptuary laws, includes instructions for stuffing hens, geese, and pigeons in his *de Agricultura* (written a year after the Lex Fannia) suggests that the restrictions on stuffed meat applied only to normal calendar days.[76] An exception would naturally be made for feasts and holidays. The Roman stereotype of the humble peasant with his flitch of salted bacon[77] assumed that, once a year at least, the fatted animal would be killed, perhaps at a festival or wedding which coincided with the start of winter. This, incidentally, helps to explain the particular connections between pigs and the Saturnalia (along with their anthropomorphic features).[78] For example, the so-called *Testamentum Porcelli*, 'Pigswill', the pet text of Roman schoolboys, looks as though it has specific connections with the Saturnalia.[79] A literate pig, M. Grunnius Corocotta, bequeaths various sections of his body to different sections of society, and asks that he may be well treated (*bene condiatis de bonis condimentis nuclei, piperis et mellis* 15), a joke on the similarities between pigs and men, spicy cooking and embalming. The imaginary date of the will is given as *XVI Kal. Lucerninas*; *XVI Kal.* is also how

[74] Macrob. 3. 13. 13: *quasi aliis inclusis animalibus gravidum, ut ille Troianus equus gravidus armatis fuit.*

[75] Plin. NH 8. 223, 36. 4. Cf. Plaut. *Fab. Incert.* fr. 31: *glirium examina.*

[76] Cato, *de Agr.* 89–90.

[77] See e.g. Juv. 11. 82–3. Cf. Hémardinquer (1979) on a similar myth of the 'family pig' under the *ancien régime.*

[78] Petr. *Sat.* 41. 4: a *porcus pilleatus* (wearing a freedman's cap) represents Saturnalian *libertas*. Lucian, *Sat.* 28: the rich glut themselves on pigs and cakes. Mart. 14. 70: a pig as a Saturnalian gift. Gell. 16. 7. 11 notes that *botulus*, sausage, appears in Laberius' mime *Saturnalia*. Cf. other mime titles: 'Hog Sick', 'Hog Well', 'The Pig', 'The Sow' (see Beare (1955: 132–4)).

[79] Buecheler, 268–9.

one would express the first day of the Saturnalia, 17 December.[80]

Extra expenditure for market days, the Saturnalia, weddings, and a few other monthly dates was always accommodated, then, within the official strictures.[81] This meant that the occasional increase in consumption was knitted into the structure even of the humble peasant's year; the fathers of the city did not begrudge the odd blow-out, within limits. If only one pig was slaughtered annually, the pork cuts listed by Pliny and Plautus could be eaten once only. But increased wealth and availability and the growth of market-shopping had turned Rome into a potentially over-consuming society: it became theoretically possible to eat several *sumina*, *altilia*, or *abdomina* in one meal. Pliny's anecdote about the mime-writer Publius illustrates the connections between comic licence and quasi-festival consumption: after getting his freedom, the mime-writer Publius never ate a meal which did not include *abdomen*, from which he acquired the nickname *Sumen*, 'sow's udder'.[82]

Varro contrasts modern luxury with the opprobrium attached in the old days to a peasant who bought bacon at the market instead of producing it on his own plot;[83] later he grumbles that there is a feast every day now in Rome, instead of every year.[84] Mary Douglas (1975) has shown us a parallel set of distinctions in our own society, too: festivals are differentiated by their amplified versions of normal food—the stuffed Christmas turkey, or the three-tier wedding-cake. And a case-study of peasant and noble food in south-west France has illustrated how one foodstuff acquires different meanings according to how often it is consumed: the *confit*, a preserve of goose-meat and fat, was ordinary for the landlord who owned many geese, but precious for the

[80] The link is reinforced by the name *Lucerninas*: *cerei*, wax candles, were a typical Saturnalian gift (see Mart. 5. 18. 2).

[81] The Lex Fannia allowed tenfold expenditure during the Saturnalia and the Roman and plebeian games (Gell. 2. 24. 3).

[82] *NH* 8. 77. 209.

[83] *RR* 2. 43; cf. Juv. 11. 79–81 on the ditcher who scoffs at vegetables and yearns for *vulva* from the cookshop.

[84] Varro, *RR* 3. 2. 16.

tenant whose supply was limited.[85] The same applies to Roman pork: one versatile substance could evoke primitive Roman living or festival excess in proportion to the amount consumed. That is why the pork catalogues in Plautus belong to the world of Carnival rather than to the frugal existence of the 'barbarians'.

It may well be the case, then, that the sumptuary laws were aimed not only at curbing displays of wealth (and, of course, canvassing for political control), but also at reinstating symbolically the traditional distinction between weekday food and amplified festival food which was being blurred by increased prosperity and availability. When moralists use the Saturnalia as an image of social disruption, the point is not that the festival itself is immoral, but that it is being celebrated all the year round. Satirists, too, use the licensed gluttony and reversals of the Saturnalia as an image of what has become ordinary in everyday life.[86]

Indeed, moralists and medical theorists used the same analogies between stuffing eaters and the fattened animals they ate. Pliny uses a fattening metaphor to describe snails bred to 'fill' human greed: *ut cochleae quoque altiles ganeam implerent*. Describing pig-fattening, he speaks of *saties* as the limit of stuffing, a metaphor from human appetite.[87] In Ammianus Marcellinus' words, the Romans themselves are given over to 'distended crammings'.[88] The perfect example of these connected images can be found in Macrobius' comments on luxury at a late-Republican pontifical banquet.[89] Among the dishes he lists are various kinds of fattened poultry: *gallinam altilem, altilia ex farina involuta*,

[85] Valéri (1977).

[86] See ch. 1, sect. 3 above.

[87] Plin. *NH* 9. 57, 8. 209. Cf. Cic. *Carm.* fr. 32. 13 on a fattened liver: *iecore opimo farta et satiata affatim*.

[88] Amm. Marc. 28. 34: *in his plerique distentioribus saginis addicti*. Cf. Sen. *Ep.* 108. 15: *edacibus et se ultra quam capiunt farcientibus* ('greedy and stuffing themselves beyond their capacity'), 119. 14: *quemadmodum non impleat ventrem, sed farciat* ('how to stuff the stomach, not fill it'); Macrob. *Sat.* 7. 5. 14: *fartus cibo stomachum vel ventrem gravatur* ('by stuffing himself with food, he weighs down his stomach or abdomen'); Jerome *Ep.* 38. 5. 2: *pingui aqualiculo farsos homines* ('men stuffed with fat bellies'; *aqualiculus* is normally used of a pig's stomach).

[89] 3. 13. 10–16.

and *altilia assa*. Then follows his conclusion: *ubi iam luxuria tunc accusaretur, quando tot rebus farta fuit cena pontificum?* ('In what context could you condemn luxury in those days, when even a pontifical dinner was stuffed with so many ingredients?') With the word *farta*, a leap has been made from the fattened birds, and battening eaters, to the dinner itself as an image of excessive consumption, stuffed out with an *antecena* worthy of many normal dinners. The train of thought continues with three examples of monstrously stuffed animals: the *porcus Troianus*, or 'Trojan Pig'; fattened hares (*saginarentur*, 14); and fattened snails (*cochleam saginatam*, 15). Later in the *Saturnalia*, Macrobius uses the example of the multiple illnesses of birds crammed for slaughter to support an argument against mixed diets for humans: *quibus, ut altiles fiant, offae conpositae et quibusdam condimentis variae farciuntur* ('birds which are fattened by being stuffed with food flavoured with different seasonings').[90]

But Plautus had already used these analogies for his own comic purposes. In his case, not only was the relationship between the fattened fowl or pig and the human beings who 'cram' themselves one of context: it was also one of metonymy. Plautus commonly confuses human eaters and stuffed or fattened animals, for example at *Most.* 65: *este, ecfercite vos, saginam caedite*, ('Go on, eat, stuff yourselves, kill the fattened calf'); or at *Truc.* 104: *de nostro saepe edunt: quod fartores, faciunt* ('They're always eating out of our resources, like cooks making stuffing'). In one of the fragments, the phrase *comesa* (or *condita*) *farte* ('when stuffing had been eaten') becomes ambiguous.[91] Is this stuffing the contents of a human stomach, or is it culinary stuffing that is then consumed? The fact that the phrase *venter suillus* ('pig's stomach') is used earlier in the fragment increases the ambiguity.[92]

[90] Macr. 7. 4. 4–5. The author of *Mul. Chir.* calls faeces in the gut of a live animal *farciminalis*, 'stuffing', like a mixed sausage that imitated the mixed intestines of live capons: see Adams (1982: 244).

[91] Fr. 23 (29; ex Fest. 333).

[92] Cf. *Most.* 169: *fartim* used of the stuffing, i.e. the body inside a woman's dress; fr. 136 (Lindsay), †*quinas*† *fartas*, again suggests a confusion between stuffed food

Comic confusions, moral ideology, and medical theory: all these helped to keep alive the associations between gluttony and the crammed food most suitable for a glutton. The constant appearance of the *gallina non altilis* and forbidden cuts of pork in the laws of and after 161 bore witness to the power of these associations. Not only did the ban help to redefine the structure of weekday and festival in the Roman year: the limited proportions of the everyday human or animal body, the meagre fowl on the table, also supplied a model for the proper limits of the Roman state's consumption.

Plautus himself does not, of course, make any symbolic attempt to conceal the expansion of the Roman body politic. His treatment of the Greek sources results in an expanded, more noticeably hybrid product. References to satiety or fat stomachs or crammed menus draw attention to the amplified nature of the plays themselves. It would be anachronistic, of course, to draw parallels between the plays as 'farces' (the word originally meant an interlude which provided 'stuffing') and the stuffed gluttons or *altilia* within them. But these blatantly crammed plays do share some of the qualities of *satura*, which, as we shall see, takes its name from a mixed dish which looked like a stuffed human stomach. Embryonic in Plautus are the linguistic confusions between womb and stomach which become more frequent in satire: in the pork-lists, for example, *abdomina, sumina*, and *vulva* reinforce the connection between abundant meat and *saturitas* in the male or female body.[93] And it is a measure of the close relationship between comedy, satire, and the protuberant shapes within

and stuffing eaters; *ecfertum fame* ('stuffed with hunger': *Capt.* 466), is an obvious paradox. *Sagina* is also used of 'fat' army pay at *Trin.* 722. Cf. *Cist.* 305: *dote altili atque opima* ('a fattened-up dowry'); *As.* 282: *maximas opimitates, gaudio exfertissumas* ('plump prosperity, stuffed with joy'). Sider (1978) reads *centones farcias* (with the MSS) instead of Lindsay's *centones sarcias*: 'go stuff someone else's quilts (with your inflated stuffing)'.

[93] Gelasimus (*Stich.* 155–6) claims to have carried 'Mother Hunger' in his stomach all his life, when she only carried him for ten months (see Diaconescu 1980: 61–2). At *Cas.* 777 women are criticized for stuffing their stomachs (out of illicit gluttony rather than acceptable pregnancy): *ventres distendant*. At *Amph.* 677 Plautus plays on the ambiguity of *saturam*, 'stuffed full' or 'pregnant'.

them that we cannot tell whether Naevius' *Satura* was a satire in the style of Ennius, or a play with a pregnant woman as its heroine.[94]

6 *Parasites and Cooks*

We can now turn to those characters in Plautus for whom food is of disproportionate importance: the parasite and the cook. These characters are usually explained simply as bumbling clowns, but their connections with Plautine comedy are more intimate than that.

First, the parasite. Although this figure was associated with Greek-style culture (cf. *pascite parasitos*, 'feed your parasites', *Most.* 23–4), the evidence suggests, as usual, that Plautus radically expanded his role. To hazard a guess, this was really because of the potential the parasite's obsession with food gave for parody of serious vocabularies and institutions, even those Roman ones which so conspicuously spoilt the integrity of the Greek original.[95] In other words, parasites contaminated ordinary dramatic language with food out of place, answering Plautus' need for comic 'mess'.

The law, the army, religion are all swallowed up in the parasite's relentless obsession with eating. At *Capt.* 159–63 troops of Rabelaisian pastry-cooks become a parasite's auxiliaries:[96]

> multis et multigeneribus opus est tibi
> militibus: primumdum opus est Pistorensibus;
> eorum sunt aliquot genera Pistorensium:
> opu' Paniceis est, opu' Placentinis quoque;
> opu' Turdetanis, opu' Ficedulensibus

You'll need a huge spread of an army. First you'll need the Battenburgers, and there are all sorts of Battenburgers: you'll need

[94] Ullman (1914: 22–3) assumes that it refers to a pregnant woman, comparing plays called *Satura* by Atta and Pomponius (he concedes that in the case of Pomponius at least there is probably word-play on *satura* in the sense of *farcimen*); Coffey (1976: 22) compares Novius' *Virgo praegnans*.

[95] Cf. puns on *esse*—to be/to eat—in connection with parasites, e.g. Caecilius 14W (16R3); Aquilius (6R3). See Wright (1974: 107).

[96] Cf. also *Curc.* 444, where a geographical catalogue is contaminated by *Perediam* and *Perbibesiam*.

the Genoans and the Madeirans; you'll need the Frankfurters and the Bolognans . . .

Here the words *multis et multigeneribus* first alert us to some unexpected similarities between a feast and an army.[97] Later in the play (906–7) the same parasite, Ergasilus, puns on *ius* (gravy, or law), and *pendere* (to hang (from a meat-rack) or to await trial), confusing cookery with the law, and implying at the same time that there was little to choose between shady legal processes and the suspicious activity of cooking.[98] The concept of legal *ius* as corruption is also brought out at *Poen.* 586, where advocates are described as *iuris coctiores*, rather than *doctiores*. At *Cist.* 472 a lover's oath, *ius iurandum*, is compared to a mixed sauce, *ius confusicium*, a joke-name for an ordered recipe, and chosen here as a typically dubious dish.[99] Similar suspicions are aroused by the butcher's trade. At *Pseud.* 197, it is compared to pimping: both deal in *malo iure*—a bad deal or shady business, or bad sauce. Religion, too, is appealed to in its lowest forms. At *Capt.* 877 Ergasilus prays to *sancta Saturitas* ('holy Satiety'). At *Cist.* 522 the gods of cooking are at the bottom end of the hierarchy: *di . . . omnes, magni minutique et etiam patellarii* ('All the gods, large, small, and even saucer-gods'). Another device of parasitic inversion is the personification of food.[100] At *Per.*

[97] Cf. Caecilius fr. 11W (13R3): *iamdudum depopulat macellum* (he's been ravaging the market-place'); Plaut. *Capt.* 153: *edendi exercitus* ('an army for eating'); *Pers.* 112: *proelium committere* ('to enter the fray'). See Wright (1974: 106–7; see also Fantham (1972: 110–11) on Plautus' use of incongruous military or legal imagery to describe food. In Greek, πολλὰ καὶ παντοδαπά is a standard phrase for the two salient features of a feast: *copia* and *varietas*. See e.g. Plut. *Lucullus* 40; cf. Hor. *Sat.* 2. 6. 104: *multaque de magna superessent fercula cena* ('there was a huge surplus of dishes from some enormous dinner').

[98] Plato (*Gorgias* 465d) had linked cooking and rhetoric as arts of cookery. *Ius* puns: Lucil. fr. 46(W) (*iura siluri*, at the trial of Lupus, whose name can also mean a fish, the bass; Ahl (1985: 92) translates '*court-bouillon*'); Cic. *Fam.* 9. 18. 3; *Verr.* 2. 1. 121 (*ius Verrinum*, a pun on 'Verrine justice' and 'pork gravy'); Petr. *Sat.* 35; Varro, *RR* 3. 17. 4 (*hos pisces nemo cocus in ius vocare audet*: 'No cook would dare summon these fish to *court-bouillon*'); Mart. 7. 51. 5; Vespa, *Iudicium coci et pistoris* (*Anth. Lat.* 1. 1. 190. 6, 29); Venantius Fortunatus, *Carm.* 6. 8. 18.

[99] Cf. *Most.* 277: *quasi quom una multa iura confudit cocus* ('just as when a cook mixes lots of sauces together'). The phrase *iura confundere* ('to confuse the laws') is used in a serious legal context at Liv. 4. 1. 2.

[100] See Fraenkel (1960: 95–104) on the personification of inanimate objects as especially Plautine. On Saturnalian inversion in Plautus, see Diaconescu (1980).

77 ff. Saturio visits last night's leftovers, to make sure that they had a good night and were well protected against kidnapping. The so-called drunken dinner (*cena ebria*)[101] also blurs the distinction between the meal and its participants.[102]

This connection between parasites and parody forces us to reconsider the implications of *para-* in the word *parasitus*. The parasite is not only the sidekick to every host; he is also an inverter or a parodist, a figure whose Saturnalian, food-centred view of the world causes him to introduce *parerga*, trivialities, into every official or technical vocabulary.

7 *Cooks and Wicked Pickles*

The cook is a character for whom Plautus seems to have a special predilection. He is another stock comic type whose role has again, as far as we can tell, been greatly expanded. A number of important studies on comic cooks, cooks in Plautus, and the cook in *Pseudolus* in particular, have appeared since the 1960s,[103] yet none of these has attempted to explain why cooks *per se* are amusing or what function their role, which is often episodic and unrelated to the main plot, is meant to perform.

One answer is that comic cooks have great 'metatheatrical' potential. The cook, like the parasite, is a parodist. His arcane activities in the kitchen (to which he gives such spurious importance), his alteration of ingredients—by mincing, chopping, mixing, flavouring, and stuffing—mimic other forms of organization and alteration, particularly the activities of the comic author himself. This dimension of the cook has been largely ignored,[104] despite the evidence of all those Greek and Latin texts, comic or literary-critical, from Aristophanes to Petronius, which draw analogies between the production of drama and poetry and the processes of cooking. The cook, an entertainer of dubious worth, was an

[101] *Cas.* 747, *Men.* 212, *Per.* 94.

[102] Cf. the confusing use of *madere/madidus*, which means both 'wet, well-oiled' and 'drunk' at *Per.* 92–4. See Segal (1987: 47–8). Cf also Ch. 4, sect. 3, below, on similar confusions at Mart. 10. 48. 12.

[103] Dohm (1964), Giannini (1960); Lowe (1985*a*; 1985*b*).

[104] Except for a brief mention by Giannini (1960: 209).

obvious parallel for the poet, and the comic poet in particular: cooks, with their ambitious claims for their untrustworthy art, deliver the same mixture of inane boasting and ironic self-deprecation. And the uncertain cultural status of gastronomy made it a good match for a dramatic form that presented itself as only a botched and rehashed version of tragedy.

Plautus' prologue to *Poenulus* links his brand of comedy with meagre peasant food, or no food at all. Explicit comparisons like this are rare, but this kind of analogy is actually widespread in Plautus, and his different ways of presenting it complement the uneven or contradictory picture he gives us of his own status as a Roman comedian. The comparisons have not generally been made much of, perhaps because author and cook are usually linked only indirectly through the internal characters who devise the hurly-burly and confusions of the plays (I shall refer to these characters as 'plotters'), confusions which are often conceived of in either 'cooking' or 'metatheatrical' terms. Plautus is making his own contribution to a long Greek tradition which links cooking and drama, and it may be helpful first to give a brief survey of this tradition and set Plautus' own work in that context. The implications are greatest for Plautus' longest cook-scene, in *Pseudolus*: this can now be read not only as an absurd debate between two opposed styles of cooking, but also as a projection of two contrary aspects of the author Plautus.

Food in Greek comedy, whether used literally or metaphorically, tends to drag the tone of the drama down to the level of the kitchen, material, ephemeral, and potentially messy. Cooks' speeches in New Comedy of course depend for their comic effect on the discrepancy between the art of cookery and the pretentious claims which are made for it.[105] Among the higher disciplines to which cooking is compared are rhetoric and poetry, and this ought to make us look again at the dramatic frame in which the comparisons are made.[106]

[105] In this tradition is Montaigne's Italian cook (*Essais* I. 51), who discourses on sauce-making as though he were talking about the government of an empire.

[106] This frame is nearly always comic. But cp. Soph. fr. 1133 (Pearson): ἐγὼ μάγειρος ἀρτύσω σοφῶς (where the absence of the direct article indicates that μάγειρος

So far, again, studies of comic cooks have neglected this kind of evidence, which in fact reveals that they have an extra dimension.[107]

The most famous comparison is in Euphron's story of King Niceratus and his cook, who concocted a false anchovy artfully (*μουσικῶς*) by slicing bits of turnip and seasoning them. At the end comes the tag *οὐδὲν ὁ μάγειρος τοῦ ποιητοῦ διαφέρει*: there is no difference between a cook and a poet.[108] Of course the point of the joke is that there *are* differences between cooks and poets, but it is worth noting that the kind of poet implied in this comparison with the scrap-mincing cook is a devious creator of false illusions: in other words, the analogy demeans poetic art.[109] Aristophanes, too, compares poets to caterers, whether he is satirizing the dainty meal provided by Crates, a forerunner of New Comedy, or his offerings to his own audience.[110] According to Astydamas, the wise poet will supply his audience with a varied dinner,[111] while Metagenes thinks the audience will be pleased by novel side-dishes.[112] Euphron, again, tells of a cook who

is being used metaphorically: 'As the cook, I shall dress this cleverly'). Pearson ad loc. thinks this is an extraordinary metaphor for a tragedy or even a satyr-play.

[107] They have been collected in surveys of ancient poetic metaphors instead: see Van Hook (1905); Bramble (1974); Taillardat (1965).

[108] Fr. 1 (3. 213K).

[109] Giannini (1960) does mention Euphron, and comments briefly on the shared obsession of cook and poet with *inventio*/*τὸ εὑρεῖν*.

[110] *Eq.* 538–9: 'The man who sent you away with an economical little breakfast, pressing out the choicest of ideas from the driest of mouths' (*ὃς ἀπὸ σμικρᾶς δαπάνης ὑμᾶς ἀριστίζων ἀπεπέμπεν ἀπὸ κραμβοτάτου στόματος μάττων ἀστειοτάτας ἐπινοίας*); *Nub.* 523: 'I thought you should be the first to taste it' (*πρώτους ἠξίωσ' ἀναγεῦσ' ὑμᾶς*).

[111] 4N2 = Com. Adesp. 1330K: *ἀλλ' ὥσπερ δείπνου γλαφυροῦ ποικιλὴν εὐωχίαν/ τὸν ποιητὴν δεῖ παρέχειν τοῖς θεαταῖς τὸν σοφόν.*

[112] 14K = K1. 708: *ὡς ἂν/ καιναῖσι παροψίσι καὶ πολλαῖς εὐωχήσω τὸ θέατρον.* See Ath. 9. 367c–f on metaphorical uses of *παροψίς*, especially denoting a *πάρεργον*, something superfluous; in a sexual context, 'a bit on the side', an adulterous lover (e.g. Aesch. *Ag.* 1447: Cassandra is a *παροψώνημα*). Cf. also Ath. 2. 60 (= K2. 392) quoting Alexis, who contrasts a splendid 'zodiac' dish with the usual hors-d'œuvres, *παροψίδες* and *λῆρος*—rubbish. Both *πάρεργον* and 'hors d'œuvre' signify superfluousness; each is outside the main body, whether of a meal or something more abstract. Ath. 9. 368a (= TGF. 2.748) quotes Achaeus' *Aethon*, a satyr-drama: 'let me have other well-stewed saucer-dishes (*παροψίδες*) served chopped in fine bits, and steaming dishes aflame on the side (*παραφλογίσματα*)' (tr. Gulick). These side-dishes can be seen as analogues to the satyric drama itself—*Aethon*, 'the burning one', served as supplement to the main tragedies.

twice stole some sacrificial meat: one trick is called a 'drama', the other a παίγνιον (Gulick translates: 'vaudeville skit').[113]

The same culinary metaphors that describe the suspicious creations of the poet or rhetorician are also used of schemes devised by plotters within drama itself. It is often this figure of the plotter that indirectly unites the mess-making cook and the manipulative dramatist. Teleclides, for example, uses 'roasting' of a new play by Mnesilochus, and Aristophanes describes speech-writing as 'kneading' in the *Birds*.[114] Yet 'kneading', προφυρᾶν, is also used of plotting at *Thesm.* 75: 'I've had a terrible mess kneaded up for me' (ἔστιν κακόν μοι μέγα τι προπεφυράμενον). Such parallels between dubious cooking, evil plots, and concocted language help to compromise the integrity of comedy.

In any case, cooks are themselves associated with devious and manipulative activity. Plato's analogy between rhetoric and cooking as arts of flattery had already been implied by the comedians. Aristophanes' sausage-seller, the lowest form of cook,[115] has a talent for mincing matters and disguising rubbish with seductive condiments that fits him perfectly for his new role as demagogue.[116] Paphlagonides accuses him of making 'cheese-pressed' plans (συντυρεύμενα) among the Boeotians.[117] And Aristophanes uses περιπέττω (to cook up or bake a crust around) of disguising wickedness: 'They crust it

[113] K3. 317 = Ath. 379df. And Alexis K2. 345 = Ath. 4. 164b relates how, when Heracles, true to form, selected a cookery book from a library, Linus told him: 'You are a philosopher, for you have picked the writings of Simus, who was an actor keen on tragedy, the best cook of all actors . . . but the worst actor of all cooks.'

[114] Ar. *Av.* 462f: 'My speech is ready-mixed and fermented: there's nothing to stop me kneading it' (καὶ προπεφῦραται λόγος εἷς μοι, / ὃν διαμάττειν οὐ κωλύει); Teleclides 1. 218K: 'Mnesilochus is roasting a new drama for Euripides, and Socrates is providing the firewood' (Μνησίλοχός ἐστ' ἐκεῖνος ὅς φρύγει τι δρᾶμα καινὸν / Εὐριπίδῃ καὶ Σωκράτης τὰ φρύγαν' ὑποτίθησι: Dindorf's reconstruction).

[115] See Dohm's extensive discussion (1964: 31–5), concluding that he is not technically a cook; and cf. the contrast Anaxippus draws between the παντοπώλης, hawker, and the genuine μάγειρος (fr. 1 = Ath. 9. 404a = K3. 296).

[116] *Eq.* 343: 'I can make speeches and pour sauce over them' (ὁτιὴ λέγειν οἷός τε κἀγὼ καὶ καρυκοποιεῖν). Cf. ibid. 215–16: 'You're always sweetening up Demus with delicious mouthfuls of rhetoric' (τὸν δῆμον ἀεὶ προσποιοῦ/ ὑπογλυκαίνων ῥηματίοις μαγειρικοῖς).

[117] ibid. 479. Cf. *Com. Fr. Adesp.* (3. 533K): 'a curdled intrigue' (παλάμημα καὶ τύρευμα); cf. ibid. 940, 998, 1173; Luc. *Asin.* 31: 'curdling up trouble for me' (κακὸν ἐμοὶ μέγα τυρεύων).

over with a new name' (ὀνόματι περιπέττουσι τὴν μοχθηρίαν).[118] There are also parallels between cooking and sophistry, the art of making the unacceptable superficially palatable. Alexis mentions a cook who got himself enrolled in the sophists' guild.[119] And in another of his plays, *Lebes*, a cook who gives instructions for remedying a dish of burned pork is told: 'You are a much better speech-writer (λογογράφος) than cook. What you say you unsay. You bring your art into disrepute.'[120] Chrysippus describes how the plain-living Athenians flogged a cook in the Lyceum for a trick of sophistry (παρασοφιζόμενον): disguising salt meat as salt fish.[121] This word παρασοφίζομαι is usually translated 'over-refine'. But the sense of παρά that implies an inadequate parallel, something analogous but also different, missing the mark—as in παρῳδός, πάρεργον—is present here as well. The cook is a parody, even of a sophist. In the cook-scene in Plautus' *Curculio*, the cook poses as a wise man, which exploits the traditional assumption that he is an expert who nevertheless falls short of genuine wisdom.[122]

Innovation is another ambition shared by cooks and poets. As well as Teleclides' 'new drama' (δρᾶμα καινόν) and Metagenes' 'novel side-dishes' (καιναῖς παρόψισι), we also have Straton's 'new words' (καινὰ ῥήματα), used of the unintelligible expressions of a sphinx-like cook.[123] A cook in Hegesippus asks: 'Can you do anything new, compared with your predecessors?'[124] Often the food itself inspires comic neologisms, for example the longest surviving word in the Greek language, the unspewable λοπαδοτεμαχοσελαχογαλεοκρανιολειψανοδριμυποτριμματοσιλφιοκαραβομελιτοκατακεχυμενοκιχλεπικοσσυφοφαττοπεριστεραλεκτρυονοπτοκεφαλλιοκιγκλοπελειολαγῳοσιραιοβαφητραγανοπτερυγών from Aristophanes'

[118] Ar. *Pl.* 159. Cf. ibid. fr. 321 (= K1. 477), *Com. Fr. Adesp.* 338 (= K3. 470), Plato, *Leg.* 886c περιπεπεμμένα; Ar. *Vesp.* 668: 'buttered up with nice words' (τοῖς ῥηματίοις περιπεφθείς). See Taillardat (1965: 235), Brotherton (1978: 46–8).

[119] Ath. 379b = K2. 351. Cf. Ath. 7. 292e: ἄλλος σοφιστής μαγειρίσκος; 9.377f: μεγὰς σοφιστής.

[120] Ath. 9. 383e = K2. 341.

[121] Ath. 9.383e = K2. 341; Ath. 4. 137f.

[122] 251–79.

[123] K 1.218; K 1.708; K3. 361 = Ath. 9. 382c.

[124] K3. 312 = Ath. 9. 405d.

Ecclesiazusae, or the more modestly glutinous sweets concocted by Philoxenus in his *Banquet*, which are outdone by the novelty of the conversation.[125] Originality, it is implied, is always a substitute for genuine quality.

Linked with this is the idea that new plays are composed out of scraps from earlier literature. Aeschylus is reputed to have termed his plays 'slices (or steaks) from the banquet of Homer'.[126] And one anonymous 'recipe' for a good tragedy (or comedy?) calls for 'a pinch of Sophocles, a pinch of Aeschylus, a whole Euripides, with salt [wit] added, real salt and not just inane chatter.'[127] New plays, this implies, whether Greek New Comedy or Roman *palliatae*, are minced-up scraps dressed up with pleasing condiments, each rehashing requiring a more potent surface flavouring.[128] Underneath, they may be tasteless, worthless, stale, or even rotten. Aristophanes talks of the lyric poets Ibycus, Anacreon, and Alcaeus 'seasoning' their works, and says that the tragedies of Sthenelus need salt and vinegar to make them edible.[129]

For all these reasons, it seems likely that, even if we account for Athenaeus' special interest in gastronomy, food was a dominant ingredient in Greek comedy. As object and metaphor, a substance which was potentially sordid and messy, it thickened the atmosphere of festive squalor in which the comic poets wallowed. Fragments of Greek comic 'shopping-lists' can be read as literary 'recipes' too.[130] Alexis, for example, presents a *κατάλογος ἡδυσμάτων* in his *Lebes*

[125] Ar. *Eccl.* 1169–75; *PLG* 3. 606 = Ath. 14. 643: *τι καινὸν ἐλέχθη κομψὸν ἀθυρμάτιον*.

[126] Ath. 8. 347e: *τεμάχη τῶν 'Ομήρου μεγάλων δείπνων*; cf. Aelian, 13. 22 on the vomit of Homer lapped up by other poets.

[127] *Com. Adesp.* 12a Demiańczuk. Gellius uses the alternative image of a botched-together patchwork when he describes Caecilius' version of a play by Menander (2. 3. 21): *consarcinantis*. See Ch. 3 n. 65 for parallels. A sewing image is used of the fabrication of lies at Plaut. *Epid.* 455: *alium quaeras quoi centones sarcias*. *Centunculus*, motley, was the costume of the Roman mime-actor: *mimi centunculus* (Apul. *Apol.* 13).

[128] Aristotle uses *ἥδυσμα* as a term for literary embellishment (*Poet.* 1450b. 16). For other examples, see below p. 100 n. 186.

[129] Ar. *Thesm.* 162: *οἵπερ ἁρμονίαν ἐχύμισαν*; Ar. fr. 151 (= K.1.429): A. *καὶ πῶς ἐγὼ Σθένελον φάγοιμ' ἂν ῥήματα*/B. *εἰς ὄξος ἐμβαπτόμενος ἢ ξηροὺς ἅλας*.

[130] e.g. Alexis K2. 343 (= Ath. 4. 170a); Antiphanes K2. 69 (= Ath. 2. 60a); Alexis K2. 362 (= Ath 4. 170a).

(The Cooking Pot), which includes mashed raisin, fennel, anise, mustard, cabbage, silphium, dried coriander, cumin, capers, marjoram, garlic, thyme, and rue.[131] The title of the play allows us to read this list as an absurdly specific 'recipe' for a play which is itself a comic hotchpotch.[132]

The connection between the list of ingredients within a play and the poet's conception of the play itself is never as easy to point out as it is here. There is one other list, however, that gives a salutary warning against drawing over-literal conclusions about food-lists or cook-scenes. A fragment of Aristophanes contains another list of sharp flavourings which are described literally as 'buggery for big meat', that is, probably, stuffing for meat that is high.[133]

> *ὀξωτά, σιλφιωτά, βολβός, τευτλίον,*
> *περίκομμα, θρῖον, ἐγκέφαλος, ὀρίγανον,*
> *καταπυγοσύνη ταῦτ' ἐστὶ πρὸς κρέας μέγα.*

Vinegar, devil's dung, onion, beet, mincemeat, figcake, brains, oregano: this is all stuffing for meat that's gone off.

If this had survived in isolation, we would assume that it was yet another ordinary shopping-list. As it is, we have a context for it. Diogenes Laertius tells us that it is a description of the style of Euripides, a poet who was pilloried for his own contamination of tragic language with the mundane: 'In the words of Aristophanes about Euripides—'the acid, pungent style', *ὀξωτὰ καὶ σιλφιωτά*—which, as the same author says, is 'strong seasoning for meat when it is high', *καταπυγοσύνη ταῦτ' ἐστὶ πρὸς κρέας μέγα.*'[134] The steaks of Aeschylus have turned into Euripides's bitter olla podrida.[135]

A final, important passage comes from Anaxippus' *Enkaluptomenos*, where a boastful cook explains the origins of his

[131] K2. 343 = Ath. 4. 170a.

[132] See Coffey (1989: 15–17) on the connections between *satura* and other dish metaphors for miscellaneous collections of material.

[133] K1. 423 = 131. Dindorf reads *ὑπότριμμα*, after Cod. Falckenburgii *υποτρικομμα*; compare Ar. *Eccl.* 1170, where *-υποτριμματο-* is sandwiched between the sharp tastes of *-δριμυ-* and *-σιλφιο*.

[134] Dio. Laert. 4. 18.

[135] Kock ad loc. is adamant that critics have been wrong to 'sniff out' an obscene meaning in the word *καταπυγοσύνη*; the recipe relates simply to Euripides' transformation of Aeschylus' style. With *κρέας μέγα* he compares Eubul. 7. 8, Philetaer. 9.

classical cooking style. He tells us that his teachers, Sophon of Acarnania and Damoxenus of Rhodes, had abolished the mortar and wiped out all the old seasonings a mere pedlar (παντοπώλης) might use: cumin, silphium, cheese, and coriander. They had started again from scratch, with oil, a new stewpan (λοπάδα καινήν) and a hot fire, and the aim of freeing their guests from all the coughs and sneezes which pungent seasonings bring on.[136] Their pupil himself dreams of leaving his own innovative writings.[137]

This fragment can also be read as a statement about writing comedy. 'Damoxenus' is of course too early to be the comic poet Damoxenus; Athenaeus refers elsewhere to Sophon as a cook rather than a poet. Nevertheless, the detailed, self-conscious emphasis on technique and innovation suggests a parallel with literary style. The pungent flavourings represent those traditional comic styles which are 'highly spiced'. As for the new school, olive oil, chief export of Attica, could be a metaphor for Atticism, the unsuperfluous style, as well as a smoothing agent;[138] λοπάδα καινήν would signify 'a new cooking pot' or, metaphorically, a new vessel for a new style of writing.[139] So Anaxippus could well be discussing a new style for New Comedy in the guise of a cooking lecture, appropriating the images once used to characterize the piquant mêlée of earlier comedy in the cause of a new restraint and simplicity. After all, Plutarch describes the humour of the comic poets in their own language as a 'wicked pickle': ὑπότριμμα μοχθηρόν.[140]

Of course, not every comic list of seasonings points so suggestively to associations with comedy itelf. But the cook

Cp. J. Henderson (1975: 129) on the obscene meaning of meat (Ar. *Eq.* 428 and schol. 484).

[136] K3. 296 = Ath. 9. 403e. Cf. Damoxenus K3. 349 = Ath. 3. 102e, where a cook complains that his contemporaries do not understand the harmonic principles of seasoning.

[137] Ath. 9. 404b: καὐτὸς φιλοσοφῶ καταλιπεῖν συγγράμματα / σπεύδων ἐμαυτοῦ καινὰ τῆς τέχνης.

[138] For unblocking as a literary metaphor, see below pp. 149–50 on Hor. *Sat.* 2. 4.

[139] See Coffey (1989: 16) on λοπάς as an equivalent metaphor to *lanx satura*; at *Anth. Pal.* 12. 95. 10 (= GP 4406–7) Meleager's παίδων Ῥωμαικὴν λοπάδα ('a Roman dish of boys') shows that the word had come to be proverbial of a miscellany.

[140] *Mor.* 68c.

and his spiced, suspicious dishes or mundane recipes always offered a model for making the comic poet's creations look ridiculous, and the extent to which Greek cook-scenes debate comic style has been greatly underestimated. The fact that the role of the cook is formalized in Menander and virtually absent in Terence indicates perhaps that the details of cooking were seen to compromise the classical 'purity' of their style of comedy. The cook Sikon in Menander's *Dyskolos*, for example, summarizes all the old comic cook clichés 'with neatness and restraint'.[141] And Terence, at *Andria* 483–5, chooses to miss out a specific prescription for four egg-yolks, losing all the flavour of Menander's original in the process.[142]

For the sceptical reader, all the links between comic poetry and the suspicious art of cookery that are made in ancient comedy are finally spelled out in an English Renaissance play, Ben Jonson's *Staple of News*. His cook, Lickfinger, is fully conscious of his classical ancestry (III, iii. 21 ff.), for it is said of him:

> He holds no man can be a poet,
> That is not a good cook, to know the palates
> And several tastes of the time. He draws all arts
> Out of the kitchen, but the art of poetry,
> Which he concludes the same with cookery.

In his masque *Neptune's Triumph*, Jonson expands this theme into a dialogue between a cook and a poet which emphasizes the similarities between them. The cook justifies his art as a form of poetry, claiming that the kitchen is where 'the art of poetry was learned and found out . . . and the same day, with the art of cookery' (70–2); (cf. 228–9); he offers the poet by way of an antimasque a 'metaphorical dish' (223), an olla podrida composed of the dregs of society.[143] These

[141] Handley on *Dyskolos* 489 ff. He concludes (1968: 200): 'Sikon could be described as a modernized and sophisticated cook, just as Chaireas could be described as a modernized and sophisticated parasite.'

[142] See Shipp (1960: 23). Arnott, reviewing Dohm (*JHS* 85 (1965), 182–4), criticizes him for not explaining why the cook plays such a small role in Terence; though cf. the relatively formalized scene where cooking parodies philosophy at *Ad.* 355–446.

[143] 'Olla podrida', which of course by this stage meant a 'miscellany' too, is also the name of Lickfinger's dish in *The Staple of News*.

antimasques are dismissed by the poet as the equivalent of ancient *parerga*:

> things so heterogeneous, to all devise,
> Mere byworks, and at best outlandish nothings. (222–3)

At the same time, he confesses that his own uses are only seasonal: 'A kind of Christmas ingine; one that is used, at least once a year, for a trifling instrument, of wit, or so'.

8 *Cooks in Plautus*

Plautus, another festive and heterogeneous poet, would probably have accepted the role of 'Christmas ingine' and creator of 'outlandish nothings' quite happily. Like his Greek predecessors, he bustles about making us see parallels between dramatic intrigue and violence, cooks' behaviour in the kitchen, and his own supreme manipulation of the plot. His characters often break the dramatic illusion to discuss with the audience their ability to determine the plot from inside. This metatheatrical dimension, which has become the focus of some interest,[144] is not just a sophisticated, self-conscious device, but also another of Plautus' methods of compromising the integrity of classical drama.[145]

The comic plotter, it has been convincingly argued, is a projection of the author in his capacity as plot-maker.[146] However, when it comes to the cook in Plautus, critics have been much slower to see him as another parallel for the comic writer. Perhaps this is because the connections are made less directly; it is the figure of the plotter who provides the missing link between cook and author. Many of the internal machinations which suggest the manipulative powers of the creator also draw on the language of cooking: violent dismemberment, transformation, evil concoctions. It becomes clear that cultural reasons for the cook's prominence—his newness in Roman society, the suspicions

[144] See Slater (1985: 13–16, 172–8); Barchiesi (1970), Knapp (1979).

[145] As Slater (1985: 146) puts it: 'the heady but anarchic mixture of script and improvisation, stock types and variations, illusory and non-illusory theatre'.

[146] Ibid. 16.

attached to cuisine—add to his metaphorical potential. Plautus is choosing to identify himself with all the messier aspects of the Graeco-Roman cultural mixture and with the peripheral and frivolous phenomenon of the festival. It is time for these analogies to be reconsidered.

It is not surprising, of course, that slapstick violence in Plautus is so often described in the language of cooking, when so much of it is perpetrated by or against cooks: 'The victim [of intrigue] is struck (*ferire, percutire, tangere*), beaten (*verberare*), boned like a fish (*exossare*), torn limb from limb (*deartuare, articulatim concidere*), mutilated (*mutilare*), lacerated (*lacerare*), bitten (*admordere*) and devoured (*devorare*).'[147] The cook's knives and spits double as butcher's weapons. Chrysalis in *Bacchides* threatens to use his spit against the soldier Cleomachus' sword, while promising to roast his master Nicobulus like a chick-pea (767) and punch him full of holes like a shrew's gut (*soricina nenia*, 889). He also compares this victim to a putrid fungus (821) and a thrush caught in a snare (792). Trachalio in *Rudens* eggs the slaves on to slice out Labrax's eyes, like a cook with a cuttlefish; Stratophanes in *Truculentus* threatens to chop the cook Cyamus ('Bean') into chunks with a meat-cleaver; and Purgopolynices cannot wait to make mincemeat of an enemy (*fartem facere*).[148]

The verb *vertere/versare*—to twist, turn over—also forges links between plotter, author, and cook. The verb is used of manipulation, confusion, translation—and of tossing or turning over in cooking.[149] In *Asinaria*, for example, the meanings of *versare* double up in a simile which compares a woman manipulating her lover to a cook preparing a fish:

> is habet sucum, is suavitatem, eum quo vis pacto condias,

[147] Brotherton (1978: 4); see Fantham (1972: 105) on *vorsare, exossare*.

[148] *Rud.* 659: *iube oculos elidere, itidem ut sepiis faciunt coqui*; *Truc.* 613: *iam hercle ego te hic hac offatim conficiam*; *Mil.* 8.

[149] Brotherton (1978: 48) argues that *Bacch.* 766 (*vorsabo ego illunc*: 'I'll toss him about') is, in its context, a metaphor from cooking. Cf. also *Cist.* 94, *Pers.* 795–6. She is in fact anticipated by Lambinus ad *Bacch.* 767, Ussing, and Fantham (1972: 105). Brotherton also compares Ar. *Ach.* 1005 τρέπειν, to turn in cooking; cf. Antiphanes K2. 105 (= Ath. 14. 623b): a cook whirls everything together like a magician with a rhombus (συντρέφοντος).

vel patinarium vel assum, verses quo pacto lubet. (179–80)

Fresh ones have the juice and the sweetness: you can stew them or roast them, season them to taste, and toss them anyhow.

Vertere/versare also expresses the manipulations, mental indecision, and reversals of identity or fortune which fill out comic plots.[150] What is remarkable is how often the word is used in a negative context, linked with torture, punishment, or a change for the worse.

Plautus is of course famous for using *vortere* as a term for his own adaptation of Greek originals: *Trin.* 19, *Philemo scripsit, Plautus vortit barbare*; *As.* 11, *Demophilus scripsit, Maccus vortit barbare.*[151] *Barbare* and *vortere* together turn the plays into second-hand, second-best perversions of Greek material. Segal sees the negative quality of the word when he writes: 'What a paradox that Horatian *versare* is a praiseworthy practice and Plautine *vortere* . . . is looked upon as a reverse alchemy which transmutes the gold of Athens into Roman dross.' It is the versatility of this word in other contexts that helps to drag Plautus' plays down to the level of the farcical confusions and suspect messes they depict.[152]

The verb *coquere*, to cook, similarly applies to both inner turmoil and comic plots. At *Trin.* 223–5, for example, it is paired with *versare* to suggest a character's bewilderment:[153]

agitare multas res simitu in meo corde vorso

[150] For example, *Per.* 795: *quo modo me hodie vorsavisti* ('The way you've messed me around today'); *Cist.* 207: *stimulor, vorsor in amoris rota* ('I'm being goaded, twisted on the wheel of love'); *Cist.* 94: *ut ego illum vorsarem* ('How I'd knock him around'); *Most.* 191: *quod malum vorsatur meae domi illud* ('The trouble that's loose in my house'); *Pseud.* 745: *scitne in re advorsa vorsari* ('When things take a twist for the worse, can he twist himself out of it?': NB the etymological twist here); *Merc.* 470: *divorsum distrahor* ('I'm being twisted and pulled apart').

[151] And it seems likely that there is some pointed connection being made between Plautus' 'twisted' translations and the twists and turns of his comic plots. See Slater (1985: 51 n. 15): 'One also wonders if the way Toxilus has twisted Dordalus around ([*Pers.*] 795) has not some relation to the Plautine process of poetic composition.'

[152] Segal (1987: 6). It could be argued that *vortere* is a metaphor from money-changing: Greek currency is exchanged for worthless cash on the foreign market. Cf. Plautus' title *Trinummus*, 'Farthing', 'exchanged' for the Greek *Thesaurus*, 'Treasure' (*Trin.* 18–20).

[153] Brotherton could have used this passage to support her argument that *vorsabo* at *Bacch.* 766 is a cooking metaphor (see above, n. 149).

> multum in agitando dolorem indipiscor:
> egomet coquo et macero et defetigo.

The number of things I'm tossing about in my head at once, and the grief it all brings me: I'm sweating and stewing, it's flaking me out.

At *Mil.* 208 it is used of a 'well-baked' idea: *incoctum non expromet, bene coctum dabit* ('He'll produce it well-cooked, not half-baked').[154] At *Per.* 52 a slave concocts a disaster for an evil pimp: *usque ero domi dum excoxero lenoni malam rem / aliquam* ('I'll stay at home until I've cooked up a big mess for that pimp'). The combination of *coquere* and *malam* gives cooking a typically black name. These scant but explicit representations of the plotter as cook add another dimension to our understanding of the cook's role itself.

There is also, as we might expect, a strong emphasis on mixing. While Terence homogenizes both style and setting, Plautus exposes and indeed highlights his imperfect blending of two cultures (for example in the satirical 'map' of Athens which seems to have the topography of Rome at *Capt.* 478 ff.). *Commisce mulsum* ('Mix the mead') may be a real command at *Per.* 87, but mixed sauces are used elsewhere metaphorically of lovers' oaths (*ius confusicium,*[155] *Cist.* 472); and of a nasty mixture of sweat and perfume: *una multa iura confudit coquos* (*Most.* 277, 'A cook mixes together lots of sauces'). *Miscere* and its synonym *conturbare*, like *coquere* and *versare/vertere*, are used to describe either external or internal confusions, again usually with some negative element.[156] Once again, cooks and parasites are held responsible not only for mixed-up priorities but for actual disturbances, riotous behaviour. A slave in *Captivi* describes

[154] See Brotherton (1978: 47) on the metaphor *calidum consilium*, 'a piping-hot plan' (*Mil.* 226—bracketed by Leo; *Epid.* 256, 284; *Most.* 665; *Poen.* 914; cf. Ar. *Ach.* 119: θερμόβουλον).

[155] A pun on the two meanings of *ius* is of course implied here.

[156] e.g. *As.* 310: *tantum adest boni . . . verum commixtum malo* ('all this good luck, mixed as it is with bad'); *Aul.* 279: *malum maerore metuo ne mixtum bibam* ('I fear I shall have to drink malice and misery mixed'); *Trin.* 285: *turbant, miscent mores mali* ('these wicked men make a muddle of our lives'); ibid. 122: *malumque ut eiius cum tuo misceres malo* ('so you mix up his knavishness with yours'). Cf. also *Most.* 416, 1032, 1053.

the havoc made in the kitchen by the parasite Ergasilus; and in *Casina* the cooks are heard offstage creating mayhem.[157]

Two occurrences of *conturbare* actually make it ambiguous whether Plautus is referring to the internal plot or to the play itself. One of these is at *Cas.* 880: *ridicula . . . ea sunt quae ego intus conturbavi* ('It's a comic brew that I've been cooking up inside').[158] The other comes from *Most.* 510: *hanc conturbant fabulam* ('They're stirring up this plot'). The idea that comedy contaminates pure tragedy is presented with a flourish in the prologue to *Amphitryon*. Mercury, god of changes, links transformation (from divine to human) within the plot (*Amphitruonis in vortit sese imaginem*: 'He's taken an Amphitryon's likeness'), with the play's metamorphosis from Greek tragedy to Roman hybrid: *faciam ut commixta sit: [sit] tragicomoedia* (59 'Let's make it a mixture: a tragicomedy'). The prologue of Plautus' *Poenulus*, which we have looked at before, also opens by referring to tragedy:

> Achillem Aristarchi mihi commentari lubet:
> ind' mihi principium capiam, ex ea tragoedia. (1–2)

I fancy basing my play on Aristarchus *Achilles*: so I'll take my opening from that tragedy.

But the analogy between the play and theatre-snacks labels it at once as comedy. Plautus is visibly 'contaminating' tragic language with comic subject-matter. We have seen how he takes self-deprecation to an extreme by advertising his play as a poor substitute for real food. Could this perhaps be leading up to an unspoken pun on *comoedia*, comedy, and *comedere*, to eat?[159] A makeshift etymology of comedy as 'communal eating' would justify the links between the festive atmosphere of comedy and the games at which it was performed, where the *populus* ate together as a body.[160] Plautus often ends his plays by regretfully excluding the

[157] *Capt.* 909–18; *Cas.* 772–5.

[158] Cf. 511–12, where the play's plot is likened to a brew which needs flavouring.

[159] See Ahl (1985: 54–60 and 60–3) for Latin puns which ignore vowel distinctions and linguistic boundaries; *oe/e* is discussed p. 58.

[160] See Veyne (1990: 208–14) on the largesse of public magistrates who paid for the games; and Hor. *Sat.* 2. 3. 182 on gifts of food distributed by the aediles: chick-peas, beans, and lupins.

audience from the revels they have shared with the cast, and sending them home to their own dinners.[161]

Plautus' native style is, as we have seen, gingered up with his own mixture of flavourings. This reflects a balance in the plots themselves, where heady flavours add enchantment to the bland everyday; women and love become symbols of ambiguity, suspicious brews of contradictory flavours—associations compounded of course by women's role as cooks or potential poisoners. They pester their husbands for money to make *condimenta*, and tend garden plots that are linked with shady practices: 'Any woman that's trouble doesn't need to go to the greengrocer: she's got a garden at home, with all the herbs she needs for her trouble-making.'[162] A love-letter brings the response: *dulce amarumque una misces mihi* (You've stirred me up a bitter-sweet mess').[163] Love in *Casina*, in the form of a cinnamon-scented girl, is extolled as the spice of life.[164]

Plautus twice suggests that his own plays are unflavoured dishes, bland barbarian pottage that requires seductive spicing. We have already seen the contrast between bland play and seasoning applause in *Poenulus*. In *Casina*, an intrigue within the play becomes a plot that too many cooks have tried to spoil, to which Chalinus adds his own complications:

[161] e.g. *Rud.* 1418–23: *spectatores, vos quoque ad cenam vocem, / ni daturus nihil sim neque sit quicquam pollucti domi, / nive adeo vocatos credam vos esse ad cenam foras. / verum si voletis, plausum fabulae huic clarum dare, / comissatum omnes venitote ad me ad annos sedecim. / vos hic hodie cenatote ambo* ('Spectators, I would ask you to dinner too, if it weren't the case that I'm not giving a dinner today and have nothing in the larder. But if you like, you can give us your loud applause and then come to dinner with me in sixteen years' time. As for you two, you'd better come today'); *Stich.* 774–5: *saltatum sati' pro vinost. / vos spectatores, plaudite atque ite ad vos comissatum* ('That's enough dancing for the amount of wine that's been consumed. Spectators, give us your applause, and go off and have a drink yourselves').

[162] *Mil.* 692; 193–4: *nam mulier holitori numquam supplicat, si quast mala: / domi habet hortum et condimenta ad omnes mores maleficos.* Brotherton (1978: 7): 'In the culinary realm the plot is spoken of not only as concocted in some picturesque way, but the whole process is entered into with a minutiae [*sic*] of detail—the cook, garden ingredients, seasoning of all varieties.' In Latin the link between women and food is already rooted in the word *mulier*, 'miller', 'grinder of flour'.

[163] *Pseud.* 63. Cf. *Cist.* 69 on love: *gestui dat dulce, amarum ad satietatem usque oggerit* ('it gives you a taste of sweetness, then piles up bitterness until you're sick of it'); *Truc.* 178: girls are a mixture of honey tongues and vinegar hearts; *Pseud.* 694.

[164] *Cas.* 217–18.

ibo intro, ut id quod alius condivit coquos
ego nunc vicissim ut alio pacto condiam,
quo id quoi paratum est ut paratum ne siet
sitque ei paratum quod paratum non erat. (511–14)

I'll be off indoors, to try my hand at changing the flavour of a mess another cook has flavoured before me: the man the mess was meant for will never receive the mess he should have had; he'll get a mess he didn't expect instead.

The metaphor is appropriate at the end of a scene where Lysidamus has mentally 'concocted' a delicious meal for Casina (490–501). In *Persa*, too, a cooking metaphor applied to plotting appears just before the entrance of the parasite Saturio: *usque ero domi dum excoxero lenoni malam rem / aliquam* ('I'll stay at home until I've cooked up a big mess for that pimp').[165] The *Casina* lines, in particular, look like another statement about the playwright's art, in the Greek comic tradition: another author has chosen the ingredients of the plot, but Plautus will determine its final flavour. To reinforce this, later in the play, a character congratulates himself on the intricacy of his own plot, which he says is more complicated than any poet's:

nec fallaciam astutiorem ullu' fecit
poeta atque ut haec est fabre facta ab nobis. (860–1)

No poet ever created a more deceitful plot than the one we've so cleverly cooked up here.

9 *The Cook in* Pseudolus

The intricacy of the connections Plautus makes between plot-making and cooking must make us take a fresh look at the cook scene in *Pseudolus*. This is Plautus' last extant play, and certainly the one that is agreed to have the strongest metatheatrical dimension. The cook-scene can be read as the final statement of Plautus' art in cooking terms. The cook is allotted the central scene, which is unrelated to the rest of the

[165] *Pers*. 52–3. This is remarked on by Chiarini (1979: 43): 'Quasi magicamente evocato dalla metafora culinaria (*excoxero*), metafora usata anche altrove, in Plauto, per indicare l'attività creativa inventiva del *servus calidus* . . . ecco apparire il più gastronomico dei personaggi, il perenne sognatore di manicaretti, il professionista della tavola: il parasito Saturio.'

plot, and he has by far the fullest role (compared with other plays, where cooks may simply be part of a procession of caterers[166]). Just as Pseudolus is the archetypal *servus callidus*, so this cook, *multiloquom, gloriosum, insulsum, inutilem* (794, 'verbose, boastful, disgusting, good-for-nothing'), is an archetypal comic cook.[167]

It has already been argued that Pseudolus, the author of plots and stage-manager, doubles for Plautus throughout.[168] But critics have been slower to acknowledge that the cook is a parallel for Plautus too. Chiarini (1979) makes the leap when he discusses *excoxero* (*Per.* 52) and *bene coctum* (*Mil.* 208) as metaphors for creative ability;[169] in a footnote he suggests tentatively that the whole cook scene of *Pseudolus* is a metaphor for the poet's ability to invent.[170] The cook, however, displays other qualities apart from novelty in this scene—ingenuity, sophistry, light fingers, bravado, and an interest in flavourings—that hint at more extensive parallels between cook, plotter, and playwright.[171] The central plotter, Pseudolus (described as *vorsutus* at 1243, the ultimate chameleon), has already made the comparisons. At 382 he is inspired by the violence of a cook boning a fish: *exossabo ego illum simulter idem ut murenam coquos*. And at 404 he considers his affinity with poets, characteristically lowering poetic art to the level of charlatanry or a pack of lies:

[166] *Pompa* of a procession of cooks: *Truc.* 549, *Cas.* 719, *Bacch.* 114, *Curc.* 2, *Stich.* 638; of a procession of food: *Bacaria* fr. 4 (*cuius haec ventri portatur pompa*: 'To whose stomach this procession is leading'); *Stich.* 683; Varro, *RR* 1. 2. 11; Petr. 60. 5; Mart. 10. 31. 4, 12. 62. 9; *Acta Fratrum Arvalium* (*more pompae*: 'in the manner of a procession'). For verbal imitation of processions, see the crocodile of festive virtues at *Capt.* 770–1: *laudem, lucrum, ludum, iocum, festivitatem, ferias / pompam, penum, potationes, saturitatem, gaudium* ('Praise, profit, playfulness, jollity, festivity, feasting, pomp, provisions, satiety, joy'). In general see Fraenkel (1960: 391), Rosati (1983).

[167] See Dohm (1964: 142–3).

[168] Wright (1975: 416) claims that Pseudolus is transformed into Plautus himself, the master of ceremonies, at the end of the play. Slater (1985) argues that Pseudolus has always been Plautus, at least since the poet simile at 401–4.

[169] See above n. 165.

[170] Chiarini (1979: 43) compares *Pseud.* 403 (*facit illud veri simile quod mendacium est*: 'he makes lies look like the truth') and 838 (*tuis istis omnibus mendaciis*: 'with your pack of lies').

[171] Slater (1985: 135) suggests that Pseudolus and the cook were played by the same actor.

> sed quasi poeta, tabulas quom cepit sibi,
> quaerit quod nusquamst gentiumst, reperit tamen,
> facit illud veri simile quod mendacium est,
> nunc ego poeta fiam.

But what about a playwright, who picks up his pen and searches high and low for something that doesn't exist and still finds it, and manages to make lies look like the truth: I'll turn playwright now myself.

Although Wright (1975) lists the cook among Pseudolus' transformations, and sees in the poet simile a premonition of the final bow of Pseudolus the master of ceremonies, he does not see any 'metatheatrical' implications in the cook-scene itself. This is despite the fact that, earlier in the play, cunning slaves had themselves been likened to caterers. As we saw, the stooge Simius' head is a spice-market (*pantopolium*) of different flavours of speech (*sapores*) for every occasion (742): a goatish *sapor* (*hircum ab alis:* 'goat in the armpits' 738), sharpness (*aceti . . . acidissimi:* 'sharpest vinegar' 739), or sweetness (*murrinam, passum, defrutum, mellam, mel:* 'sultana-juice, raisin-wine, grape-syrup, mead, honey' 741), a good description of Plautus' own cosmopolitan bag of tricks. *Pantopolium*, the free-for-all bazaar, associated with shady deals and lucky finds,[172] reminds us of παντοπώλης, the word used scathingly of herb-pedlars by Anaxippus.[173] Pseudolus is another of Plautus' slave-caterers: *condus promus sum, procurator peni* (608, 'I'm butler, I'm steward and pantryman'). And he describes his prolific supply of words in the language of largesse: *largitu's dictis dapsilis* (396, 'You have showered him with talk'). Already, then, even before the cook-scene, the links between the cook, the plotter, and the poet–caterer have been strengthened. And once again, the author who pretends to be mean

[172] See Stallybrass and White (1986).

[173] See above p. 85. *Panto-* is commonly associated with 'low' forms of entertainment (e.g. 'pantomime', 'pantaloon'), as though, however motley or sordid, the entertainment in question covers an entire spectrum of life, a panorama. Cf. Ath. 1. 20b on Pantaleon, a famous buffoon. The misshapen body of the pantaloon (see the description of Pseudolus quoted above) contains a variety of different extremes. Compare the body of the African Scybale in the *Moretum* (32–5, *tota patriam testante figura*: 'Her entire body betrayed her native land'), a figure for the poem itself, which embraces a universe in microcosm. See above p. 47.

and insular also provides us with parallels for himself from the generous and cosmopolitan end of the scale.

Apart from Chiarini's note, the cook-scene has up to now been read literally. Dohm and Lowe concentrate on isolating Plautine additions to the original Greek material, with the result that up to half the scene is now ascribed to Plautus. This includes the extensive alliteration, hyperbole, verbal play, the unusual emphasis on the cook as pilferer, and the allusion to *mortuis cenam* (796, 'funeral feast'), a Roman institution.[174] The scene owes many of its stock themes to Greek comedy, but it has clearly been expanded and reshaped by the new author. We need to explain some of these new emphases.

The cook has been hired by the grasping pimp Ballio to prepare his birthday dinner. The scene is a duel of verbal resourcefulness, with the cook devising sophistic or hyperbolic arguments to prove his own superiority to the rest of the profession, and Ballio countering with rebuttals which voice the audience's suspicions of all culinary pretension. A distinction is drawn between 'other cooks', who are cheaper and who pile whole meadows of seasonings on to their dishes, so piquant that they virtually disembowel the guests and scrape out their eyes like screech-owls, and this exceptional cook, who caters for Jupiter, and has discovered his own unique seasonings which confer immortality:

> non ego item cenam condio ut alii coqui
> qui mihi condita prata in patinis proferunt,
> boves qui convivas faciunt herbasque oggerunt,
> eas herbas herbis aliis porro condiunt:
> indunt coriandrum, feniculum, alium, atrum holus,
> apponunt rumicem, brassicam, betam, blitum,
> eo laserpici libram pondo diluont,
> teritur sinapis scelera, quae illis qui terunt
> priu' quam triverunt oculi ut exstillent facit.
> ei homines cenas ubi coquont, quom condiunt,
> non condimentis condiunt, sed strigibus
> vivis convivis intestina quae exediunt. (810–21)

[174] Dohm (1964: 139–54); Lowe (1985*b*: 415).

I don't flavour my dinners as other cooks do, who just serve up plates of seasoned grass, and make dinner-guests into cattle with their mounds of herbs, and then season those herbs with more herbs: they pile on coriander, fennel, garlic, and horse-parsley, then add dock, cabbage, beet, and spinach; they dilute the mixture with a pound of devil's dung, grind in some vicious mustard, which makes the grinders' eyes drop out before they've finished. These people don't use seasonings, they use screech-owls which disembowel the guests alive.

The debate is usually seen as a satire on some contemporary vogue for vegetarian cooking.[175] That ignores the similarity between the phrase *non ego item cenam condio ut alii coqui* (810) and the *Casina* speech we have just looked at:

> ibo intro, ut id quod alius condivit cocus,
> ego nunc vicissim alio pacto condiam. (511–12)

I'll be off indoors, to try my hand at changing the flavour of a dish another cook has flavoured before me.

The whole scene can also be read as a debate about styles of dramatic 'seasoning', in the tradition of Greek comedy. But which, in that case, is Plautus' own cooking style? Is it the bovine diet of vegetables overpowered by condiments, which is treated with such contempt? Or is it the celestial but illusory alternative which is so hyperbolically but suspiciously advertised? A case can be made for either, which calls into question the identify of the *alii coqui*.

Among the ingredients of the *alii coqui*, Lowe thinks that Dohm distinguishes too rigidly between *Gewürze* and *Gemüse*, herbs and vegetables. He also argues that Plautus must be conflating two separate points in the description of the 'other' cooking school: these cooks are vegetarian rather than omnivorous, and they use too many herbs.[176] In fact, the conflation makes more sense if we look at the passage in more general cultural terms. There *is* a distinction being made between between coriander, fennel, garlic, and *atrum holus* (814), and *rumex* (dock), *brassica* (cabbage), *beta* (beet), and *blitum* (spinach): it is meant to indicate the upside-down

[175] See Dohm (1964); Lowe (1985*b*).

[176] Lowe (1985*b*: 413): 'The emphasis on the *condimenta* of cheap cooks . . . obscures the fact of their vegetarianism.'

proportions of this style of cuisine. Herbs, normally the superfluous garnish of a dish, are being made its basis, while flavourless staples are added as trimmings (*apponunt*, 815). A whole pound of silphium (when usually only a drop of this virulent juice was needed[177]) and 'vicious mustard' (*sinapis scelera*) are poured on in profusion. The cooking style is an absurd inversion of normal good taste.[178]

Apart from that, we have seen many of these ingredients before, associated with the uncivilized and antisocial margins of the Roman world, ingredients whose agreeableness is compromised by their acrid or stinking qualities. Garlic, *alium*, appears twice in Plautus: once as the reek of a rustic serf (*Most.* 39), and once as the diet of galley-slaves, the dregs of Roman society (*Poen.* 1314).[179] Silphium, or devil's dung, is the crop grown by the uncultivated fishermen of Cyrene in *Rudens* (630), the foul-smelling complement to their harsh lives.[180] Mustard lends its bite to a hostile stare at *Truc.* 310: *si ecastor hic homo sinapi victitet, non censeam / tam esse tristem.*[181] In the *Pseudolus* passage, the metaphor becomes literal: Plautus describes mustard as a plant so acerbic that it stings out the eyes of anyone who grates it (817–18).[182]

As for the vegetables 'added' to these herbs, at least two of them, beets and spinach, had a reputation for cheapness and

177 See Apic. 1.10.

178 Willcock's solution to this is to reorder the lines logically, with vegetables preceding herbs; he presumably understands *apponunt* as 'serve up' rather than 'add'.

179 See below, Ch 5, on the role of garlic in Hor. *Epod.* 3.

180 See Hunter (1983: 105) on silphium in Eubulus fr. 7.

181 Cf. Ar. *Eq.* 631; *Ran.* 603: βλέποντ' ὀρίγανον; *Vesp.* 455: βλεπόντων κάρδαμα; *Ach.* 254: βλέπουσα θυμβροφάγον. All four plants, mustard, origanum, cardamom, and thyme are listed by Theoph. *Hist. plant.* 1.12.1 among plants with a bitter taste.

182 The passage clearly belongs to a comic tradition that ascribes violent properties to innocent vegetables, e.g. Naev. *Apella* 18–19W, cursing the inventor of the onion: '*ut illum di perdant, qui primum holitor protulit/ caepam!*'; Cic. *Fam.* 7. 26. 2 after an attack of diarrhoea: *a malva et a beta deceptus sum* ('I was deceived by Messrs mallow and beet': abl. + *a* signifies that the vegetables are personified); Hor. *Epod.* 3, a hyperbolic tirade against the indigestibility of garlic; *Mor.* 83. an etymological description of watercress, the nose-twister: *quaeque trahunt acri vultus nasturtia morsu* (*nasus* + *torqueo*); *Anth. Lat.* 1. 1. 281 on the onion: *mordeo mordentes.* Compare also the satirical tradition of including human reactions to a foodstuff among the plant's own attributes: Enn. *Sat.* 12–13W *triste sinapi . . . caepe maestrum*; Lucil. 194M = 216W *flebile cepe.*

blandness.[183] *Blitum*, occupying an emphatic position in the line, stands in Plautus, as we have seen, for everything insipid and oafish. *Barbaricum bliteum* is contrasted with delicatessen at *Cas*. 748; *bliteus* describes a mucky prostitute at *Truc*. 854: *blitea et lutea meretrix*. Spinach has become synonymous with 'rubbish'.[184]

Another comic inversion, which Fraenkel, Dohm, and Lowe regard as Plautine, is the transformation of the eaters into cattle (812), unthinking ruminants. This takes us back to another passage in Plautus that we looked at earlier, where a slave compares his lupin toy-money (*aurum comicum*) to cattle food, specifically Italian cattle food: *macerato hoc pingues fiunt auro in barbaria boves* (*Poen*. 597). The images of fatness and pulp (*pingues, macerato*) connect this passage with Plautus' name Maccus (cf. *pultiphagonides*) and with the notion of the play as an amplified, mashed-up version of the Greek original.

In other words, the *Pseudolus* passage probably has very little to do with any specific vegetarian craze. With garlic, spinach, and cattle food surfacing again, it makes much more sense as part of Plautus' ironic picture of Roman culture as a whole, insipid and worthless in itself, pepped up with a pungent mixture of native and exotic flavourings. The *alii coqui* are another manifestation of Plautus' 'barbarians', seen again through Greek eyes as alien and uncivilized foreigners. And underlying Plautus' parody of Roman culture is a parody of his own comic style, presented as a product of provincial, backwoods culture, and concealed in the culinary metaphors of the comic tradition. Crude and uncultured native farce is made piquant, strong-smelling, and indigestible with a surfeit of indelicate *condimenta*, which disembowel the 'guests' like screech-owls. The comic reversals of normal culinary proportions, flavourings outweighing the base, suggest not only technical incompetence,[185] but also

[183] The two are connected at Isid. 17. 10. 15: *blitum genus holeris saporis evanidi, quasi 'vilis beta'* (Spinach is a tasteless vegetable, like 'the vile beet'). See above n. 30.

[184] Cf. Laberius *Tucca* 92: *bipedem bliteam beluam*.

[185] Cf. Damoxenus (K3. 349 = Ath. 3. 102e): a cook laments that modern cooks do not know the proper rules and limits of seasoning.

cultural worthlessness: insipid material has to be camouflaged with immediately arresting embellishments, jokes, pleasurable language, and so on.[186]

Both this passage and Anaxippus' description of a new cookery school may seem on the surface to be polemics for *nouvelle cuisine*, but they are also advertisements for a 'new' style of comedy to replace the 'wicked pickle' of the highly flavoured old-fashioned version. There is one important difference, though, between the two. With Anaxippus, we can reasonably assume that the *new* cuisine is the one which we are to identify with this author's 'New Comedy'. In Plautus' case, we have become so used to the author's ironic perspective on his own provincial style that it is no longer clear what the distance is between himself and the bovine *alii coqui*.[187]

If we turn to the alternative, Ballio's cook himself, we find that he is a parallel for certain other aspects of Plautus, which begins to confuse matters. For a start, his claims for his own superior style of cooking are constantly being regarded with suspicion by Ballio, which gives the scene something of a metatheatrical dimension anyway. Ballio speaks for the audience, withholding his trust in the dramatic illusion, while the cook's hyperbolical claims stretch credulity to its limits. The emphasis on charlatanry links the cook with the deceitful poet conjured up earlier in the play. Ballio lumps the cook's flavourings and his lies together in one curse:

At te Iuppiter
dique omnes perdant cum condimentis tuis
cumque tuis istis omnibus mendaciis. (836–8)

The victims of culinary illusionism are treated as dupes; when Ballio asks how much he needs to pay the cook to teach

[186] Cf. the literary-critical use of ἥδυσμα to characterize literary trimmings: Ar. *rhet.* 3. 3. 3. 1406a. 18–19 (epithets); *AP* 1450b. 16: μελοποιία; 1449b. 28 (rhythm and harmony; Lucas ad loc. comments: 'the point of ἥδυσμα is that one does not need much of it'); Plut. *Mor.* 54f (jokes); Quint. 9. 3. 4 (*sal*, wit); 14. 35 (*voluptates*); 9. 3. 27: figures of speech are like bitter but pleasant food.

[187] Dohm (1964) points out that the criticism of strong flavourings is very similar to the Anaxippus passage discussed above; but he connects the two passages on grounds of subject-matter only, without seeing their common metaphorical potential.

him how to protect himself from his thieving hands, the reply is: *si credis, nummo; si non, ne mina quidem* (877, 'if you believe me, sixpence; if you don't, not even a farthing'). The cook's words echo those of the slave Pseudolus: *edico . . . ut caveant, ne credant mihi* (127–9 'I'll tell them to watch out, I'm not to be trusted').[188] And, when the cook leaves, Ballio's thoughts turn to Pseudolus again: *ne quis quicquam credat Pseudolo* (904, 'Pseudolus is not to be trusted').

Cooking as a dangerous or grotesque form of seduction seems to be the idea behind the cook's playful suggestion to Ballio (881–4) that he invite all his enemies to dinner, since the food will be so delicious that the guests will want to gnaw their fingers off.[189]

> nam ego ita convivis cenam conditam dabo
> hodie atque ita suavi suavitate condiam:
> ut quisque quidque conditum gustaverit,
> ipsus sibi faciam ut digitos praerodat suos.

I shall serve your guests such a spicy dinner today, spiced with such savoury succulence, that I'll make whoever tastes the spicy mess want to gnaw off his own fingers.

The indulgent alliterative style is reminiscent of the cloying speeches of Plautus' pimps,[190] which recalls Plato's parallels between rhetoric and cooking as branches of the art of flattery, and exploits the common links between the two sensual pleasures of eating and sex.

Another prominent feature is the emphasis on the cook as a thief, an element substantially more developed here than it is in any Greek play.[191] This can also be explained on a literary level as a playful reference to a sensitive issue: whether or not Roman literature was parasitic on Greek. Terence uses the word *fur* to describe a plagiarist in his

[188] Pseudolus as liar: cf. 581: *industria et malitia fraudulenta*; 671–2: a letter presents opportunities for manifold deceptions: *cornu copiaest . . . hic doli, hic fallaciae omnes, hic sunt sycophantiae*; 943: *mera iam mendacia fundes*.

[189] See Lowe (1985*b*: 415): this is a common Greek metaphor twisted and made literal.

[190] Though Ballio is comically uningratiating in this scene.

[191] See Dohm (1964: 142), Lowe (1985*b*: 413–14. Cf. Chiarini plant. (1979: 209) on thieving as an element of Greek cook scenes: Euphr. 1K (= Ath. 9. 379f), 10K (= Ath. 377d). See also Plaut. *Aul.* 322, 326).

Eunuchus apology (pref. 23–4: *exclamant furem, non poetam fabulam / dedisse*); Plautus, in the person of this cook, seems to be presenting the use of Greek material as petty theft.

Self-awareness is also, perhaps, behind another dialogue in this scene, the allusion to the myth of the archetypal evil cook, Medea, and her experiment with Pelias. It has always been thought surprising that the cook uses this primeval botched soup as a parallel for his own restoring broths:

> quia sorbitione faciam ego hodie te mea,
> item ut Medea Peliam concoxit senem,
> quem medicamento et suis venenis dicitur
> fecisse rursus ex sene adulescentulum,
> item ego te faciam. (868–72)

Because I shall have the same effect on you with my soup today as Medea had when she cooked up the old man Pelias. They say she rejuvenated him with her drugs and potions: that's what I'll do to you.

Surely he means Aeson, who was successfully rejuvenated, not Pelias, who ended up in the soup? Is this yet another case of Plautus making a mess of his original material?[192]

It is more likely, of course, that the mess is deliberate. After all, Ballio reacts (872) by asking: *es veneficu's?*[193] And our suspicions should be aroused anyway by the use of the verb *concoquere*. In other Latin accounts of the Medea myth, the word used is *recoquere*,[194] while *concoquere*, on the other hand, suggests boiling or cooking together,[195] or fusing together several ingredients,[196] always suggesting an irreversible process.[197] So the cook distorts the myth and ends up exposing the deceitfulness of his profession. In any case, he uses the ambiguous word *venenum*, poison or potion: *[senem] quem medicamento et suis venenis dicitur / fecisse*

[192] As argued by Fraenkel (1960: 77–8).

[193] See Forehand (1972).

[194] For example at Cic. *Sen.* 83, where the confusion between Aeson and Pelias is also made. *Recoquere* is also used metaphorically of any new beginning. Cat. 54. 6, Hor. *Sat.* 2. 5. 55, Petr. fr. 21, Quint. 12. 6. 7.

[195] e.g. Sen. *Ep.* 95. 28: *ostrea, echini, spondyli, mulli perturbati concoctique ponantur*.

[196] Varro, *Men.* 428: *humanae quandam gentem stirpis concoquit*.

[197] *Concoqui*, proposed by Ussing for *consequi* at *Amph.* 1055, would also imply irreversible confusion.

rursus ex sene adulescentulum (870–1). Ballio is justified, then, in asking whether the cook is a poisoner. The cook replies: *vero hominum servator* (873, 'No, I'm the saviour of mankind'). As a whole, Plautus' travesty of the Greek myth suggests his own self-deprecating view of his restoration of Greek drama, the *senem* to his *adulescentulum*.[198] Medea ('the cunning one'), an archetypal plotter and deceiver, is the perfect model for Plautus' clever slaves.[199] Further suspicions are awakened by a pun on *conquiniscet* (864, squat, or defecate) and *coquinare* (cook) a few lines later (874).[200] The art of cookery is contaminated by the mention of its end-product.

As for the cook's own fantastical repertoire (compare *multiloquom*), it revives all the Greek comic parallels between food and language as joint systems of largesse, invention, and manipulation. The resourcefulness of the cunning slaves, Plautus' doubles within the play, has already been described in culinary terms (742, 608, 396). Now the metaphor is realized. Instead of demanding olive oil and a new saucepan, Ballio's cook advertises the preposterous virtues of his own unique herbs (which can make a man live for two hundred years), unique in more than one sense of course, as they appear to be imaginary. The cook's comment on the 'other cooks'' condiments, *formidulosas dictu, non essu modo* (824), anticipates the far-fetched names of his own herbs.

As we might expect in the case of neologisms, there are many textual possibilities involved,[201] and the problem is compounded by the fact that we do not know whether to understand them as Latin or as Latin transliterations of Greek. Here again is the problem of Plautine originality in a nutshell. In the case of the first herb, *cocilendrum*, the ending seems to be Greek -ενδρον, but *coci-* suggests either an

198 NB the satirical diminutive. For rehashing, cf. Juvenal's *crambe repetita* (7. 154).

199 Petrone (1983: 52): 'Medea, come i servi Plautini, è un' eroina della μηχανή, ovvero delle *machinae*, cioè delle *fallaciae*, infine degli inganni.'

200 See Willcock (1987) ad loc.

201 *Cocilendrum* (*cocilindrum* A *cicilendrum* P); *cepolendrum* (A; *si polindrum* P); *secaptidem* (A; *sauccaptidem* P); *maccidem*; *cicimalindro* (Ussing; *ciomalindro* A *cicmandro* P); *hapalocopide* (A); *hapalopside* P); *cataractria* (AP; *cactaractria* B *catactri* C) (Leo's text and apparatus).

obvious connection with Latin *cocus* (cook), or perhaps with Greek κόκκος (berry, or pomegranate seed, used for dye; hence also scarlet), or again with Latin *co(c)cetum*, from Greek κυκεών, the mixed drink of honey and barley (from κυκᾶν, to mix).[202] If κόκκος (a seed), is intended, this would suggest links both with food and with the sort of paltriness suggested in Latin by *frit* and *nugae*.[203] *Cepolendrum* suggests both Greek κῆπος (garden), and Latin *cepa* (onion). *Secaptidem* (Leo) suggests *secare*, to cut;[204] *hapalocopide* (Leo) combines Greek ἁπαλός (delicate, luxurious) with κόπτω/κοπίς (chop/chopper), a tight compression of the paradoxes involved in the cook's job, where violent acts are performed for cultivated ends. The double sense of κόπτω in Greek, to chop and to bore to death, is a favourite joke in Greek cook-scenes.[205] Greek κόπις can also mean prater, liar; a pun on this in the mouth of the *coquus gloriosus* may be intended. *Hapalopside* (Lindsay) combines ἁπαλός with ὄψον, a cooked dish. *Cicimalindro* may be from Greek κίκι, the castor-oil tree. A connection is also suggested with apples, via Greek μάλινος. *Cataractria*, the last-named herb, suggests, as Leo observes, a contrast with *hapalocopide*: one a gentle pounding, the other a violent chopping. *Cataractria* is uncompromisingly violent, from καταράσσω, to strike, beat, or perhaps ῥαχίζω, to cleave, or, to play the braggart—again, like *-kopide*, suggesting links between the violent activities of the cook[206] and his wearisome talk (*multiloquom* 794).[207]

In short, there are a number of plausible meanings, but everything is extremely speculative. One can guess that, if the original text had survived, these nonsense words would turn out to be composed of intelligible and appropriate

[202] Ernout-Meillet s.v.: *cocetum* is linked with the same root as *coquo*. So associations with mixing in the kitchen may come across from either language.

[203] See above p. 59 on *nugae*.

[204] I cannot see the significance of *saucaptidem* (in Lindsay's text).

[205] See Handley (1968) ad Men. *Dyskolos* 397 f. (comparing 410; *Samia* 68–70, 285; Anaxippus 1K 23 (ἐμὲ κατακόψεις, οὐχ ὃ θύειν μέλλομεν; Alexis 173K 12; Sosipater 1K 20); cf. also Plat. Com. 35K; Gomme and Sandbach (1973) on *Samia* 285.

[206] Cf. *Rud.* 659: *oculos elidere, itidem ut sepiis faciunt coqui; Pseud.* 382: *exossabo ego illum simulter itidem ut murenam coquos.*

[207] *Insulsum* (794) also paradoxically contrasts the cook's repertoire of spices with his boring discussion of them.

elements, compact versions of the monster words of Aristophanes or Philoxenus. It is difficult, of course, to judge the balance of Greek and Latin elements, but the words would undoubtably have had a 'nonsensical' or 'unplaceable' flavour. We can at least note the frequency of reduplicated syllables, *coci-*, *cici-*, and rhyming word-endings, *maccidem, secaptidem, hapalocopide; cocilendrum, cepolendrum, cicimalindro*. These are the *scitamenta*, titbits, of a facetious writer, and they are especially appropriate in a hyperbolical, self-advertising description of cooking. I have deliberately omitted *maccidem* until now, because it seems to be the most telling of all these imaginary *condimenta*. It cannot be a coincidence that it sounds so similar to Plautus' name Maccus or Maccius, which he uses to link his own works detrimentally with the peasant diet of barbarian society and its animals. But it is Maccius with a difference. *Maccidem*, with its Greek 'patronymic' ending, lifts the Latin patronymic on to a pseudo-Greek level appropriate to the cook's pretensions.[208] The word appears in this list as a kind of signature for Plautus himself: Plautus *pergraecans*.

There is one last posture at the end of this scene to convince us that it has a metatheatrical dimension. Ballio, in frustration, banishes the cook to make the dinner: *intro abi et cenam coque* (890). Almost immediately (891–2), a slave appears and asks Ballio to take his seat before the dinner is spoiled: *iam corrumpitur cena*. This leap in time might be explained as an inadvertent slip or a textual omission; or *cena* could refer to the provisions ordered by Ballio rather than to the cooked dinner itself.[209] But all those parallels we have seen between cooking and other kinds of intrigue (of which the prime example is *Cas.* 511: *ibo intro, ut id quod alius condivit cocus / . . . condiam*) suggest that two kinds of dinner are being alluded to here: the real dinner, which the cook and Ballio have been discussing, and a metaphorical dinner, the pantryman Pseudolus' concocted plan, which has suffered a

[208] Cf. the patronymic *pultiphagonides* (*Poen.* 54) which links mundane diet with the heroic past. It may be significant that the list of herbs begins with *cocilendrum*, when the only extant occurrence of the word *maccum* is as a gloss on Greek κοκκολάχανον.

[209] Lowe (1985*b*: 415) thinks it is an inconsistency similar in style to *Men.* 225.

delay because of this interlude and is growing stale. Plautus has another similar image of a 'piping hot' plan (perhaps a pun on *calidus*, hot, and *callidus*, cunning?).[210] The word *corrumpitur* now leaves a nasty taste in the mouth: the dinner will indeed need to be seasoned.

Many different factors, the self-conscious character of the play, the gestures towards traditional comic metaphors, and the emphasis on thieving, charlatanry, and invention, seem to encourage a double reading of this scene. Above all, the cook is depicted as a parodist.[211] His pompous periphrases for 'meat' and 'fish' (*Neptuni pecudes, terrestris pecudes*, 834, 835) resemble one of Pseudolus' grandiloquent speeches, which at 707 provokes the response: *ut paratragoedat carnufex* ('How the rascal strikes his tragic attitudes!'). The cook, who is of course literally a *carnufex*, is introduced to parody the making of comedy itself. By serving mixed, flavoured soups of dubious worth or substance, he reinforces links between comic drama and the peripheral or trivial spheres of the kitchen or festival. He is not just 'para-tragic', but also 'para-comic', a parallel to, or parody of, the comic author.

Our reading of the scene needs to be a double one in other ways too. If cooking styles are comic styles too, which cooking style is supposed to represent Plautus: the incompetent 'barbarian' or 'other' style, or that of the purveyor of pleasure, drenched in all the most outlandish and Graecized flavours of Roman civilization—a manifesto signed, with a flourish, *maccidem*? However contradictory it sounds, neither ought to rule out the other: we need to be able to see-saw from one to the other in order to understand the complete Plautus. The 'barbarian' stereotypes are all too familiar. However, if Plautus is the 'civilized' cook, advertising his own unique flavourings, the *alii coqui* will have to be identified with other poets, either Greek or Roman, who

[210] *Epid.* 256; cf. *Most.* 665. See Brotherton (1978: 47). Cf. also Ath. 8. 354: 'Because of this long feast of words (*διὰ τὴν πολλὴν τῶν λόγων ἑστίασιν*), somebody ordered the cooks to see that the dishes should not get cold.'

[211] Slater (1985: 135): 'The cook is not yet capable of complete dramatic self-creation; his theatrical thinking is parodistic, borrowing earlier forms (the Medea story) and converting them to his own purpose'. Cf. Milazzo (1982) for a discussion of parody in Vespa's *Iudicium coci et pistoris* (*Anth. Lat.* 1. 1. 190. 76–93), where the cook borrows metamorphosis myths to exalt his own art.

have made a botched job of flavouring comedy. That would makes this one of those rare occasions when Plautus advertises his own skill as superior to that of his predecessors.[212] But the very fact that his boasts are voiced by a cook drags them into the sphere of the absurd, and cultural suspicions towards the *coquus gloriosus* come into play.

We have seen that many of the parodic aspects of Plautus' cook are common to Greek comedy as well. Nevertheless, a new cultural setting, the hybrid Graeco-Roman city, where comedy was one more stage removed from the 'pure' origins of drama, must have added a new dimension to the analogy between mixed food and mixed literary forms. Plautus 'tampers' with the shape and flavour of his originals,[213] sharpening parallels with *imbroglio* within the plot, and with Rome itself, a mixed civilization whose boundaries with the rest of the world are not easily defined. He uses the dual nationality of his plays to identify himself with two different peripheries of Roman culture. Seen through Greek eyes, he is a crude barbarian; seen through Roman eyes, he is a revelling feaster. Perhaps what emerges in the end from Plautus' plays is not so much an 'identity crisis', for Plautus or for Rome, as an ironic double perspective which makes distant some of the polarities that were by now *internal* to Roman culture.

Chapter appendix on *nenia*

Nenia commonly means 'trifle' or 'funeral dirge' (*OLD* s.v.), but ancient etymologists identify it with an edible part of the gut: Paulus (161M) calls it *extremum intestinum* (the end of the gut); Arnobius (*Adv. Nat.* 7. 24) includes it in a list of

[212] Plautus parodies the clichés of Greek comedy at *Pseud.* 1081–2: *nugas theatri, verba quae in comoediis / solent lenoni dici, quae pueri sciunt* ('theatrical rubbish, boys' slang, the words they say to pimps in comedies'); *Cas.* 860–1 (dramatic intrigues): *nec fallaciam astutiorem ullu' fecit / poeta atque ut haec est fabre facta ab nobis* ('no poet ever created a more deceitful plot than the one we've so cleverly cooked up here'); *Bacch.* 649: *non mihi isti placent Parmenones, Syri* ('I haven't any time for these Parmenos and Syruses'), a joke at Menander's expense: Handley's papyrus of *Dis Exapaton* reveals that the slave in Menander's play was called Syrus.

[213] Gratwick (1982: 105): 'He [Plautus] cuts, stretches, squashes and amplifies.'

sausages, describing it as an intestinal waste-duct; it is listed as a pork dish in the *Notae Tironianae*. Heller (1939) argues that this late sense of the word arose out of a misunderstanding of Plaut. *Bacch.* 889, or from a false connection with Greek νήτη, the lowest chord of the lyre (hence Paulus' definition as *ultima cordarum*, where *corda* can mean both a lyre-string and a gut). He argues (1943) that the general meaning of *nenia* is the same as Greek παίγνιον, 'a plaything', easily transferable to other meanings—a trifle, literary or otherwise—and a funeral dirge (since impromptu mimes were often performed at funerals). Since the slave has been using culinary similes throughout this scene (cf. *frictum cicer*, 767, *turdus*, 792, *putidus fungus*, 821), *soricina nenia* must, therefore, mean a morsel of food which the shrew has gnawed holes in and then saved to play with. However, Heller's objection that a shrew's gut would not cover much of a sausage is irrelevant, as these other examples are very paltry kinds of food. Whether or not Paulus' etymology is right (if it were, *nenia* the funeral dirge could be explained by analogy with 'coda'), and whether or not he based it on this passage of Plautus, it is still revealing that he and others think a connection between a trifle or funeral dirge and the lowest part of the intestine is plausible. For the link between a sausage and a trifle, cf. Varro, *LL* 5. 111, who derives the word *hilum*, 'the least bit, a jot', from the word *hila*, 'a small intestine, a sausage'. Varro's catalogue of Roman gods began with Janus and ended, appropriately, with Nenia: *clausit ad Neniam deam* (Aug. *Civ. Dei* 6. 9). For associations between offal and rubbish, cf. James (1982) on the Northern dialect word 'kets', which originally meant 'offal' and is now used of children's joke-sweets.

3
Black Pudding
ROMAN SATIRE

Introduction

The cook pulled back his tunic, snatched up his knife, and proceeded to slash the pig's stomach all over, his hand trembling. At once, out of the wounds stretched by the slithering weight of the pig, came tumbling mounds of blood puddings and sausages.[1]

This famous passage from the *Cena Trimalchionis*, which describes the mock-gutting of a pig, fuses two essential impulses of ancient satire: primeval *frisson* and comic relief. Full-blooded satire was, in Horace's words, the kind that stripped the skin off society, revealing the rot beneath the shining exterior.[2] Here, Petronius only scrapes the surface: the moment of exposure is snatched away at the last minute, and civilization is restored in its more grotesque and risky form. But often satirical food goes closer to the bone; it tends towards the putrid, disgusting, or taboo, exposing the tenuous and often arbitrary divide between what is considered edible and what inedible, embodying corruption in the midst of civilization. Food is in the guts of Roman satire, not just because of its laughable qualities and its capacity for grabbing at our most basic instincts of revulsion: the very name, 'satire', is culinary by origin, and Petronius' grotesque sausages are an internal image of the form itself. Any ancient satire is, by virtue of its name, its author's own black and

[1] Petr. *Sat.* 49. 9–10: *recepta cocus tunica cultrum arripuit porcique ventrem hinc atque illinc timida manu secuit. nec mora, ex plagis ponderis inclinatione crescentibus tomacula cum botulis effusa sunt.*

[2] *Sat.* 2. 1. 64–5.

putrid offering, stuffed for the table with an acrid and dubious blend of spices, giving a sour or salty kind of pleasure to its recipients.[3]

Our best source for the origins of the word *satura* is the fourth-century AD grammarian Diomedes, who offers four possible etymologies: (1) from *satyri*, satyrs; (2) from *lanx satura*, literally, a full dish; (3, and related) a kind of forcemeat or stuffing, a *farcimen*; (4) the name of a political bill with many headings, the *lex satura*, or, more accurately, *lex per saturam*.[4] Diomedes gives first place to 'satyrs', and explains the connection as irreverent or naughty humour: *res ridiculae et pudendae*. But it is now generally thought that at least one of the culinary etymologies is correct. *Lanx satura* was a dish of harvest fruits, from *satur*, 'full up'; in other words, the dish was personified as though it had eaten its fill. The sausage sounds like a culinary imitation of this raw dish. Varro, apparently, had a recipe for it: raisins, polenta, pine-nuts in honeyed wine, and, if liked, pomegranate seeds.[5] Again, this was a sort of personification: a stuffing imitating the contents of a full human gut. So *satura* was originally some kind of mixed dish named by analogy with a person or his stomach, mixed with a great variety of things and bursting at the seams.[6]

The fourth possibility, the bill with many heads, is clearly a case of putting the cart before the horse: the bill was named by analogy with medleys, not vice versa. When we find the phrase *per saturam* on its own, it means something like muddled or messy (its opposite is *generatim*, classified by type).[7] It was an insulting name given to controversial bills at the time when Ennius was producing the first *saturae*: the

[3] Quintilian on figures of speech which give the same pleasure as bitter food: 'They have a certain attraction because they are almost defects, like bitter foods, for which you can still acquire a taste' (*habent quendam ex illa vitii similitudine gratiam, ut in cibis interim acor ipse iucundus est*, 9. 3. 27).

[4] Diomedes, *Grammatici Latini* (Keil) 1. 485.

[5] '*Uva passa et polenta et nuclei pini ex mulso consparsa*'. *Ad haec alii addunt et de malo punico grana*.

[6] See Van Rooy (1965: 2–20).

[7] Fronto (212N): 'If we categorize arguments by type, rather than piling them up indiscriminately, like some hotchpotch (*non sparsa nec sine discrimine aggerata, ut ea quae per saturam feruntur*).'

kinds of suspicions attached to these bills shed some light on the self-consciously dubious qualities of satire itself. By the time of Diomedes, the satyr etymology seems to have taken over; this shift occurred when satire came to mean any kind of burlesque or ribaldry, rather than a specific miscellany. 'Satyric' humour had always been a part of satire, and the confusion is understandable. Even so, it is the culinary origins of the word that will be of most interest here.

The equation of writing with cooking goes back to comedy, as we saw in the last chapter. There is an obvious overlap between comedy and satire:[8] both genres were easily identified with the dubious art of cookery because they shared a similarly humble position in the generic hierarchy. As we saw, some comedies had culinary titles, for example Alexis' *Lebes* (The Cooking Pot) or Plautus' *Aulularia* (related to *olla*, 'pot'). There were many other possible names for miscellanies in Greece and Rome. Pliny and Gellius both give lists of them: bunches of flowers, bundles of clothes, cornucopias, and so on, and Pliny comments on their amazing aptness (*mira felicitas*).[9] But to name a work after a dish of food, or, going further back, after a bulging human body, is a particularly undignified advertisement, which sounds more grotesque and messy than enticing. The connection between literary miscellanies and culinary terms survives of course today in words such as farce, olio, gallimaufry, pot-pourri, olla podrida (both meaning 'rotten pot'), hotchpotch, mish-mash, macedonia, mélanges, pastiche, and macaronic poetry (the Renaissance nonsense verse that mixed Latin and vernacular languages).[10] 'Miscellany' itself, as we shall see from Juvenal 11, was originally a dubious plate of leftovers eaten by gladiators.[11]

[8] Noted by Diomedes, who says that satire was originally written in the style of ancient comedy: *archaeae comoediae charactere compositum*. Cf. Hor. *Sat.* 1. 4. 1–2.

[9] Plin. *NH* pref; Gell. *AN* pref. Some of the titles in this list, for example κέρας Ἀμαλθείας, 'cornucopia', and πάγκαρπον, 'basket of fruit', have obvious connections with the harvest dish *lanx satura*.

[10] Cf. John Gay, 'On a Miscellany' (from *Lintott's Miscellany*, 1712), lines 1–10: a mixed poem as 'the Muses' *Olio*', which is 'compounded of all kinds of poetry'.

[11] 11. 20. Tertullian, in a list of proverbial mixtures (*Adv. Val.* 12), includes *Acci Patinam*, 'Accius' dish', and *Nestoris Coccetum*, 'Nestor's barley-drink', though whether these refer to ingredients or titles we shall never know. *Coccetum* translates

This connection between mixed literary forms and their subject-matter is obviously not confined to satire. We have already seen its appearance in Plautus. And the *Moretum*, the pseudo-Virgilian work which describes the concoction of a herb-flavoured cheese (the solid version of Nestor's mixed drink), is such a mixture of styles and levels that a cosmogony out of warring elements seems to be taking place in one tiny mixing-bowl.[12] Here, there is a very close fit between subject and form: the cheese and the poem are created simultaneously. Similarly, Apuleius' *Metamorphoses*, a patchwork of *variae fabulae*, has at its centre an image of its own stuffed contortions. A baroque punishment is devised for the ass and his damsel in distress (6. 31): he is to be gutted, dried in the sun, and 'stuffed' with the girl he has rescued, transformed into a stuffed sausage like those that often seduce him on his journey: *insiciatum et fartilem asinum*.[13] The ass stuffed with a girl is the structure of the book in miniature: the rich and pleasurable *Golden Ass* stuffed with the story of Psyche.[14] Centuries later, Martianus Capella describes his *Marriage of Mercury and Philology*, influenced by the style of Apuleius, as a stuffed mixture of useful and worthless ingredients: *miscilla . . . docta indoctis adgerans, / fandis tacenda farcinat immiscuit* (this idea of satire as a curate's egg is a constant theme).[15] Pungent with herbs or rich and cloying? Satire is neither: its flavour, as we shall see, is more vinegary, saltier, and black to the core.

Abundance and variety were the two main qualities implied in the term *satura*. According to Diomedes, the word signified abundance and fullness: *a copia ac saturitate rei*. While there are no surviving instances of the phrase *lanx satura* in ancient literature, a passage from Virgil cited by

Greek κυκεών, which we have already seen used to exemplify a self-sufficient whole (see above p. 47 n. 212).

[12] e.g. 115: *circuit inque globum distantia contrahit unum*. See p. 47, above.

[13] Cf. 10. 16: the ass stuffs himself with food: *iam bellule suffarcinatus* ('now I was nicely stuffed').

[14] There are also parallels between the narrator Lucius and the narrator of *Cupid and Psyche*, an old woman who combines the roles of story-teller, comforter, and cook.

[15] 9. 997–8.

Diomedes includes the phrase *lancibus pandis*, which conveys the bulging qualities suggested by *satura*.[16] Diomedes' own definitions of *lanx satura* and *satura* as *farcimen* also record the idea of physical stuffing implied in their names: 'stuffed with a huge variety of first-fruits' (*referta variis multisque primitiis*); 'stuffed with a variety of ingredients' (*multis rebus refertum*). Meanwhile, Quintilian stresses the *mixed* nature of the genre: 'not just a varied mixture of songs' (*non sola carminum varietate mixtum*, 10. 1. 95).[17] Of course, the notions of fullness and mixture are often different ways of looking at the same phenomenon: different parts coexisting create an impression of abundance in the whole.[18] The mixed dish provides a model in ancient culture for any coexistence of parts and whole, from the cosmological gnome[19] to the sympotic riddle.[20]

We cannot entirely rule out the other etymological possibilities for *satura*—the 'satire' law and the satyr—because there are traces of their influence throughout Roman satire.

[16] *Georg.* 2. 194. Cf. schol. on *pandis*: 'either shallow, or bent by the weight of the entrails.' Van Rooy (1965: 12) suggests that Virgil is also paraphrasing *satura* in the phrase *oneratis lancibus* ('loaded dishes': *Aen.* 2. 215, 8. 284). It is interesting that these *lances* hold entrails rather than fruit: another link with the stuffed *farcimen*?

[17] Cf. Fest., Lindsay, 416: '*satura* is a kind of food flavoured with a variety of seasonings' (*cibi genus . . . ex variis rebus conditum*); Isid. 20. 2. 8: 'composed of a variety of ingredients' (*vario alimentorum adparatu compositum*).

[18] Isidorus' definition of *farcimen* (20. 2. 8), for example, includes both minced-up meat and the capacity of the new whole to 'fill out' a gut: 'forcemeat is minced meat so-called because it is used to stuff a gut, that is, fill it, with a mixture of other ingredients' (*farcimen caro concisa et minuta quod ea intestinum farciatur, hoc est impleatur, cum aliarum rerum commixtione*). Van Rooy (1965: 4–5) notes that the capacity of *satura* to embrace both fullness and miscellaneity had already been exploited by Plautus at *Amph.* 667: the body of Alcmena is described as *saturam*, simultaneously 'full of food' and 'pregnant', and, since she is doubly pregnant or over-full with the two children of different fathers, 'full of a miscellany': her stomach holds a mixed stuffing. On the ambiguous title *Satura*, of works by Naevius, Atta, and Pomponius, see: Van Rooy (1965: 45 n. 5); Ullman (1914: 22–3). *Gastron*, 'The Pot-Belly', is the title of a play by Antiphanes (Ath. 10. 448e).

[19] Heraclitus fr. 125DK likens the world to a κυκεών, a mixed drink, which would separate into its constituent parts if it stopped being stirred: καὶ ὁ κυκεὼν διίσταται [μὴ] κινούμενος. See above p. 47 n. 212.

[20] See *Anth. Lat.* for sympotic riddles exploiting the dual or triple nature of mixed or stuffed dishes: e.g. 281. 82: on *conditum*, a three-in-one mixed drink: *tres olim fuimus qui nomine iungimur uno. / ex tribus unus, et tres miscentur in uno. / quisque bonus per se, melior qui continet omnes*. Cf. Apicius' recipe for *conditum paradoxum* (1. 1. 1); and cf. other paradoxical mixtures in *Anth. Lat.*: 217 (mincemeat); 222 (a mixed dish); 165 (a goose stuffed with the rest of the lunch).

Laws and lawbreaking are a perpetual concern of the satirist, who alternately plays moral censor and insatiable voyeur. Both Horace and Juvenal refer to a 'law' of satire (Horace at the beginning of Book 2, where he claims that his critics have slated Book 1 for exceeding the 'law', and Juvenal in *Satire* 6 where he soars off into flights of transgressive sublimity).[21] Clearly the law is there to be broken, and that is part of the joke. Satire is the one Roman genre that really has no defined boundaries, the exception that proves the rules about other genres, which are normally based on principles of decorum. We tend to talk of mainstream and off-beat satire, as though satire followed or deviated from rules: with Menippean satire as a sideline, Ennius as a red herring, Persius as a cul-de-sac. But to call a satirist 'anarchic' is something of a tautology.[22] To this extent, references in satire to rules should be taken with a pinch of salt.

At the same time, a teasing reference to 'satire laws' must be involved as well. These were not so much real laws as a bad name given to any bill that was suspected of hiding unpopular measures among the popular ones.[23] It is curious that this insulting name began to be bandied around at around the same time as Ennius was writing the first *saturae*.[24] The idea of a legal hotchpotch, a mixture of good and bad measures all wrapped up with deceitful promises, exploits old cultural prejudices against mixed dishes. The similarities emerge in an ancient commentary on Cic. *Mil.* 14:

hoc autem solebat accidere cum videbatur aliquis per saturam de multis rebus unam sententiam dixisse. et habebat nonnumquam conexio huiusmodi rerum multarum fraudulentas captiones, ut rebus aequis res improbae miscerentur atque ita blandimentis

[21] Hor. *Sat.* 2. 1. 1–2: *et ultra / legem tendere opus*; Juv. 6. 635: *et finem egressi legemque priorem.*

[22] As Bramble (1982: 599) calls Juvenal.

[23] Diomedes (1. 485 *GLK*) says it was named after the satire law, which included many items under one heading: *lege satura quae uno rogatu multa simul comprehendat, quod scilicet et satura carmine multa simul poemata comprehenduntur* (cf. Isid. 5. 16: *satura vero est lex quae de pluribus simul rebus eloquitur*).

[24] The bills were eventually banned by the *Lex Caecilia Didia* of 98 BC: *ne quis per saturam ferret.*

quibusdam obreperunt ad optinenda ea quae si per se singulariter proponerentur, displicere deberent.

It was often the case that a mixture of ingredients was presented in the form of one bill. Sometimes a number of suspicious clauses were included under this heading, a mixture of unreasonable demands with reasonable ones: by this devious method, motions were passed which would have been rejected had they been presented on their own.

Like the sausage-seller with his spiced donkey-meat in the *Knights*, the author of the composite bill is suspected of seducing his victims with an uneven, unidentifiable mixture.[25] Martianus Capella used the same idea to describe the uneven mixture of his own all-inclusive work: piling learned ingredients on to unlearned ones, stuffing unspeakable words with utterable ones.[26] Take it or leave it, pot luck, seems to be the challenge with which satire provokes its audience: the whole truth or no truth.

Culinary confusions and satire laws may well be being linked in two adjacent fragments of Lucilius. One fragment seems to be referring to a *lex per saturam*.[27] Another, describing the trial of a fishy character called Lupus (= bass) and other small fry, nicely ties together the connections between legal and culinary brews and the genre named after them:[28] *occidunt, Lupe, saperdae te et iura siluri* ('Wolf [-fish]! salt mullet and sheat-fish will serve you in court-bouillon'). It has been remarked of fr. 47, 'Lucilius, fighting for the conservative cause, would naturally attack a *lex per saturam*.'[29] But Lucilius, ironically enough, was himself writing a work *per saturam*, a medley of good and bad elements. The fact that the development of *satura* was contemporaneous with 'satire laws' suggests that writers consciously identified themselves with a political *bête noire*:

[25] Cf. Justin. *Praef. dig.* p. xv: a hotchpotch mixing useful and useless ingredients (*[opus] passim et quasi per saturam collectum et utile cum inutilibus mixtum*).

[26] *Docta indoctis adgerans, / fandis tacenda farcinat, immiscuit* (9. 998–9).

[27] Fr. 47W.

[28] See Ahl (1985: 97); this is his translation. NB the pun on *ius*, sauce/law, which is frequent in satire.

[29] Ullman (1913: 180).

the moral conservatism of satire coexisted, in a perverse way, with the practice of generic anarchy. Even within the *œuvre* of Ennius and Seneca, authors who also wrote in conventional genres, the satire represented a supplement or a rag-bag of indeterminate worth.

Similarly, although the culinary etymology may be right, it would be naïve to think that, to a bilingual Roman, the word *satura* did not also conjure up satyrs. In late antiquity the adjective *satiricus* came to mean 'satirical' in our sense, as well as 'pertaining to a satyr'.[30] Satire is spelled *satira* in Horace' *Satire* 2. 1, justifiably, as he goes on to present himself as a very pathetic drooping satyr, Horatius *Flaccus*, who has lost his nerve (*vires / deficiunt*, 12–13), and there is almost certainly a reference to the alternative root when he presents himself as a tamed Priapus in *Satire* 1. 8. It is also hard to believe that Petronius was not punning on the similarities in his title *Satyrica*.[31]

Any direct connection between satire and satyr-plays has been pretty much ruled out.[32] But the two genres clearly share the element of burlesque. Satyr-drama, tacked on to the end of the tragedy, was a supplementary parody of the 'pure' form. It could even have been the confusion of the two words *satura* and *satyrus* that made satire develop along mocking and in our sense satirical lines.[33] The evidence for satyr-plays at Rome may be scanty,[34] but satyr-dancers performed an equivalent role in the procession in the circus (*pompa circensis*), where they would prance behind the serious actors who were impersonating the dead or triumphing man, and imitate their movements to ward off divine

[30] Van Rooy (1965: 136): Greek σατυρικός had come to mean 'mocking' or 'derisive', 'satirical' in our sense, by 200 BC, i.e. by the time Ennius started writing satires.

[31] Van Rooy (1965: 155) rejects this, on the grounds that the *Satyrica* lacks the moral function of Varro's Menippean satires.

[32] Though Van Rooy detects what we would call 'satirical' elements in some late Greek satyr-dramas, the Menedemus of Lycophron (a satire on a meagre meal) and Sositheus' *Daphnis or Lityerses*, dealing with a glutton.

[33] Indeed, Van Rooy (1965: 156) maintains that it was because of the potential for confusion in the adjective *satiricus* that it did not come to be applied to satire until the genre was nearly extinct.

[34] See Wiseman (1988).

nemesis.[35] So the satyr is a good analogy for the satiric persona: well-fleshed, earthy, and irreverent.[36]

Another similar-sounding word is Saturnalia, the December festival of misrule, at which gambling, feasting, and freedom for slaves were licensed. The festival gave much greater scope for free speech: jokes, puns, pillorying of great men, and so on. It was associated either with *more* food than usual (the sumptuary laws allowed extra expenditure over this period) or with food that was unusual in other ways, made of mud or other such inedible substances: in other words, boundary-crossing or joke food.[37] This method of expanding or transforming food is associated with satirical descriptions too, which suggests that the Saturnalia acted as an invisible context for much of Roman satire, particularly when it was connected with botched or preposterous feasts. Again, there is no strict etymological connection between the two words (*Saturnalia* is from *Saturnus* rather than *satur*), but the similarity seems to have helped to cement together all the other associations that satyrs, Saturnalia, and satire shared: carnival licence, parody, uninhibited mockery, and excessive consumption.

Petronius' *Satyrica*, indeed, sets every possible etymology in play at once. The farcical *Cena Trimalchionis*, a Saturnalian gathering of ex-slaves, is composed of stuffed or pregnant dishes, which are complemented by the literary pastiches served up as interludes. The menu includes two 'satyrical' dishes: a pastry Priapus propping up a cornucopia of fruit, and peeling wineskins representing the flayed satyr Marsyas.[38] The narrator is metaphorically satiated before he

[35] Dion. Hal. *AR* 7. 71–2 (citing Q. Fabius Pictor).

[36] It is perhaps because of this that the ultimate, absurd stage in the fusing of *satura* and *satyri*, the scholiasts' etymology of the word from *saturi*, 'fat people', was ever regarded as plausible (e.g. the scholiast on Persius claims that satyrs were so called because they were full of food and wine). Compare the pun in the baker's speech in Vespa's *Iudicium coci et pistoris* (*Anth. Lat.* 1. 1. 190): *Thyrsiten[ens] Satyros: facio et saturos ego plures* (44: 'The thyrsus-bearer leads the satyrs: I make more people satiated'). Hor. *AP* 224 calls the audience of Athenian satyr-plays *exlex*, temporarily released from laws.

[37] Petr. 69. 9; also *SHA Elagabalus* 26. 7, 27. 1–6.

[38] Stuffed dishes: 33. 5; 36. 1; 40. 5; 49. 10; 60. 6; 69. 6: pastry thrushes stuffed with nuts and raisins (similar to Varro's *satura* recipe?). Saturnalia: 44. 3: *istae*

has even reached the table, filled with pleasures, sated with astonishment,[39] and his disgust steadily mounts to the inevitable conclusion: *res ibat ad summam nauseam* (78. 5, 'things reached the height of nausea').[40]

Even so, the persistent link between *satura* and cooking has tended to be underestimated. Satire began as a medley, any mixture of verse and prose, Greek and Latin, speech and singing, or different levels of diction; by the time of the classical satirists, Horace, Persius, and Juvenal, it had lost this meaning and been transformed into something approaching its modern form, a malicious exposé of vice and pretension. This is the usual theory, and it corresponds with the view of Diomedes, who says that satire was 'a kind of Roman poetry which nowadays is abusive and composed to censure human vices in the manner of Old Comedy, the kind written by Lucilius, Horace, and Persius; formerly, it was just a mixture of different poems, as written by Pacuvius and Ennius'.

There is no reason, however, why the two sorts of poetry should have been mutually exclusive.[41] Unevenness and ludicrous bulging, as formal devices, actually help to serve the purposes of abusive or caricaturing poetry; in other words, mixture and misproportion have always been a vital part of satirical technique. One of the best ways of deflating pretension is through parody, which essentially involves juxtaposing two disparate ingredients, a grand style with a mundane subject, or classifying something below its proper station, for example Seneca coining *Apocolocyntosis*, 'Gourdification', instead of the expected *Apotheosis*, 'Deification',

grandes maxillae Saturnalia semper agunt ('for those big jaws it's one long Saturnalia'); 58. 2: *io Saturnalia, rogo, mensis december est?* ('Happy Saturnalia! Is it really December again?'); 69. 9: *vidi Romae Saturnalibus eiusmodi cenarum imaginem fieri* ('I've seen this sort of mock-dinner at Rome at the Saturnalia'); Priapus: 60. 4; Marsyas: 36. 3.

[39] 30. 5: *his repleti voluptatibus*; 28. 6 *admiratione iam saturi*.

[40] Cf. 64. 6: a retching puppy is stuffed with half-eaten bread: *nausea recusantem saginabat*.

[41] Even the diatribe, which is thought to have given satire its moralizing focus, was a hybrid form: Van Rooy (1965: 4): 'a motley compilation of elements derived from the philosophical treatise, the oration, the school lecture, and the dialogue'.

for his satire on the death of Claudius.[42] Caricature works by exaggerating a prominent feature out of all proportion, so that it is almost grafted on to the body from outside: for example, Persius uses *aqualiculus*, a word for a pig's stomach, to describe human pot-bellies, which at the same time exposes the animal aspects of the human body.[43]

The culinary etymology in fact played an essential part in both the theory and the practice of satire. The title of Varro's lost treatise, *de Compositione saturarum*, is nicely ambiguous, suggesting either a satirical 'Poetics' or a recipe-book.[44] And when Varro appears as a character in a dialogue of Cicero, he uses culinary images to describe his own Menippean satires:

in illis veteribus nostris, quae Menippum imitati, non interpretati, quadam hilaritate conspersimus, multa admixta ex intima philosophia, multa dicta dialectice. (*Top. Acad.* 1. 2. 8)

In my early works, which were imitations, rather than interpretations of Menippus, I sprinkled in plenty of humour and added a good dash of serious philosophy and intellectual discussion.

The words *conspersimus, admixta*, and *invitati* suggest the concoction, seasoning, and convivial sampling of a dish. And this 'recipe', with its ill-assorted mixture of ingredients, has something in common with Varro's other recipe, as it survives in Diomedes, for the forcemeat *satura: uva passa et polenta et nuclei pini ex mulso consparsa* (raisins, polenta, pine-nuts doused with honeyed wine).[45] In Livy's account of dramatic *saturae*, which is probably adapted from Varro, the phrase *impletas modis*, 'filled with tunes' (7. 2), also seems to be an allusion to satire's culinary origins.

[42] The image of the gourd, proverbial for empty swelling (see Eden 1984: 4), is a humiliating comparison for Claudius' swollen head, typical of the satirist's techniques of inflation and deflation.

[43] Pers. 1. 57. See Bramble (1974: 111): Isid. 11. 1. 136 (pig's stomach); Vegetius, *Mulomedicina* 40 (horse's stomach).

[44] *Compositio* can refer to literary or culinary composition. Titles of recipe-books: Ath. 14. 643e πλακουντοποιικὰ συγγράμματα; Pollux, *Onom.* 6. 70 ὀψοποιικὰ συγγράμματα. See in general *RE* s.v. *Kochbücher*.

[45] This is attributed by Diomedes to Varro's *Quaestiones Plautinae*. The word *satura* may have been used in a culinary sense in Plautus, which would give further support for the argument that there is a relationship between food in Plautus' plays, images of fat bodies, and his own amplified version of the Greek original. However, as Van Rooy (1965: 16) points out, Varro's note could simply be incidental to his comments on e.g. *Amph.* 667. The juxtaposition none the less draws attention to connections between the stuffed dish or stuffing and the stuffed body.

What is more, Horace, Persius, and Juvenal themselves seize on culinary metaphors to describe their own variations on the satirical 'recipe'. Horace, it will be argued later, uses cooks and recipes to exemplify his own restrained use of mixed ingredients. Persius offers *aliquid decoctius* (1. 125, 'something more concentrated'), a metaphor from boiling down liquids,[46] and serves his own *plebeia prandia* (1. 18, 'plebeian lunches'), suggesting his scorn for pretension (the metaphor is not entirely suitable, as will become clear later). Juvenal refers to the *farrago* (mixed mash) which feeds his book (1. 86), and there are other, less explicit allusions that can be pointed out in his work.

A fragment of Ennius, which describes a greedy parasite, has been singled out for showing that there was a moralizing element embryonic in satire right from the start.[47] In fact, it can also be used to show how important links with culinary stuffing still were. One phrase that the passage includes, *inferctis malis*, 'with stuffed jaws', recalls the original 'stuffed sausage' definition of satire (with perhaps a pun on *mala*, jaw, and *malum*, evil, thrown in). If moralizing was there at the beginning, notions of mixture and stuffing were equally persistent.

This all suggests that, when food crops up elsewhere within satire, its meaning is not merely literal. It is wrong, then, to separate obvious allusions to the origins of *satura* from other references to food within the satirical text. Mixed dishes, or mixed combinations of dishes, need to be seen not only as symptoms of moral frailty and corruption, but also as internal metaphors for the generic form, particularly if they are the focus of 'set pieces', such as a meal, a tableau, an offering, or a recipe. Since *satura* is a bodily metaphor, the

[46] Cf. Cic. *de Or*. 3. 26. 103; Quint. 2. 4. 7. Bramble (1974: 139) quotes Anderson (in the preface to Merwin's (1961) translation of Persius: 'His purpose has been to boil down ideas to their minimum.' *Coccetum* and *conditum*, which are both used, like *satura*, as metaphors for paradoxically disparate wholes, are also liquids (though Festus defines *coccetum* as a kind of foodstuff).

[47] Van Rooy (1965: 32 n. 19); fr. 14–19 Vahlen (= 14–19W), beginning: *quippe sine cura laetus lautus cum advenis / inferctis malis expedito bracchio* ('since you bounce in without a care in the world, your jaws stuffed full, your sleeves rolled up'). He categorizes it as one of those 'satirical pieces in which the poet describes and mocks or criticizes harmful types in society'.

same goes for swollen or protuberant bodies: foodstuff and body are often confused in satire. The kind of food that is prominent in satire tends to blur the distinction between men and animals, or between food and the inedible: bulging stuffed entrails in the place of real guts, dubious stews with human-looking limbs in them, rotten or excremental messes, which are exhibited to us as disgusting or morally objectionable symptoms of the society to which they belong. As in Plautus, the stuffed stomach and womb (*vulva*) are paramount among satirical dishes. This is not only because of their capacity to straddle both alimentary and physical spheres, and exemplify moral excess as well: they are also perfect metaphors for *satura* inside the text.[48]

This brings us to one of the most freakish properties of satire, which has never been properly acknowledged: the genre which exhibits grotesque or distended bodies as images of excess exemplifies the same deformities itself. Food in satire is usually only considered as a moralizing focus. A vital contribution of John Bramble has been to point out another important aspect: its role as a literary metaphor. According to the long metaphorical tradition we have seen, bloated or cloyingly rich food embodies the excesses of contemporary style, while a simple and wholesome diet stands for literary restraint.[49] So, for both Horace and Persius, rich food is simultaneously the object of moral and aesthetic repugnance.[50]

By uncovering this extra metaphorical dimension, Bramble has in fact opened a can of worms. Nowhere in his discussion of the contrast between *plebeia prandia* and bloated feasts[51] as metaphors for literary 'consumption' (45–59) does he mention how apt these images are within *satura*, which is itself a bodily or culinary metaphor. Nor does he seem to be aware of the complications that this

[48] Stomach or gut: Pers. 1. 57: *aqualiculus*; Juv. 4. 107: *venter*; Pers. 2. 42: *tucceta . . . crassa*; Pers. 6. 74: *illi tremat omento popa venter*; Juv. 10. 355: *exta et candiduli tomacula porci*; Petr. *Sat.* 49. 10: *tomacula cum botulis*; Petr. *Sat.* 66: *botulo . . . sangunculum et gizeria . . . cordae frusta et hepatia.* Womb: Juv. 11. 81: *vulva*; cf. Hor. *Ep.* 1. 15. 41: *nil vulva pulchrius ampla.*

[49] See Bramble (1974: 45–59).

[50] Ibid. 46–7, 54–6, 111–12, 114–15.

[51] e.g. tragic *mensae* (5. 17) or the 'stewpot of Thyestes' (5. 8).

involves. The fact that *satura* was a bodily metaphor automatically gave it moral and aesthetic connotations. Fullness and variety might often be desirable characteristics, but there is a thin line between these virtues and their corresponding vices of fatness and messiness, which were grotesque deviations from a classical bodily ideal.[52] *Ubi uber, ibi tuber*, 'where there is abundance, there is also malignancy', according to one Latin proverb.[53] When it was not a sign of pregnancy, the swollen stomach was a physical and moral aberration.[54] Obesity embodied moral excess,[55] it was also an image of stylistic misproportion for Callimachus and his followers,[56] for whom the fat woman, originally Antimachus' Lyde, embodied the bulky text, the παχὺ γράμμα καί οὐ τορόν (fr. 398 Pf.).[57] Fat women continue to represent literary expansiveness in later European works, and the 'grotesque body' for the Renaissance, with none of its humps or orifices suppressed, is also a successor in this tradition.[58] Mixture or

[52] Van Rooy (1965: 17) shows how *satur* could be used in a favourable sense, of fertility and abundance: e.g. Virg. *Georg.* 4. 335, Sen. *QN* 1. 5. 12: of colour; *Aetna* 12: of crops. Offerings of *lances saturae* to the gods were presumably symbols of the earth's fertility, microcosms of the harvest. *Copia* is similarly a rhetorical virtue, within limits (see Cave 1979: 5: its antitheses are '*inopia*, poverty of diction . . . empty prolixity (*loquacitas*), *copia* without *varietas*, or Asiatic over-elaboration'; he refers to Cic. *Brut.* 13. 51; *Orat.* 69. 231; *Opt. gen.* 3–4; Quint. 2. 4. 5; 8. 2. 17; 10. 1. 62, 3. 2, 5. 22).

[53] Quoted by Apuleius, *Flor.* 18. The conflict is exemplified in his *Metamorphoses*, where the narrative is both a retrospective account of the perils of *voluptas* and the product of newly-learned rhetorical *ubertas*. Stuffed or rich dishes in the text can be seen as figures for cloying pleasure *and* for the literary *copia* to which Apuleius aspires: e.g. 1. 4, 2. 7, 5. 15, 7. 11, 9. 22.

[54] Plaut. *Cas.* 677: women are criticized for stuffing their stomachs with food (*ventres distendant*), an action implicitly opposed to the more desirable form of female distension, pregnancy. Cf. puns on two forms of abdominal distension, from eating and pregnancy, at Ath. 6. 246.

[55] *Excedere* was originally a physical metaphor, as we can see, for example, from Pliny's description of expanding cucumbers at *NH* 19. 65: *cum magnitudine excessere (cucumeres), pepones vocantur*. For gourds as images of the distended stomach, cf. Virg. *Georg.* 4. 122: *cresceret in ventrem cucumis*; *Mor.* 76: *in latum dimissa cucurbita ventrem*; Prop. 4. 2. 43: *tumidoque cucurbita ventre*; Col. 10. 380: *praegnans cucurbita*. See above pp. 13–14 on *luxuria* as a physical metaphor.

[56] See Bramble (1974: 35–8) on the history of bodily metaphors used of disease or distortion of the *corpus orationis*; cf. 56–8 on the associations of fatness, literary and moral.

[57] See below, Ch. 4 pp. 233–4 on Catullus' Quintia.

[58] Cf. Bakhtin (1968); Parker (1987: 8–35), apparently unaware of the Callimachean precedent.

messiness is similarly ambivalent. In the culinary sphere, dishes made of mincemeat or scraps smack of speciousness or the second-rate. Their jokey names—*catillum concacatum* or *penthiacum*—suggest the excluded zones of messy slaughter or rotten excrement, like the modern olla podrida or pot-pourri.[59] Even descriptions of more luxurious mixed dishes, like Seneca's *patina* of mixed seafood which resembles half-digested pulp or vomit, are blighted with suspicion.[60]

This prejudice against composite food, which we have seen in accounts of *leges per saturam*, also emerges from the 'culinary' history of Greek literature charted by the dramatists. 'Pure' drama undergoes a progressive deterioration, from Homer's grand feasts, to Aeschylus' offcuts (τεμάχη), to the meretricious pickle used by Euripides to disguise Aeschylus' rotting meat.[61] Demagogic rhetoric in the *Knights* is likened to sausages made of minced-up dog or donkey meat disguised with a piquant sauce (1399; cf. 343, 215–16). Like over-fat works, literary 'scraps' were also regarded with contempt.[62]

[59] *Penthiacum* (Petr. *Sat.* 47. 10; cf. *pentheus*, Vespa *Anth. Lat.* 1. 1. 190. 77); Greek πνίγω 'to throttle, choke', is used of cooking in a covered vessel (cf. Eng. 'sweat'), e.g. at Ar. *V.* 511; cf. Ath. 9. 396a–b. Cf. ὀνθυλεύω, 'to stuff with forcemeat' (from ὄνθος dung); *catillum concacatum* ('shitty stew': Burman's conjecture for *catillum concagatum* at Petr. *Sat.* 66. 7).

[60] Sen. *Ep.* 95. 26–9. In Roman *haute cuisine*, mixed dishes were often juxtaposed with plain ones for the sake of variety. Macrobius' menu from a pontifical feast includes simple dishes followed by their stewed version: oysters and mussels are followed by oyster stew and mussel stew; sow's udder by sow's udder stew. Sidonius, *Ep.* 2. 9. 6 describes the 'senatorial' style of serving abundant food on a few dishes, achieving variety by having some stewed, some roasted. A dish known to Pliny as *tripatinium*, which he describes as the height of culinary elegance, consisted of lampreys, bass, and mixed fish (*NH* 35. 162). Cf. the debate in Macrobius (7. 4. 3 – 7. 5. 32) on the virtues of mixed versus plain food.

[61] Ath. 8. 347e; Ar. fr. 130.

[62] See Bramble (1974: 143–6) on *esca, frustum, globulus*, and other pejorative images of food which denote meretricious details or superficial scraps of information. Cf. two other examples of disciples 'gleaning' from their masters: Sen. *Ep.* 27. 7 (*analecta*, the slave who picks up scraps from the table); Philostratus, *Life of Apollonius* 1. 19: Damis, A.'s biographer, accused of being like a dog who picks up scraps from a feast, retorts: 'The gods have attendants to pick up ambrosia from their feasts.' For scraps from feasts of words, cf. Ath. 6. 223d: 'We restore to you the morsels (λείψανα) left by the dinner-sophists'; *Love's Labour's Lost* V. i. 39, Moth (on the macaronic mixture of Latin and English spoken by Holofernes and Nathaniel): 'They have been at a great feast of language and stol'n the scraps.'

Where, in that case, does satire itself stand in the debate between bloated and mixed food and pure and simple food taking place within its own pages? It seems to deviate from the ideal physical form in two ways, being both over-stuffed and heterogeneous. If the muse of elegiac poetry was the beautiful and well-proportioned body of the poet's mistress,[63] the muse of satire was none other than the bloated bodies, protuberant guts, and messy stews that are the chief objects of the satirist's abuse. As we shall see, the contradiction is resolved with some difficulty by the Callimacheans, Horace and Persius. Their satires are still crammed with images of excess, and display all kinds of internal inconsistencies, stylistic and otherwise. The contradiction seems to be one of the many odd properties of satire, a form which cannot expose corruption without wallowing in it as well, and cannot represent the ugliest and filthiest aspects of Roman life without reeking of them itself.

In the light of all this evidence for the ambiguous status of composite dishes, bills, and literary works, Quintilian's proud claim for 'Roman' satire, *satura quidem tota nostra est* ('Satire at least is exclusively Roman'), begins to look a bit ragged.[64] Miscellaneous, and wallowing in the muddy and messy areas of life, satire drags itself straight down to the bottom of the Roman literary hierarchy, which has room for epic, tragedy, and history, unified, important genres, at the top. From the earliest times, it was a rag-bag which held all the aspects of Roman life for which there was no room anywhere else: autobiography, jokes, daily conversation, miscellaneous lists. And food, a subject which provokes blushes and apologies, a subject which, however civilized, is inextricably linked to bodily functions, belongs there too. Satire is Rome's own humble pie, composed of the offal of every other genre. What sort of a novelty was it? Simply a patchwork of all the other genres grafted on to Roman

[63] See Wyke (1984).

[64] Quint. 10. 1. 95.

culture from the Greeks.[65] The only purely Roman genre was its most impure, a medley of styles and subjects.[66]

In its way, then, satire is a peculiarly suitable by-product of Roman culture. Satire food overlaps with the negative food images that the Romans used in other contexts to represent their corrupt civilization: a plate of mushrooms mixed with poison and destined for one of the emperors, a dish of shellfish so mashed up that it resembled vomit (cited by Seneca as a symptom of topsy-turvy morals), or a huge platter of delicacies that exhausted the whole Roman empire (Vitellius' so-called Shield of Minerva).[67] And a curious vindication of Quintilian's 'Roman' satire can be found in an erotic epigram by Meleager, a catalogue of pederastic pleasures which ends: 'If the gods granted you these delights, you lucky man, what a Roman dish of boys you would be garnishing!'[68] For *lanx satura* to be translated as Ῥωμαικὴν λοπάδα, 'Roman dish', suggests that mixed dishes were already associated with Rome, the supreme example of a hybrid civilization. Today, we still use culinary mixtures as convenient images of mixed cultures, not just literary miscellanies: for example, the gumbos of Creole cuisine as an image of the Southern melting-pot. Quintilian gives satire the status of a tattered carnival king, an emperor with no

[65] The metaphor of the variegated patchwork is often an alternative to mixed food created out of miscellaneous ingredients: e.g. *cento*, 'rhapsode'; and see above Ch. 2 n. 127. The farce-title *Pannuceati* suggests a connection between the hybrid, 'third-rate' status of the mime and the costumes of its actors, ancestors of the harlequin's suit and the fool's motley ('harlequin' was actually the name of a stew made of scraps in 19th-c. Paris: see J. Brown (1984: 96 ff.)). Cf. Sid. *Ep.* 2. 2. 6 on the multicoloured costumes of comic actors. Eratosthenes uses the image of multicoloured clothing to designate the adulteration of philosophy: 'Bion was the first to dress philosophy in flowery clothes' (Diog. Laert. 4. 52); cf. flower images used of anthologies, e.g. Ἴα, violets, Λειμών, meadow, Ἀνθηρῶν, flowers (Plin. *NH* pref. 24); also Στρωματεῖς, tapestry hangings, bedspread (Gell. *NA* pref. 6). The *cento* also appears in the picaresque novel (Petr. *Sat.* 7. 2; Apul. *Met.* 7. 5) suggesting a connection with its disparate episodes, cf. Apul. *Met.* 1. 1: *conseram*. Cf. Lucilius' patchwork images: fr. 841W; fr. 1026W. The fact that this one image can be applied across so many genres, the picaresque novel, comedy, satire, mime, suggests that they share an identity: all of them are constructed in opposition to 'first-rate', 'first-hand' canonic writing.

[66] Cf. Horace *Sat.* 1. 10. 66: satire is paradoxically the only literary form 'unsullied' by the Greeks.

[67] Tac. *Ann.* 12. 67; Sen. *Ep.* 95. 26–9; Suet. *Vitell.* 13. 2.

[68] *Anth. Pal.* 12. 95. 9–10.

clothes. Ironically enough, the Romans' contribution to literature is the verse-form that lays bare the real Rome; satire and its native city are both infinitely adulterated.

Horace

1 *Introduction*

Horace breaks all the rules for satire that I have just laid out. He is neat, not messy, subtle, not crude, and reticent, not outspoken. The satirical recipe of the Republic has been drastically changed. This is partly, Horace tells us, a sign of the times, a response to new threats of censorship and prosecution.[69] But he is not just making a virtue out of necessity. Lucilius' loose-tongued, sprawling mess is also aesthetically undesirable. Satire, for the neo-Callimachean, must also be art; Lucilius' *παχὺ γράμμα* 'fat book', must be slimmed down.[70] The result is virtually an oxymoron: *tenuis satura*, 'slim fat dish'. Horace reconciles these opposites by bringing out notions of 'satisfaction' rather than excess. His satires are an exercise in knowing limits—literary, political, moral—with *iam satis est* ('that's quite enough') as their universal slogan.

What is the shape and flavour of Horatian satire? Later we shall look at two poems which seem to provide the answer: *Satires* 2. 4 and 2. 8. Both of them deal with gastronomy: one is a list of instructions, the other describes a disastrous dinner party. They are normally looked at in isolation, and their subject is regarded as nothing more than a contemporary phenomenon ripe for satirical treatment.[71] In fact, the social meal and the capacious stomach are prominent images right from the beginning of Book 1, and this is because Horace is still aware of the culinary origins of *satura*. They

[69] On the question of satire and libel laws, see LaFleur (1981).

[70] Cf. Euripides 'slimming' Aeschylean tragedy at Ar. *Frogs* 940. On Callimachus' influence on Horace see Wimmel (1960); Clausen (1964); Cody (1976). On the opposition between *pinguis* and *tenuis* as literary terms, see Bramble (1974: 56–8, 156–8).

[71] e.g. by Rudd (1966) and Classen (1978). For a more sophisticated discussion along these lines, see Hudson (1989).

are images that link the literary form with its subject-matter: they cover questions about living in society and the place of the writer in that society, and the right flavour and size of a literary text. Horace's move towards gastronomy in Book 2 is, it will become clear, an ironic expedient; in an imperfect, post-Republican society, the writer has to compromise his own ideals. Above all, gastronomy in Horace is a metaphor for the necessary transformation of satire.

In Book 1 the satirical recipe is still in transition. Republican satire, acerbic, uncontrolled, and vicious, is being toned down into something more subtle, disciplined, and innocuous. Nostalgic traces of Republican *libertas* remain: the lost freedom of Lucilius (4, 10), primitive ritual clowning and abuse (5), Priapic obscenity (2, 8), squabbling Pompeians (7), rustic Italian invective (7), all of which, Horace says explicitly, make their audiences laugh. But sharp satire is not always palatable for its victims, and it is being used here as a foil for Horace's own more soothing brew of irony, disarming apology, personal modesty, self-criticism, and reticence. Horace's jokes, like sweet biscuits offered by indulgent schoolmasters to children learning their ABC, neutralize the bitter poison of Republican satire: the black ink of the cuttlefish or undiluted copper-rust, as he calls it, or bruised and biting savagery.[72] Lucilian wit becomes the spiteful *invidia* of the outrageous dinner-guest, who slanders even his host; free speech is perversely reduced to drunken abuse.[73] This priggishness on Horace's part of course needs to be taken with a pinch of his own black salt. Fundanius in 2. 8, as we shall see, merrily makes mincemeat of an entire dinner party, and the host's efforts in particular.

Another 'recipe' for Republican satire is presented in the form of a lawsuit in Pompey's camp. *Satire* 1. 7 is a spitting cauldron of abuse made up of acerbic or inedible flavours: *pus atque venenum* (1, 'pus and poison'), *sermonis amari* (7, 'bitter speech'), *salso* (28, 'salty'), *Italo . . . aceto* (32, 'Italian vinegar'), boiled together (*compositum*, 20) into one acrid ketchup (*in ius / acres procurrunt*). The idea of a trial as a brew

[72] 1. 4. 100: *hic nigrae lolliginis, haec est / aerugo mera*; 1. 4. 93: *lividus et mordax*.
[73] With a pun on Liber, Bacchus.

of culinary ingredients revives the '*court bouillon*' of Lentulus Lupus described by Lucilius: satire's favourite pun, on *ius* (court-case, or gravy), brands litigation as a nasty pickle, drawing on the analogy between suspiciously mixed dishes and composite bills implied in the term *lex per saturam*. Horace, it seems, is not so much describing a duel between the flavours of Horatian and Lucilian satire,[74] as scrutinizing the whole process of Republican abuse and disowning it. In this, however, he has little choice, and it is doubtful whether Horace is not in fact nostalgic for his native sauce, 'Italian vinegar', abuse which could once be heard in every hedgerow.[75] These acrid tastes, after all, inject their flavour into his satires.

Even so, Horace's final recipe is new and sophisticated, made from a tighter use of resources. Crude laughter is no longer enough (*non satis est risu diducere rictum*, 10. 7):

> est brevitate opus, ut currat sententia, neu se
> impediat verbis lassas onerantibus auris;
> et sermone opus est modo tristi, saepe iocoso,
> defendente vicem modo rhetoris atque poetae,
> interdum urbani, parcentis viribus atque
> extenuantis eas consulto. ridiculum acri
> fortius et melius magnas plerumque secat res. (10. 9–16)

You need brevity, to make the train of thought flow, and not block up the ears with a weight of verbiage. And you need a style that moves from solemn to joking, playing orator and poet by turns, and making room for wit too, the kind that can judge when to spare its strength and when to spin it out. Humour packs more of a punch than spite, and is better at solving knotty problems.

Humour is still an ingredient, but with its sting taken out. Mixture is still there, but toned down and patterned. To mix Latin and Greek is a gaffe, like mixing Falernian and Chian wine.[76] To use a modern analogy, Horace's satire is not

[74] As argued by Perini (1975). The fact that both Persius and Rupilius Rex use 'swollen' and 'gushing' language suggests that they are both guilty of muddy Lucilian excess (*tumidus*, 7, *multoque fluenti*, 28).

[75] 1. 7. 30–1.

[76] 1. 10. 23: '*at sermo lingua concinnus utraque / suavior, ut Chio nota si commixta Falerni est.*'

'saturated', but 'polyunsaturated', less clogging, more metabolically efficient. The list of qualities is worth remembering, as it surfaces again in Book 2 in an unexpected form.

As for the shape of satire, Horace uses the human appetite in Book 1 as his model for the proper recognition of limits, moral and literary. In Satire 1. 1 images of satisfaction and surfeit flesh out a homily on the theme of dissatisfaction with one's lot:[77] the natural boundaries of appetite (50), the plot of land (50), the granary (53), moneybags (70), the human stomach (46), all overflowing with superfluous greed.[78] This Epicurean tirade is also a literary polemic against the excessive consumption of words. Horace sets a good example to the prolix Stoics with his own concern for the jaded reader: *ne te morer* (14, 'to keep it brief'), *nec longa est fabula* (95, 'the story is not a long one'), *iam satis est*, (120, 'that's quite enough'). Later in the *Satires*, garrulous Lucilius takes their place as the antithesis of Callimachean restraint.[79] The final caution (*iam satis est*), and the simile of a man leaving life like a satisfied dinner-guest (*uti conviva satur* 119[80]), stamp satire with the image of sufficiency rather than distension, and set up parallels between life and the *convivium* that will have further implications for *Satires* 2.

Snatches of autobiography in *Satire* 6 add to this impression of Horatian compactness. His modest diet (6. 115) and perfectly controlled appetite: *pransus non avide, quantum interpellet inani / ventre diem durare* (127–8, 'after a light lunch, enough to satisfy the stomach for the day') match his economy with words: *pauca locutus, / infans namque pudor prohibebat plura profari* (56–7, 'I could barely speak, the embarrassment stopped me stuttering more than a few words'). By contrast, Lucilius' meals are part of a continuous

[77] This is really only a generalized philosophical extension of *invidia*, the traditional impulse of satire: see Hubbard (1981: 319).

[78] Tantalus, insatiable appetite personified (68), echoes in the sound of his name the endless quantifying of unlimited amounts which surrounds him: *tantundem* (52), *tantuli quantum* (59), *tanti quantum* (62). See ibid. 312.

[79] The image of the ambitious man swept away in the waters of the river Aufidus (54–8) prepares us for the Callimachean muddy-river imagery transferred to Lucilian satire at 1. 4. 11.

[80] Adapted from Bion, via Lucretius. See Glazewski (1971) on the history of the simile.

flow with his literary output: *amet scripsisse ducentos / ante cibum versus, totidem cenatus* (10. 60–1, 'he liked to have written two hundred lines before dinner and the same again afterwards'). Above all, the promise of *Satires* 1 is modest hospitality. Like the inn on the road to Brundisium, stuffed (*differtum*, 5. 4) with sailors, landlords, and rhetoricians, it is a convivial tight squeeze of filth and learning.[81]

In *Satires* 2, the restrictions close in. Horace is advised by a legal expert that satire is too dangerous, he suffers from writing blocks, and is left out of dinner parties. Reticence is taken to its logical conclusion. Horace relinquishes the platform to a series of self-appointed pundits, recording them with Socratic irony: his *sermones* belong to other people (*quiescas*, 2. 1. 5: 'be silent'; *nec meus hic sermo*, 2. 2. 2: 'these are not my words'; *celabitur auctor*, 2. 4. 11: 'the author must remain a secret'). His voice only returns when he escapes to Sabine self-sufficiency (*sermo oritur*, 71). For Fraenkel, the apparent petering out of the satires (there are only eight poems instead of ten) was itself an indication that Horace recognized the limits of his satirical powers. In fact, the two books are the same length: the extra material in Book 2 has been absorbed by 2. 3, a monster Stoic sermon,[82] on which Horace eventually imposes a limit: *iam desine* (323, 'it's time to stop'); *teneas . . . tuis te* (324, 'keep to your limits'). The book ends with a feast at which nothing is eaten. It seems more likely that Horace is playing with our expectations of 'satisfaction' from his satires.

Horace lays out the problems of writing satisfying post-Republican satire in the first lines of the book, mixing all the traditional etymologies of satire in the process—satyrs, laws, and mixed dishes:

> Sunt quibus in satira videar nimis acer et ultra
> legem tendere opus; sine nervis altera quidquid
> composui pars esse putat, similisque mearum

[81] The clientele (*differtum nautis, cauponibus atque malignis*) are the same picaresque scum who made up the mad hurly-burly of *Satire* 1: *perfidus hic caupo, miles nautaeque* (1. 129).

[82] This is the ironic solution to Horace's writing block: *sic raro scribis* (1 'you write so rarely'); *nil est* (6 'the page is blank').

mille die versus deduci posse.

Some people think my satires are too keen and break all the rules; others think everything I write lacks sinew and verses like mine can be squeezed out at the rate of a thousand a day.

He is caught between two contradictory complaints: his writing is either too sharp and aggressive, or too flabby and insipid. If he is a satyr,[83] he has been both too keen and too drooping.[84] He has also been accused of infringing 'the law', an indeterminate amalgam of real libel laws and so-called *leges per saturam*.[85] The legal consultation which follows becomes the right context for a discussion of contemporary *satura*. Literary-critical and legal verdicts are eventually merged: *ius iudiciumque* are visited upon *mala carmina*, bad poems, or malicious ones (82–3). Finally, a culinary metaphor is involved too: *acer* also means 'sharp to the taste', and reminds us of the outspoken combatants in the legal pickle of 1. 7 (*in ius / acres procurrunt*). Now, Horace intimates, the satirical recipe will have to be toned down and made tighter.[86]

The ambiguous words *acer* and *ius* revive the links made in *Satires* 1 between sharp-flavoured satire and its impact on a sensitive regime. They are a sign that eating is going to continue to provide parallels for the consumption of words, both in writing and in conversation. But the changes are all too obvious. For a start, food is now a theme in its own right: the fashionable science of gastronomy has taken over the lives of the élite and become a sinister instrument of power and exclusion. The book still has a convivial frame: the *convivium* is the occasion *par excellence* for *sermo*,[87] and Horace follows Cicero in using it as a model for 'living

[83] He posed as a municipal Priapus in 1. 8.

[84] Horatius Flaccus is shortly to plead literary impotence: *cupidum . . . vires deficiunt* (12, 'the desire is there, but I've lost my nerve').

[85] See Van Rooy (1965: 69) on Horace's reform of Lucilius' *lex saturae* (see Lucil. fr. 47W).

[86] There is no obvious culinary sense for *sine nervis*, but I shall argue later that Horace does produce one.

[87] *Sat.* 2. 6. 71: *sermo oritur*. Quint. 6. 3. 14, 6. 3. 28 thinks dinner parties and daily chit-chat the right occasions for jokes.

together', 'life in society'.[88] The number of poems is the same as the number of guests at Nasidienus' party, and the model of Socratic dialogue revives the ideal Platonic form of the *Symposium*. But *Satires* 2 is a *convivium* gone wrong. The imbalance of speech between pundit and listener in each *sermo* is not only the antithesis of the ideal dialogue, but also of the ideal *convivium*, from which Varro, in his satire *The Night is Young*, banished *loquaces . . . convivas [et] mutos* ('garrulous guests and dumb ones').[89] Horace is eager to dine out, but by the end of the book he is still looking for a companion (*quaerenti convivam*).[90] He has rejected all the philosophies of living he has encountered, and can find no *conviva* in the widest sense of the word too: someone who is socially compatible. In many ways, the final fiasco of a dinner party is the expected metaphor for the breakdown of social communication.

Nimis acer is the first of many culinary puns spilling over from the gastronomic passages into moral, literary, or social contexts. Once again, we are reminded that the vocabulary of taste, intelligence, pleasure, wholesomeness and corruption, adequacy and excess has a culinary base. As well as *acer*, Horace uses *acerbus* ('sharp'), *dulcis* ('sweet') and *amarus* ('bitter') of the flavour of writing and the pleasures and pains of experience.[91] *Sapiens* means 'tasty or savoury', 'tasteful' and 'wise'.[92] *Putidus, integer, vitiatus* are used of wholesome or rotten food or judgement. *Vitium* is both a moral and a literary fault, a breach of etiquette, or putrefaction in food.[93]

[88] Cic. *Fam.* 9. 24. 3; *de Sen.* 13. 45.

[89] *Nemo scit quid vesper serus vehat*: *Men.* 330.

[90] 2. 8. 2.

[91] See above p. 41. *Acer*: 2. 1. 1. In 1. 7, *amarus* (7), *acetum* (32), *acres* (21) are applied to bitter speech. At 2. 7. 107, *inamarescunt epulae sine fine petitae* ('feasts without end lose their savour') puns on the sensory and abstract meanings of *amarus*.

[92] Taste (*sapor*): 2. 4. 36, 54; tasteful (*sapiens*): 2. 4. 44; wise (*sapiens, sapere*): 2. 1. 17, 2. 2. 3, 63, 2. 3 *passim*, 2. 7. 83, 2. 8. 60. On *sapor, sapere* as literary metaphors, see Bramble (1974: 50 ff.). For a pun on the culinary and literary senses, see Petr. *Sat.* 2. 1, 3. 1; on the culinary and moral senses at Cic. *Fin.* 2. 25 (praising Laelius' wise indifference to gastronomy; Horace alludes to this at 2. 1. 72: *mitis sapientia Laeli*).

[93] *Putidus*: 2. 3. 75 (of judgement), 2. 4. 66 (of food); *integer*: 2. 1. 85 (of salvation), 2. 4. 54 (of food). *Vitium/vitiatus/vitiosus*: 2. 2. 21, 78 (moral fault,

Bonus and *malus* can be transferred across gastronomic, moral and literary codes. *Sal* is both 'salt' and 'wit'.[94] *Ius* is both 'sauce, gravy' and 'the law'. The word appears in its legal sense at 2. 1. 82, of the juridical proceedings which threaten any aggressive modern satirist. Its return to the culinary sphere later in the book (2. 4. 38, 63) makes a parallel between the codification of the law and gastronomy, and also marks the law as a kind of institutionalized mess akin to food.[95]

The confusion created by these puns is licensed by another implied frame, a disruptive one: the Saturnalia. The festival is named as the background for two satires (2. 3, 2. 7), where a mad Stoic and a cheeky slave assume free speech to attack authority in the shape of the author: the satirist turns on himself. Horace suppresses them too late, only after they have given us a new and unflattering picture of him as a consumer. Horace is a split soul who plays town and country mouse at whim. He claimed to be satisfied with his lot: *modus agri non ita magnus* (2. 6. 1, 'a plot of land, not too large'); *nil amplius oro* (4, 'I ask no more'); *uncta satis . . . holuscula* (64, 'greens with just enough dressing'), but turns out to have an appetite out of all proportion to his size. He only praises country vegetables when he is not loitering outside urban dinner parties; he secretly prefers *epulae sine fine petitae* to *tenuis victus*. He is also a pretentious, over-spending dwarf, liable to inflate himself like an ambitious frog.[96] The body which knew its limits perfectly in Book 1, then, is both stunted and inflated in Book 2.[97]

suggesting putrefaction), 2. 4. 76 (breach of etiquette), 2. 3. 92, 307 (moral fault), 1. 4. 9, *AP* 31 (literary fault).

[94] 2. 2. 17, 2. 4. 74, 2. 8. 87 (salt); 1. 10. 3 (Lucilius' wit commended); cf. Quint. 10. 1. 94 on Lucilius: *acerbitas et abunde salis*; *Ep.* 2. 2. 60, *sal* is used of the *Sermones* as part of a dinner-party metaphor for Horace's writing: *Bioneis sermonibus et sale nigro* ('the coarse wit of Bion's diatribes'). See Bramble (1974: 53) for other examples of *sal* as salt or wit.

[95] Cf. 2. 8. 45, 69. See p. 77 above.

[96] 2. 7. 29–32; 107; 2. 3. 307–25. Horace becomes part of the panorama of dissatisfied human beings he surveyed in *Satire* 1. 1.

[97] 2. 3. 308: *ab imo / ad summum totus moduli bipedalis* ('from top to toe all of two feet'); cf. *Ep.* 1. 20. 24: *corporis exigui* ('a tiny body'). Bramble (1974: 159–61) sees these descriptions as complements to Horace's stylistic consumption: 'Horace the moralist depicted himself as the incarnation of humility and simplicity.' But cf. *Ep.*

What are we to make of these double meanings and unsettling contradictions? The trivial and serious uses of Horace's words inevitably compromise each other. On the one hand, they remind us that questions of diet had always been on the fringes of moral philosophy. For Ofellus ('Little Morsel'), the old countryman who speaks in 2. 2, principles of diet (*vivere parvo* and *victus*) are part of a whole way of life. On the other hand, gastronomy *per se*, food as more than just a fuel for the body, is a different kettle of fish. Rehashing the terms of Greek philosophy in a new gastronomic context, or, alternatively, spattering philosophy with the terms of the kitchen, is Horace's way of exposing Roman culture as materialist, and parasitic on Greece.

Saturnalian puns and confusions also make havoc, as we shall see, of the neat parallels between good writing, simple food, and the ideal society that appear in Book 1, and later in the *Odes*.[98] These parallels may have stayed intact in the countryside, but the satirist is now urban and sophisticated, like the society he depicts. We are still aware of Horace's sense of proportion, and excessive eating is unequivocally damned. But the relaxed suppers of the Republic, where the food was rancid or makeshift, and the eaters were truly *sapientes* (wise, if not tasteful), have been replaced by perfect dinner parties with stifled conversation, an uneasy parallel for Horace's correct but hampered writing. In a social context, Horace appears to be nostalgic for the freedom of the *ancien régime*. In a literary context, however, control has

1. 4. 15–16, where Horace becomes a ridiculous pig: *me pinguem et nitidum bene curata cute vises / cum ridere voles Epicuri de grege porcum* ('you can visit me when you want a laugh, a fat and shiny piece of bacon from Epicurus' herd'); and the *Vita Horati*: *habitu corporis fuit brevis atque obesus qualis et a semet ipso saturis describitur et ab Augusto* ('in shape he was short and fat, according to his own satires and the evidence of Augustus'). The letter from Augustus compares Horace's stomach to a pint-pot, or the *volumen* on which his poems were written: *sed tibi statura deest, corpusculum non deest* ('you may lack stature, but you don't lack girth').

[98] See Mette (1961) on *mensa tenuis* and the *genus tenue* in the *Odes*. Occasionally in *Satires* 2, food and speech become direct opposites. Damasippus and Ofellus compensate for being *sobrius* (2. 3. 5: 'sober') and *impransus* (2. 2. 7: 'going without lunch') by regaling Horace with words. Damasippus compares his own conversion to that of Polemon, who, when drunk (*potus*, 2. 3. 225), was taught temperance by Xenocrates (*impransi magistri*, 257). The irony is of course that Damasippus has not yet learned verbal temperance.

become not only necessary but tasteful too. The concentration of meaning in Horace's words for taste and sense shows what the threat to Republican *libertas* has lost and gained for the satirist.[99]

2 *Horace*, Satire *2. 4*

Satire 2. 4 is stuffed with food; at the same time it is oddly desiccated, a puzzling and unconvivial poem.[100] Horace persuades a man named Catius to let him into a secret: a list of *nova praecepta* (given to him by their unnamed *auctor*) good enough to supersede those of Pythagoras, Plato, and Socrates. It comes as a shock to find that these precepts are gastronomic ones. Catius starts plodding through a deadpan, miscellaneous catalogue of gnomic culinary advice: how to choose eggs, make a tough hen tender, cure constipation, make sauce, clean a dining-room, and so on. We realize that the poem is a parody—most obviously, as its frame suggests, of didactic philosophy.[101]

Our sense of surprise alerts us again to the ambivalent status of food-discussion. Diet had always been a philosophical concern, and in one sense Catius can reasonably suggest that he is following in the steps of the Greek philosophers. However, his *nouvelle cuisine* is an end in itself: what it teaches is trivial. Ovid sheds some light on the poem when he includes rules for banquets and entertaining (*epulis leges hospitioque*) in a list of subjects for Saturnalian mock-didactic poems. It belongs, then, with the *Lex Tappula* and other *leges convivales* as a seasonal skit; it invests convivial activity with spurious importance.[102] Yet, as we have seen, the comic inversions licensed by the Saturnalia were often a

[99] One conclusion of the discussion in Plato's *Timaeus*, whose opening words are parodied at 2. 8. 4–5, had been that the state guardians should live together as a community (ξυνδιαιτωμένους, 18b).

[100] Coffey (1976: 85): 'To assess the tone and intention of this work is unusually difficult.'

[101] *Ede hominis nomen* (10, 'yield the name of the man') and *canam* (11, 'I shall sing') parody epic. Other gastronomic parodies of epic are thought to have influenced Horace: Archestratus' *Hedyphagetica* (and Ennius' imitation) and Varro's Περὶ Ἐδεσμάτων.

[102] Ov. *Trist.* 2. 488. See above p. 29 on surviving convivial *leges*.

means of commenting on the strangeness of normality, and this parody has its serious aspects too. It is really a symptom of the differences between Plato's society and Horace's: the *convivium*, epitome of all social intercourse, now has as its focus food, not conversation, while even conversation, *sermo*, is saturated with gastronomy, not philosophy. Rome is now the centre of a materialist world, and *hospitium* has become a calculated art.

There is an extra layer of meaning in this poem, but that will be revealed later. First, there is more to be added to the usual interpretation of the poem (see Rudd 1966; Classen 1978), which considers the paradoxical overlap in vocabulary between cuisine and philosophy.

What is immediately striking is that Catius' gastronomic system has no place in the simple/luxurious opposition so common in Latin moralizing (illustrated, for example, by Ofellus' speech in 2. 2). Some of his ingredients would have been expensive: African snails (58–9), Falernian wine (19, 24, 55), boar (40–3), hare (44), prawns (58), Lucrine shellfish (32), Tyrian coverings (84). Others are cheap and simple: eggs (12), cabbage (15), unrefined salt (74), cheap shellfish (28), ham and sausages (60–2). What seems to unite this miscellany instead is a system of decorum based on knowledge, timing, economy, variety, and elegance, where the comfort of the guest, his palate, jaded appetite, digestion, and surroundings are of greatest concern.[103] It is a system poles apart, for example, from the luxurious *cena dubia* satirized at 2. 273 f:

> at simul assis
> miscueris elixa, simul conchylia turdis,
> dulcia se in bilem vertent stomachoque tumultum
> lenta feret pituita.

But you mix roast and boiled food, thrushes and shellfish, all together: sweetmeats are turned to bile, and sluggish phlegm pumps havoc down to the stomach.

[103] Cf. cooks in Greek comedy: Sosipater (Ath. 9. 377f. = K3. 314); Anaxippus (Ath. 9. 403e = K3. 296); Dionysius (Ath. 9. 404e = K3. 325). On harmony in seasoning cf. Damoxenus (Ath. 3. 102a = K3. 349); Machon (Ath. 8. 346a); Plut. *Qu. Symp.* 657d, e.

A fastidious selection is made from the vast range of foodstuffs flooding into the emporia of Rome, from Africa, Byzantium, Cilicia, the Italian countryside, and the local greasy cookshops. We shall never know what relation Catius' precepts bore to contemporary doctrine, as we would in any case expect cookery, like medicine, to be disputed by different schools. However, they were at least plausible enough for Pliny to ascribe the doctrine that long eggs have a better flavour than round ones (12–14) to Horace himself.[104] All we can assume is that the recipes realistically parodied the idiom of pseudo-scientific dietary lore.

Something, even so, is clearly amiss. Although the precepts themselves avoid excess, Catius' obsessive perfectionism shows a lack of balance or proportion. For example, the master's fine differentiation between shapes of eggs, a controversy in antiquity from Aristotle onwards,[105] is preposterous when we know that the Latin for 'as like as two peas' is 'as like as two eggs'.[106] Horace's puns focus on the parallels, and the discrepancies, between gastronomic and philosophical knowledge: *doctus* (19, learned, expert), *sapiens* (44, tasty, wise), *ignarus* (38, ignorant), *pernoscere naturam* (63–4, to understand the nature of), *ingenium* (47, talent), *cura* (8, 48, 85, care, anxiety), *reprehendi iustius* (86, more reprehensible), *male credere* (21, mistrust), *mendosus* (25, erroneous), *flagitium ingens* (82, a monstrous crime), *rectius* (72, more correct), *decet* (26, cf. *decebit* 65, it is appropriate), *bonus* (5, good), *melius* (27, better), *malus* (42, 49, bad), *satis* (37, 48, enough), *laborare* (49, to labour), *sectari* (44, to subscribe to), *ratio* (36, system).[107] Metaphors of flavour and wholesomeness used earlier in the book in a moral sense[108] are restored

[104] Plin. *NH* 10. 54. 145: *quae oblonga sint ova gratioris saporis putat Horatius Flaccus*.

[105] See Palmer (1925) ad loc.

[106] See Cic. *Luc.* 57: *videsne ut in proverbio sit ovorum inter se similitudo*; Sen. *Apoc.* 11. 5: *hominem tam similem sibi quam ovo ovum*; Quint. 5. 11. 30.

[107] See Rudd (1966: 213), Classen (1978: 336, 339).

[108] e.g. in 2. 3: *sapiens* (wise), 35, 46, 56, 97, 265, 296; *sanus* (healthy, sane), 128, 138, 158, 160, 218, 241, 246, 275, 284, 302, 322; *insanus* (unhealthy, insane), 32, 40, 48, 74, 102, 120, 130, 159, 197, 201, 221, 225, 271, 298, 302, 305, 326; *desipere* (to lose one's reason), 47, 211; *putidus* (addled; cf. *putescere*) 194, 75; *integer* (whole, in one's right mind), 220; *satis* (enough), 127, 178.

to their organic meaning: *putescere* (66, to rot), *vitiatus* (54, high), *sapor* (54, flavour), *sapiens* (44, tasty, tasteful), *salubris* (21, wholesome). In 2. 2 Ofellus had shown that gastronomy is no mirror of morals. The Romans' ancestors had served rancid boar, but in this they paradoxically displayed moral integrity, in so far as they valued ease of hospitality over cuisine. In 2. 4, the more the two vocabularies of cooking and philosophy overlap, the more perceptible the distance between gourmet and sage becomes. Someone who thinks a man is discerning (*sapiens*, 44) if he prefers the wings of a female hare or that it is a cardinal sin (*immane vitium*, 76) to serve fish on too small a plate has no proper sense of proportion; he does not know what *satis est*.

More can be made of the Platonic allusions at the beginning of the poem. Fraenkel (1957: 136–7) points out that all three food-centred poems (2, 4, 8) open with quotations from Platonic dialogues. The first lines of 2. 4 parody the opening of *Phaedrus*.[109]

> unde et quo Catius? non est mihi tempus aventi
> ponere signa novis praeceptis, qualia vincent
> Pythagoran Anytique reum doctumque Platona.

Where have you come from, Catius, and where have you been? I haven't the time, I'm gasping to note down these new rules, which will supersede those of Pythagoras, the condemned Socrates, and learned Plato.

The allusion has a hidden point. Socrates goes on to ask Phaedrus about the speech of Lysias he has just heard (227b): *Τίς οὖν δὴ ἦν ἡ διατρίβη; ἢ δῆλον ὅτι τῶν λόγων ὑμᾶς Λυσίας εἱστία;* ('What was the discussion about? Or was it the case that Lysias feasted you with his words?'). The image of feasting someone with words recurs at *Timaeus* (27b), and Fraenkel points to this as a distorted allusion in *Sat*. 2. 8, where Fundanius is asked to recall, not a feast of words, but a real feast. He fails to point out, however, that the same feasting metaphor is concealed in the allusion to *Phaedrus* in

[109] *Phaedrus* 227a: 'Dear Phaedrus, where have you come from and where have you been?' '. . . he went to learn the speech by heart . . . he met someone dying to hear it too' (''Ὦ φίλε Φαῖδρε, ποῖ δὴ καὶ πόθεν;' . . . 228b '. . . ἐξεπιστάμενος τὸν λόγον . . . ἀπαντήσας δὲ τῷ νοσοῦντι περὶ λόγων ἀκόην κτλ' 227a).

2. 4.[110] This cannot be a coincidence. In both poems, allusions to Platonic word-feast metaphors stress the discrepancy between philosophical meetings and Horace's representations of 'real' meals, where food rather than philosophy is the focus of attention.[111] *Aventi* ('starving, gasping' 1) suggests real as well as philosophical hunger. And Horace's metaphor at the end, *ut / . . . haurire queam vitae praecepta beatae* ('to drink deep of the secrets of the good life', 94–5), is comically appropriate in the context of gastronomic knowledge.[112]

To take the analogy with *Phaedrus* further, it is worth mentioning the detail that Phaedrus heard the speech in question at someone's house—which is what made it appropriate to call Lysias the 'host' of a feast of words. It is normally assumed that the precepts were delivered as a lecture.[113] But it is more likely, considering the analogy with Plato, that Catius heard them at a real feast in someone's house. That is, after all, how Nasidienus delivers his own precepts in 2. 8, to a captive audience of guests. Catius' gastronomic knowledge, it is implied, will give him access to a privileged and exclusive social group: such is the prestige of culinary expertise. That may be one reason why Catius conceals the name of the originator of the precepts (*celabitur auctor*, 11). By expressing enthusiasm at the end, and asking to be taken to see the speaker, Horace plays the social climber (like the bore in *Sat.* 1. 9). *Vitae beatae* (95) becomes

[110] Rudd (1966: 301) points out that *unde et quo* could refer to several Platonic dialogues, but this ignores the special significance of the feast metaphor in *Phaedrus*.

[111] See p. 41 n. 185 on the philosophical 'feast of words'. The anecdote told at Ath. 9. 381–382a of a host who used to make his cooks learn the dialogues of Plato and say as they brought in the dishes 'One, two, three; but where, my dear Timaeus, is the fourth of our guests from yesterday, who are our hosts today?' also exploits the discrepancy between Platonic food metaphors and their re-enactment on the 'real' level.

[112] Philosophical thirst is opposed to the real thirst appropriate to a drinking party (cf. Plat. *Symp.* 175e); see also Fantham (1972: 161) on *haurire* in Cic. *de Or.* 1. 12, 193, 203; 3. 123.

[113] Palmer (1925: 316); Rudd (1966: 208); Classen (1978: 339). See as evidence for cooking-schools Col. 1 pr. 5: *contemptissimorum vitiorum officinas* ('despicable dens of iniquity'); Juv. 11. 136–41; cf. Hor. *Sat.* 2. 5. 80: young Romans are keener on cuisine than on romance (*nec tantum Veneris, quantum studiosa culinae*).

not only the philosophical good life, but also 'the life of the rich' (cf. *beatus vidisse*, 92; *Nasidieni . . . beati*, 2. 8. 1).

There is another Platonic allusion in *ede hominis nomen, simul et Romanus an hospes* (10, 'yield the name of the man: was he a Roman or a foreigner?').[114] But *hospes*—foreigner—has another meaning, 'guest', and the word is used in this sense at 17: *vespertinus hospes*. So the question could mean 'whether he was a Roman or a guest', that is, whether the speaker was the host or a guest at the dinner, in which case the satire is also advice on how to get on in society.[115] The precepts seem biased towards the distinction between appearance and inner worth (*facies*, 12; *suci melioris*, 13; cf. *suco . . . facie*, 70–1) or trustworthiness and deceit (*male creditur*, 21), and concern for the guest's well-being.[116] Horace appeals to Catius *per amicitiam divosque* (88), putting the ties of *amicitia* on a level with the sanctions of the gods. Behind the parody is another serious implication: that *hospitium* was succeeding to rhetoric as the route to distinction and power. Cicero implies this when he describes giving lessons in rhetoric in return for lessons in cookery and puns on the two meanings of *ius* (*Fam.* 9. 18. 3): *tu istic te Hateriano iure delectas, ego me hic Hirtiano* ('While you were enjoying Haterius' laws, I was enjoying Hirtius' sauce'). In Horace's world the law is 'cooked' as much as dishes are. A recipe for double sauce: *est operae pretium duplicis pernoscere iuris / naturam* (63–4, 'you ought to know the recipe for double sauce') conceals another layer of advice: 'it is worth knowing the workings of the two-faced legal system.'[117] Alternatively, since *haute cuisine* was largely Greek in influence, the question might concern Greeks and Romans, a symptom of the Romans' understandable insecurity about their cultural debt to Greece.[118]

[114] Rudd (1966: 301 n. 21): *Protagoras* 309c.

[115] Davus in 2. 7 parodies Horace's own eagerness to be Maecenas' *conviva* (30–5); Horace is Maecenas' *convictor* at 1. 6. 47.

[116] Presents of food are legacy-hunters' bait at 2. 5. 10–14.

[117] At 2. 2. 131 Ofellus suffers by underestimating the subtlety of the law (*vafri inscitia iuris*).

[118] Horace's speakers frequently admit that their knowledge is acquired at second or third hand: e.g. 2. 3. 33, 2. 7. 45.

Classen argues that a very specific form of parody is involved: parody of false Epicureanism. The emphasis this doctrine placed on eating ('the beginning and the root of all good is the stomach', fr. 409 Usener = Ath. 12. 546 f), had already been distorted into something falsely sensual, an impression which Epicurus was at pains to correct. He assured his followers that 'it is not continuous drinking and revellings, nor the satisfaction of lusts, nor the enjoyment of fish and other luxuries of the wealthy table which produce a pleasant life, but sober reason.'[119] Classen's theory is that Catius is Catius the Insubrian, described by Cassius as one of the 'false interpreters of Epicurus' (*mali verborum [sc. Epicuri] interpretes*: Cic. *Fam.* 15. 19. 2). Undeniably, much of his language does parody Lucretius' Epicurean *de Rerum natura*.[120] But to identify Catius so firmly with a historical figure is to restrict the possibilities of his name. One of the scholiasts says that he is Catius Miltiades, who wrote a book on baking. And Palmer thinks the name is a pseudonym for C. Matius, author of works on cooking, fish-cookery, and pickle-making.[121] So the name Catius may have suggested both cooks and philosophers to a reader, and the ambiguity is important. Perhaps Horace is implying that Catius would be a Matius if he were alive today. It is also worth considering Catius as a typically Horatian type-name.[122] A pun on *catus* (shrewd, clever) would be very apt in the context of misapplied expertise (cf. *vafer*, 55, *ignarum*, 38, *mendose*, 25, *doctus*, 19, 88, *sapiens*, 44).[123] And there must also be a pun

[119] *Ep. Men.* 132, quoted from Classen (1978: 342).

[120] See Classen (1978: 342 (and Rudd (1966: 209, 211)) for a list of parallel passages. Catius is also named as an Epicurean philosopher by Porphyrion on 2. 4. 1.

[121] Schol. ad 2. 4. 47, p. 166 Keller. Matius: see Col. 12. 4. 2: *C. Matius [sc. et ceteri] quibus studium fuit pistoris et coqui nec minus cellarii diligentiam suis praeceptis instruere* ('C. Matius, who left advice on the economics of baking, cooking, and cellar-keeping'); cf. 12. 46. 1: *illi enim propositum fuit urbanas mensas et lauta convivia instruere. libros tres edidit, quos inscripsit nominibus Coci, et Cetarii, et Salgamarii* ('He wrote on the subject of polite dinner parties and elegant entertaining. He published three books: *The Cook, The Fish-Cook*, and *The Pickle-Maker*').

[122] An idea rejected by Classen: 'The name Catius can be classed neither as significant nor as denoting a type character' (1978: 335).

[123] Suggested to Classen by a member of the Cambridge audience to whom he gave the original paper. Catius is also an antitype of the old Roman hero Cato, whose didactic writing on rural food in *De agricultura* bears some resemblance to the quaint precepts of this poem, but who of course abhorred gastronomy. Cato's

on the words for pots and pans that actually appear in this poem: *catillis* (75), *catino* (77). A play on *catus* and *catinus*/*catillis* would perfectly encapsulate the confusion of philosophy and gastronomic knowledge.

Yet more can be made of the pseudo-philosophical aspect of the poem. The outer frame parodies the excesses of the moral system implied in the precepts. Catius applies his standards of punctiliousness to social exchange: *non est mihi tempus* (1, 'I haven't the time') can be compared with his obsession with the right time for everything.[124] Horace parodies this punctiliousness himself: *peccatum fateor, cum te sic tempore laevo / interpellarim* (45–6 'I know it's a *faux pas* to interrupt you at such a bad time').[125] *Interpellarim* (5), *interciderit* (6) suggest lapses or interpolations in a perfectly organized system. Catius says that he has trouble (*curae*, 8) remembering everything (despite his obsessive care in every branch of gastronomy: *nequaquam satis in re una consumere curam* (48, 'it is not enough to eat up all one's energy in one area'). Horace parodies this at the end: *at mihi cura / non mediocris inest* (93–4, 'but I have an insatiable anxiety'). He emphasizes ironically that Catius, by showing un-Epicurean concern for *mediocritas* in cooking, can no longer be called *mediocris*.[126] Misapplication of a philosophy designed to relieve care actually produces greater anxiety, as we see from the dangerous life of the town mouse (2. 6. 100–15); the sedulous host who sets fire to his own house when trying to roast some skinny thrushes (1. 5. 71–2); and Nasidienus, torn apart by anxiety (*sollicitudine districtum*, 2. 8. 68) in his dinner preparations.

own name was linked with the word *catus*: 'Afterwards he got the surname of Cato for his great abilities. The Romans call a man who is wise and prudent *catus*' (Plut. *Cato Maior* 1. 2).

[124] Cf. 23: *ante gravem quae legerit arbore solem* ('which should be picked before the sun grows too oppressive'); 45: *quae natura et foret aetas* ('the true nature and proper age'). Cf. the cook (μεγὰς σοφιστής) in Sosipater, *The False Accuser* (K3. 314 = Ath. 9. 377f) on the importance of knowing the proper order and timing of each dish.

[125] *Peccatum* anticipates *immane vitium* (76) and *flagitium ingens* (82).

[126] Cf. Ath. 4. 158a exploiting similarities between the two systems of philosophy and gastronomy: 'It is a Stoic belief that the wise man will do everything properly, right down to the wise seasoning of lentil soup.'

So is it enough to leave the poem there? Even if Horace is using the Socratic technique[127] to make the reader think *iam satis est* before the meal has even begun, terms which are common to cooking and philosophy are actually quite thin on the ground. Of course, it could be argued that the rest of the poem is *meant* to be a waste of words, mere material padding, which needs to seem trivial and boring to provoke the right moral response in the reader. *Satires* 2 is as a whole a collection of wasted words, for which Horace disclaims responsibility, and the joke behind this poem is that it is a 'feast' of words that is not in the least entertaining.[128]

There is one coincidence, however, that has been overlooked. Catius' gastronomic system encodes the same aesthetics of decorum, purity, and economy, uses the same terms of seasoning, scale, and texture that Horace himself prescribes for writing satire. Catius' summary of the prescriptions—*res tenuis tenui sermone peractas* (9, 'a subtle matter described in subtle language')—cannot fail to remind us of Horace's own Callimachean principles. Perhaps the most compelling piece of evidence comes, retrospectively, from the *Epistles*, where Horace compares himself explicitly to a cook or dinner-party host trying to satisfy guests or readers with different tastes:[129]

> denique non omnes eadem mirantur amantque:
> carmine tu gaudes, hic delectatur iambis.
> ille Bioneis sermonibus et sale nigro.
> tres mihi convivae prope dissentire videntur,
> poscentes vario multum diversa palato.
> quid dem? quid non dem? renuis tu, quod iubet alter;
> quod petis, id sane est invisum acidumque duobus.
> (*Ep.* 2. 2. 58–64)

[127] See Anderson (1963: esp. 33–4).

[128] Horace despises the verbosity of the Stoics (1. 1. 13–14; 1. 1. 120–1; 1. 4. 14–16), yet 2. 3, a speech of monster length which is ostensibly recorded during Horace's own writing block, is an imitation of this verbosity.

[129] See Brink (1982: 299–302) on this passage. *Sal niger*—coarse salt or coarse wit—fuses the literal and the metaphorical. The epithet 'black' gives 'the specific notion of polemic satire'. Cf. Pliny *Ep.* 3. 21. 1 on Martial's *sal* and *fel*. For *sal* in rhetorical theory cf. Hor. *AP* 271 with Brink's note. *Acidum* (64, 'acidic') and *palatum* (62, 'palate') can also be taken literally and metaphorically: for *acidus* Brink notes Plaut. *Pseud.* 739, Petr. *Sat.* 31. 6: *acido cantico*. For *palatum* cf. Quint. 6. 3 19 comparing critical judgement to the palate. For the writer/host analogy cf. Plin. *Ep.* 2. 5. 8; Mart. 10. 59, 9. 81.

In short, one man's meat is another man's poison: you like my odes, he likes my epodes, a third likes my satires with their caustic [lit. black] salt. They're just like squabbling dinner-guests, all asking for different food to suit their individual tastes. What am I to serve them? You'll just reject what *he* orders; and what you choose yourself is bound to be distasteful to the others.

One reader likes Horace's odes, another his epodes, and a third his *sermones* with their coarse wit (*sale nigro*).

We have seen that to characterize writing in cooking terms was a feature of ancient comedy, and it is in the tradition of 'meta-literary' cooks' speeches that Horace's satire belongs. Of course, the notion of a 'recipe' for writing is all the more appropriate in connection with the 'mixed dish', satire. In 2. 1 Horace had claimed that his recipe for satire so far had not been to the critics' taste. Like the readers set up as dinner-guests in *Ep*. 2. 2, their reactions were contradictory: some found it too sharp (*nimis acer*), others too insipid (*sine nervis*).[130] This was despite Horace's attempt in Book 1 to eliminate Lucilian verbiage. In his disregard for superfluities and strong flavourings, Catius' master recalls Anaxippus' cook, whose chaste *nouvelle cuisine* seemed to be a programme for New Comedy.[131] This looks like Horace's 'recipe' for New Satire, a toned-down, tightened-up system palatable to the post-Republican regime and its newly fastidious critics.[132]

Horace's *res tenuis tenui sermone peractas* are clearly a far cry from the *mensa tenuis*, the simple rustic food that nourishes Horace in the *Odes* and complements his writing style, the Callimachean *genus tenue*.[133] That sort of diet is

[130] See pp. 130–31 above.

[131] K3. 296 = Ath. 9. 403e. See pp. 84–5 above.

[132] Horace makes explicit parallels between old and new comedy and old and new satire: 1. 4. 1 ff., 2. 3. 11–12. The reference to legal restrictions on Athenian comedy at *AP* 283 is clearly meant as a precedent for those that threaten post-Republican satire in 2. 1. Owen (1977: 78–9) sees parallels between 18th-c. culinary aesthetics and changing tastes in general, e.g. between Collingwood and Woollams' rules for pared-down, economical cuisine (1797) and the *Lyrical Ballads*.

[133] See Mette (1961). Cf. *Od*. 2. 16. 14: *mensa tenui*; *Od*. 1. 31. 15–16: Horace's diet of olives, chicory, and light mallows; *Od*. 3. 29. 14: *mundaeque parvo sub lare pauperum / cenae* ('neat dinners under a pauper's roof'). Compare *Ep*. 1. 5, an invitation poem in which Horace sums up his poetic offerings in humble terms:

extolled by Ofellus as *tenuis victus* in 2. 2. His ideas, we are told there, are out of date and rustic, phrased in language that is the opposite of Callimachean finesse (*rusticus abnormis sapiens crassaque Minerva*, 2. 2. 3). Unlike the *Odes*, the *Satires* and *Epistles* give us conflicting and often unflattering impressions of Horace as a consumer: we are told that he idealizes *securum holus* ('uncomplicated vegetables') only as a form of sour grapes when he has no urban dinner party to go to.[134] Besides, however perfect, a metaphor of simple and innocent food would be inappropriate for satire, the mess-making form, particularly when the society that *Satires* 2 evokes is an imperfect one, urban and full of guile. Catius' *tenuis ratio saporum* (36, 'subtle system of flavours'), gastronomic subtlety, is a more sophisticated culinary transformation of the *genus tenue*, morally questionable and exploiting all the absurdities of a contemporary fashion.

Many of Catius' culinary principles can now be set alongside Horace's programmatic writings in *Satires* 1. 4, 1. 10, and the *Ars Poetica* to show how literary metaphors are restored to their physical origins. The sloppy cooks chastised by Catius' master should be seen as transformations of Lucilius, the disorganized creator of messy *saturae*. The parallels are laid out in table-form, to allow the jaded reader to skim.

(i) *lutulentus/purus* In *Sat*. 1.4 Horace compares Lucilius' writing to a muddy river full of debris: *cum flueret lutulentus, erat quod tollere velles* (11, 'His flow was so muddy, there was plenty to dredge out'). The source of this image is Callimachus' own turbid Euphrates:

> Ἀσσυρίου ποταμοῖο μέγας ῥόος, ἀλλὰ τὰ πολλὰ
> λύματα γῆς καὶ πολλὸν ἐφ' ὕδατι συρφετὸν ἕλκει.
> (*Hymn to Apollo* 108–9)

The Assyrian river is huge, but it carries vast amounts of mud and other debris on its waters.

modica holus omne patella (2, 'the whole meal of vegetables on a small plate'); see Race (1978), and Bramble (1974: 162–3) on Mette.

[134] *Sat*. 2. 7. 30–2.

Removing dirt and impurity is also central to Catius' scheme. Sediment (*limus*—also Latin for mud) is to be removed from Surrentine wine by adding Falernian lees and the yolk of a dove's egg:

> Surrentina vafer qui miscet faece Falerna
> vina columbino limum bene colligit ovo,
> quatenus ima petit volvens aliena vitellus.
> (2. 4. 55–7)

The expert will mix Surrentine wine with Falernian lees and collect the sediment with a dove's egg: when the yolk touches the bottom it carries all the dross with it.

Impurities (*aliena*) are analogous to superfluous or inappropriate expressions in writing.[135]

Heavy sediment (*gravis limus*) reappears in line 80, this time sticking to an old bowl (*veteri craterae*). This looks like an image of the old-fashioned conviviality of Lucilius' writing, with all its attendant dross.[136] Horace emphasizes the contrasting novelty of his own precepts (*novis praeceptis*, 2).[137] The same image is used in a moral context at 1. 3. 55–6: *at nos virtutes ipsas invertimus atque / sincerum cupimus vas incrustare* ('but we turn our virtues upside down, and insist on besmirching a clean jar'). The dirty bowl turns the stomach of a fastidious guest: *magna movent stomacho fastidia* (78). This only reinforces the literary analogy. *Stomachus* is used by Martial as a metaphor of literary taste; *fastidium* can

[135] See Van Hook (1905: 29) on *aliena* as a literary term (cf. Gk. ἔκφυλος—'alien', 'foreign'). *Ima petit*: cf. *AP* 378, of a poem with faults: *si paulum summo decessit, vergit ad imum* ('if it falls at all short of the heights, it plunges at once to the depths').

[136] Anaxippus' cook had demanded only a new saucepan (λοπάδα καινήν), and olive oil to replace the heavy seasonings of old comedy (see above pp. 84–5).

[137] Horace also uses an image of new wine for his Neoteric *Odes*: *de patera novum / fundens liquorem* (*Od.* 1. 31. 2–3: 'pouring new wine from its jar'). Cf. the image of storing Italian wine in a Greek jar at *Od.* 1. 20. 1–3, which Commager (1962) and Race (1978) read as a metaphor for Horace's Latin transformation of a Greek literary form. Nisbet and Hubbard (1975), ad loc., reject this kind of interpretation, but that seems unreasonable given the long-standing association between wine and poetry, together with the popularity of 'container' names for literary collections (see Plin. *NH* pref.; Gell. *AN* pref.). Pindar clearly equates poetry and wine at *Ol.* 7. 1–10, and at *Ol.* 6. 91 compares the poet to a mixing-bowl. See Bramble (1974: 48–50) on *sobrius*, *ebrius*, etc., used of style, and on the 'water-drinkers versus wine-drinkers' debate (see also Hor. *Ep.* 1. 19; Commager (1962: 29 n. 59); Wimmel (1960: 225).

be aesthetic as well as physical disgust.[138] And purity is also the guiding principle behind Horace's dinner in *Ep.* 1. 5: *ne turpe toral, ne sordida mappa / corruget naris* ('no squalid couch or dirty napkin will wrinkle up the nostrils').[139]

At 83 Catius uses *lutulentus* of a dirty broom, saying that it is a crime to use it on a mosaic floor (*varios lapides*). Again we can see a criticism of Lucilius' sloppiness.[140] By contrast Catius' cook serves food on spotless dishes (*puris catillis*, 75), which recalls the pure spring Callimachus opposed to the muddy river (*Hymn to Apollo* 111–12).

Excessive wateriness in food is another transformation of Callimachus' flooding rivers.[141] Cabbage, Catius says, must be picked from dry fields: *cole suburbano qui siccis crevit in agris / dulcior* (15–16, 'Cabbage that grows in dry fields is sweeter than the suburban kind'). Nothing is more insipid (*elutius*) than a watery garden (*irriguo horto*, 16). The same principle is behind the hint at 59 that lettuce is to be discouraged after a certain stage because it swims (*innatat*, 59) on the acid or irritable stomach (*acri stomacho*).

(ii) *pinguis/tenuis* Catius rejects a Laurentian river boar for being *carnem inertem* (41, 'tasteless meat'), fat with sedge and reeds (*ulvis et harundine pinguis*, 42). Here is a parallel for Horace's *scriptor iners*, literally, 'artless writer', at *Ep.* 2. 2. 126. *Pinguis* (fat, and, in literary terms, turgid or bombastic) is of course a vital metaphor for the neoterics, together with its opposite, *tenuis* / λεπτός (slim, finely-spun). The distinction was epitomized in physical terms by Antimachus' 'fat book' (παχὺ γράμμα, Callim. fr. 398 Pf.) and Callimachus' own 'slender Muse' (*Μοῦσαν λεπταλέην, Aetia* prol. 24).[142] When

[138] *Stomachus*: cf. Mart. 10. 45. 6, 13. 3. 8; *fastidium*: cf. Cic. *Opt. gen.* 12; Hor. *Ep.* 2. 1. 215; Plin *Ep.* 2. 5. 4; Quint. 8. 3. 23.

[139] Where *naris* suggests *nasus*, metaphor for good taste.

[140] Lucilius (fr. 84–5M) had used an image of mosaics to describe an intricate style: *quam lepide lexis conpostae ut tesserulae omnes / arte pavimento atque emblemate vermiculato* ('How charmingly composed his remarks are, like all the little mosaics in a decorated floor or inlaid slab'). Horace ironically suggests that Lucilius managed to spoil his own intricate designs.

[141] See Van Hook (1905: 12) on literary metaphors of wetness and dryness: e.g. θολοῦσθαι—to be turbid; ὑδαρής—watery, weak.

[142] See Wimmel (1960); Bramble (1974: 56–8). Call. *Aet.* 1 fr. 1 Pf. 23–4 is echoed at Virg. *Ecl.* 6. 4–5: *pinguis / pascere oportet ovis, deductum dicere carmen*

pinguis or *tenuis* appear in physical contexts, they are often also programmatic metaphors for style. The *pinguis aper* with its clogged-up diet is the antithesis of fine poetic material.[143] *Unctis manibus* (78–9) is another example of grease and superfluity.[144]

At 1. 10. 13–14 Horace had mentioned the poet's need to 'spin out' his resources: *parcentis viribus atque / extenuantis eas consulto*. The literary metaphor *tenuis* is principally derived from fine-textured weaving, yet the word can also be used of the thin juices of plants or liquids.[145] An image of a thinned liquid appears at 2. 4. 52, where Catius recommends leaving Massic wine outside in the night air to rarefy its flavour: *si quid crassi est tenuabitur aura* ('The air will thin down any thickness in its texture'). *Crassus*, used of Ofellus' style in 2. 2 (*crassa Minerva*, 3), is, ironically, the opposite of the *tenuis victus* he is advocating. Catius, by contrast, requires stylistic finesse appropriate for the subtlety of his *nouvelle cuisine*. Rustic *tenuis victus* has been transformed into a sophisticated science: *tenuis ratio saporum* (36).[146] In

('You should pasture fat sheep, but keep your poems slender'); Hor *Sat.* 2. 6. 14–15: *pingue pecus domino facias et cetera praeter / ingenium* ('fatten the flocks for their master, and fatten everything else but my talent').

[143] See Gourévitch (1974), who cites this passage as an instance of the ancient belief that a foodstuff absorbed the qualities of its environment. Classen (1978: 337) quotes Martial 10. 45. 4 as evidence that Catius has chosen the wrong kind of boar: *ilia Laurentis cum tibi demus apri* ('when I serve you the haunches of a Laurentian boar'). In fact, the boar is being used here as an ironic and negative image for Martial's honorific, flattering poems. These are described as mild, sweet, and pleasant (*lene, dulce, blanda*); the reader finds them rich (*pingue*) and not to his taste (*ad stomachum*). Martial is in effect disowning these poems. The fact that he chose Laurentian boar as an image for turgid poetry suggests that he knew *Sat.* 2. 4 and recognized its poetic meaning. At 2. 5. 40 Horace portrays the turgid epic poet Furius Bibaculus distended with rich tripe: *pingui tentus omaso* (see Bramble 1974: 64–6).

[144] For greasiness as a figure of rhetoric, cf. *Catalepton* 5. 4: *scholasticorum natio madens pingui* ('a nation of schoolboys dripping with rich grease').

[145] e.g. Col. 7. 8. 1; Plin. *NH* 14. 80.

[146] *Sapor* and *sapiens* not only confuse gastronomic and philosophical wisdom: they also include aesthetic taste. *Sapor*: Cic. *Brut.* 172: *nescio quo sapore vernaculo* ('with a certain native flavour'); Quint. 6. 3. 107: *Athenarum proprium saporem* ('the peculiar flavour of Athens'). Cf. the culinary-aesthetic pun at Petr. *Sat.* 2. 1, 3. 1. See Bramble (1974: 50); Onians (1954: 61 ff.). *Sucus*, 'juice' (*suci melioris*, 13, *suco*, 70), is also a metaphor for style, suggesting vitality or energy: e.g. Cic. *Brut.* 36: *sucus ille et sanguis incorruptus [sc. orationis]* ('the juice and uncorrupted blood (of

this recipe, the *sapor*, flavour, exuded by poetry is of a rarefied Callimachean kind: *aura* (52) suggests the finest form of vapour. But later Catius says that to strain wine through linen destroys its original flavour: *integrum perdunt lino vitiata saporem* (54). The writer must strike a balance between coarse flavouring and flavourless over-refinement. *Vitiata* looks ahead to *immane vitium* (76): the moral term is first restored to its physical origin—'spoilt', 'rotten'—and then becomes an aesthetic defect.

(iii) *durus/lenis* At 1. 4. 8 Horace calls Lucilius *durus*—rugged or robust in his writing. But the adjective can also be pejorative. At *AP* 445–8 the critic watches for lines that are insipid (*inertis*), wooden (*duros*), or untidy (*incomptis*).[147] At 2. 4. 18 ff. the cook gives instructions for removing culinary *duritas* by boiling a hen in Falernian wine:

ne gallina malum responset dura palato
doctus eris vivam mixto mersare Falerno;
hoc teneram faciet.

To stop a tough hen deadening the taste-buds, the clever thing is to plunge it alive into Falernian juice; that will make it tender'.

Palatum is another word that can be used metaphorically of critical discrimination;[148] *doctus* is used specifically of neoteric learning or subtlety.[149] So, in literary terms, this instruction reads: you will be learned in Callimachean subtleties if you exclude harshness from your writing to please the discriminating writer. Old-style *tenuis victus* often involved rancid or hastily prepared food; here host and guest are exercising high critical standards.

Durus is also the word used at 27 of constipated bowels: *si durus morabitur alvus* ('if the stomach is sluggish with constipation'). The image merges with the ones that recall clogged-up rivers. Cheap shellfish and Laelius' short,

speech'); cf. Quint. 1. pref. 24; Cic. *de Or.* 2. 93; *Att.* 4. 18. 2 (16. 10); see also Onians (1954: 62–5).

[147] Cf. *carnem inertem*; at Nasidienus' feast, which I shall argue has a literary dimension, the page-boys are described as *compti* ('well-groomed').

[148] Cf. 46: *ante meum nulli patuit quaesita palatum* ('the quest yielded nothing until I started to taste'); see above n. 129.

[149] See Kenney (1970).

humble sorrel (*lapathi brevis herba*, 29) are prescribed to remove the blockage (*pellent obstantia*, 28), together with white Coan wine (Horace's *satura* is composed of a mixture of common and recherché ingredients). The principle of unblocking internal passages is again strongly reminiscent of Anaxippus' reforming cook, who promised to free eaters from the coughs and sneezes caused by strong flavourings.[150] Similarly, Horace's own prescription for fast-moving (cf. *brevis herba*) and varied *satura* at 1. 10. 9–11 uses an image of ears blocked up by a heavy mass of verbiage:

est brevitate opus, ut currat sententia, neu se
impediat verbis lassas onerantibus auris

You need brevity, to make the train of thought flow, and not block up the ears with a mass of verbiage.

Catius also advocates smoothness. At 2. 4. 24 he mentions a figure named Aufidius who mistakenly made *mulsum* with strong Falernian wine:

Aufidius forti miscebat mella Falerno,
mendose . . .

It may be that Horace is referring to the M. Aufidius Lurco mentioned by Pliny as the first Roman to sell fattened peacocks.[151] However, the name also recalls the passage in *Sat.* 1. 1 where a man tries to fill his jar from the raging river Aufidus, transferring Callimachus' river-image to the moral sphere:

eo fit
plenior ut si quos delectet copia iusto,
cum ripa simul avulsos ferat Aufidus acer.
at qui tantuli eget quanto est opus, is neque limo
turbatam haurit aquam, neque vitam amittit in undis.
(1. 1. 56–60)

That is how those who like more than their fair share get swept away by the raging Aufidus, with the bank under their feet. But the man who demands only what he needs avoids drinking water cloudy with mud and escapes death in the torrent.

[150] K3. 296 = Ath. 9. 403e.
[151] As Palmer (1925) suggests ad loc: Plin. *NH* 10. 20. 45.

Aufidius sounds like the kind of ambitious poet who plunges his readers into words that are not easily absorbed.[152] By contrast, Catius insists that the guest must only have smooth wine to begin with.[153]

> vacuis committere venis
> nil nisi lene decet; leni praecordia mulso
> prolueris melius. (25–7)

Nothing but mild wine should be poured into empty veins: it is best to wash out the insides with mild honeyed wine.

(iv) *magnus/angustus* Lucilius' chief defect, as Horace so often tells us, was the sheer quantity of his verses.[154]

> nam fuit hoc vitiosus; in hora saepe ducentos,
> ut magnum, versus dictabat stans pede in uno.

This was his chief flaw; he often dictated two hundred lines an hour standing on one foot: what an achievement!

The sloppy cook, similarly, spoils expensive fish by serving them on a plate too small for them:

> immane est vitium dare milia terna macello
> angustoque vagos piscis urgere catino.

It is a cardinal sin to pay a fortune at the market, and then squeeze floundering fish into a narrow dish.

In the light of satire's 'mixed dish' origins, the tiny plate, *angustus catinus*, looks like an appropriate image for Horace's small-scale satire. The *lanx satura*, he is suggesting, is too small a form to bear the load of Lucilius' vague and rambling

[152] Reminiscences of the peacock-fattener would reinforce the name's associations with indigestibility.

[153] Onians (1954: 42) argues from 2. 4. 26 and other passages that *praecordia* are to be understood as the lungs, and that the ancients believed that wine was absorbed directly by the lungs. He argues p. 64 that the origin of the poetic image of *vena* as the poet's mind (Hor. *AP* 409: *divite vena*; *Od* 2. 18. 10: *ingeni benigna vena*; cf. Ov. *Ex Pont*. 2. 5. 21; *Trist* 3. 14. 34) lies in the meaning of *vena* as a pipe or vessel containing native liquid. We can see from this passage (2. 4. 25) that *vena* is also the mind of the reader or recipient of poetry. The liquid of poetry flows into his veins (cf. Petr. 4. 3 on students watered with a heavy dose of reading: *ut studiosi iuvenes lectione severa irrigarentur*).

[154] A sort of verbal diarrhoea: cf. 1. 10. 60–1: *amet scripsisse ducentos / ante cibum versus, totidem cenatus*. *Vitium* and *vitiosus* are used by Horace of literary as well as moral faults: e.g. at *AP* 31.

verses.[155] In addition, the large amount of money spent, *milia terna*, recalls Callimachus' league images for large-scale poetry (*Aetia* 1 prol. fr. 1 Pf. 4, *πολλαῖς χιλιάσιν*, 'many thousands'; 17–18, *σχοίνῳ Περσίδι*, 'the Persian league'). Horace is trying to deflect the criticism recorded at 2. 1. 3 that verses like his could be churned out at the rate of a thousand a day.[156] By contrast, Callimachean exclusiveness is suggested at 2. 4. 31: *sed non omne mare est generosae fertile testae* ('Not every sea is rich in classy shellfish'). This recalls Callimachus' *Hymn to Apollo* 110: 'Not from every spring do the Melissae carry water to Demeter.'

(v) *pigritudo/labor* Another of Lucilius' faults was laziness: *piger scribendi ferre laborem, / scribendi recte* (1. 4. 12, 'He had a lazy approach to the hard work of writing, writing correctly'). For Horace poetry is *labor* and needs to be approached in the correct way (*recte*).[157] Similarly the cook labours at 2. 4. 49 (*laboret*) and is required to follow a recipe properly (*rectius*, 72). Poetic *cura* finds a parallel in the *cura* of entertaining (85). This care, Catius says, needs to be distributed over every detail. It is not enough to devote all one's talents to making novel cakes:

> sunt quorum ingenium nova tantum crustula promit.
> nequaquam satis in re una consumere curam. (47–8)

Some people only use their talent to create new cakes: but it is not enough to use up all your energy in one department.

The choice of example can be explained by another cake-image, the one Horace used at 1. 1. 25: just as schoolmasters

[155] At *Ep.* 1. 5. 2 a small plate (*modica patella*) is offered by Horace to his reader. Martial and Juvenal also use the plate as an internal image of the poetic sphere they have chosen: see below pp. 209–10, 251.

[156] Compare his criticism of the Stoics' prolixity: e.g. 1. 1. 120–1, *iam satis est. ne me Crispini scrinia lippi / compilasse putes, verbum non amplius addam* ('That's quite enough: in case you think I've pinched the writing-cases of bloodshot Crispinus, I shan't add another word'); and cf. *Sat.* 2. 3, a Stoic sermon three times as long as any other poem in the book; the bore in *Sat.* 1. 9 reveals his vulgarity by boasting that he can write verse fast: *nam quis me scribere pluris / aut citius possit versus?* (23–4).

[157] See Brink (1963: 59) on *labor* as an Alexandrian ideal. *Rectius*: cf. *AP* 129: *rectius Iliacum carmen deducis in actus* ('It is better to draw plays from the Trojan legend').

coax children with cakes (*crustula*) into learning their ABC, so satirists use jokes to hold their readers' attention. Deciphered, Catius' message means: it is not enough simply to amuse one's readers.[158] Horace stated this plainly at 1. 10. 7–8: *ergo non satis est risu diducere rictum / auditoris* ('So it is not enough to make your reader split his sides').

(vi) *variatio* Horace's chief principle of satire-writing in 1. 10 was constant variety:

> et sermone opus est modo tristi, saepe iocoso,
> defendente vicem modo rhetoris atque poetae,
> interdum urbani, parcentis viribus atque
> extenuantis eas consulto. (11–14)

And you need a style that moves from solemn to joking, playing orator and poet by turns, and making room for wit too, the kind that can judge when to spare its strength and when to spin it out.

Variatio for the neoterics was a virtue that vindicated their small-scale writing.[159] Catius' ideal is similar, and is actually schematized in the structure of the lines that call for variety: *nigris . . . moris* (22, 'dark blackberries') occupies the same position in the line as *albo . . . Coo* (29, 'white Coan wine'); *immundis* (62, dirty) is in the same position as *puris* (75, spotless); white pepper and black salt are juxtaposed at 74. Like the satirist, the cook needs to present an ever-varying series of flavours. This means that the constantly changing state of the guest overrides general principles. For example, in his prescription *nil nisi lene decet* (26, 'only mild is appropriate'), Horace seemed to be steering away from the criticism (2. 1. 1) that he was *nimis acer*. Yet by 59 the stomach has become *acri* with wine, and piquant ham and sausages are more suitable than watery lettuce. The flavour of food, it seems, depends on the stage of the meal and the state of the guest's appetite. This reflects the satirist's need to vary *ridiculum* with *acre*, jokes with pointed criticism,

[158] NB the puns on eating and satisfaction (*satis, consumere*) used here in an abstract sense.

[159] See Wimmel (1960: 8 n. 1, 78 n. 2, 84, 155). The country mouse compensates for poverty with variety: *cupiens varia fastidia cena / vincere* (2. 6. 86).

according to context.[160] At the same time, the instructions are still precise enough to seem part of an ordered scheme, suggesting that direct parallels are being made with the carefully variegated nature of Horatian satire.[161]

The master prides himself on having invented particular mixtures, and it is here that we should look for the most appropriately 'satirical' recipes. The claim *ego primus* (73, 'I was the first') is of course typical of comic cooks.[162] But it is also a boast of neoteric poets (compare Horace's claim for the *Epodes* at *Ep.* 1. 19. 23–4: *Parios ego primus iambos / ostendi Latio*: 'I was the first to show Parian iambics to Latium'). There is an emphasis on pursuing unknown branches of gastronomic knowledge: *piscibus atque avibus quae natura et foret aetas / ante meum nulli patuit quaesita palatum* (45–6, 'the quest for knowledge about the properties and proper age of fish and birds eluded everyone before I started tasting'); *at mihi cura / non mediocris inest, fontis ut adire remotos* (93–4, 'But I have an insatiable anxiety to reach those far-flung fountains') which recalls Lucretius' step into the unknown:

quia Pieridum peragro loca nullius ante
trita solo iuvat integros accedere fontes. (*RN* 1.926–7)

Since I cover untrodden tracts of poetry, I prefer to stumble on untapped fountains.

But even in these lines we recognize Callimachus' untrodden paths (κελεύθους ἀτρίπτους: *Aet.* 1 fr. 1 Pf. 27–8) and pure

[160] Cf. 1. 1. 23–7. One obvious discrepancy between a meal and a poem emerges at *AP* 147–8. Horace recommends plunging *in medias res* when dealing with a hackneyed subject like the Trojan War, rather than starting with the obvious beginning, the birth of Helen: *nec gemino bellum Troianum orditur ab ovo* ('the Trojan War does not begin with the twin egg'). *Ab ovo*, we recall, was used of the beginning of a meal at 1. 3. 6, and in deference to this convention Catius begins with eggs. Horace may have chosen a story that begins with an egg because of its connections with the beginning of a meal (Varro, *RR* 1. 2. 11 compares the egg that marks the end of a chariot race to the one that heads the procession at a meal); conversely, he may have chosen the phrase *ab ovo* in 1. 3 to draw an analogy between the meal and other systems that begin with an egg.

[161] Not all mixtures, of course, are desirable: at 1. 10. 23–4 Horace compares mixing Greek and Latin in poetic diction to adulterating Falernian wine with Chian.

[162] Cf. Alexis (K2. 360 = Ath. 12. 516d): a recipe for Lydian stew is θαυμαστὸν ἐμὸν εὕρημα ('my own miraculous invention').

spring (*καθαρή τε καὶ ἀχράαντος . . . λιβάς*: *Hymn to Apollo* 111–12).

Among these novel mixtures is the combination of white pepper and black salt:

> primus et invenior piper album cum sale nigro
> incretum puris circumposuisse catillis. (74–5)

I was the first to sprinkle white pepper and black salt round the edges of spotless plates.

Sal, salt or wit, is a prime ingredient in alimentary/literary parallels; *sal niger* is coarse or unrefined salt, black wit. Here is the single most suggestive hint that 2.4 is an encoded recipe for *satura*. The same ingredient is used in just this double sense in Horace's description of himself as a literary host at *Ep.* 2. 2. 58 ff., and it stands there for his satires: *Bioneis sermonibus et sale nigro* (60, 'the caustic black wit of my satires'). Horace had praised Lucilius for his wit (*sale multo*, 1. 10. 3), but this was no compensation for his lack of finesse. In Horace's satire, unrefined salt is sifted (*incretum*) on to spotless plates (*puris catillis*, 75): coarse wit is channelled into an uncontaminated form.[163] The ingredients of all those artful mixtures also suggest the (paradoxically) desirable element of 'corruption' in both *haute cuisine* and perfect satire: *malis* (73), apples, punning on *malis*, evils; *faecem* (73), sediment or dregs; and the stench of *garum* (*putuit*, 66).[164] Again, the surprising inclusion of dirty cookshops (*immundis popinis*, 62) along with *puris catillis* (75) suggests the necessary contrast between the suspect origins and the neat presentation of satirical material.

Finally, the most detailed recipe is for a sauce, and here we should expect to discover the flavour of the new satire:

> est operae pretium duplicis pernoscere iuris
> naturam. simplex e dulci constat olivo,
> quod pingui miscere mero muriaque decebit,
> non alia quam qua Byzantia putuit orca.
> hoc ubi confusum sectis inferbuit herbis
> Corycioque croco sparsum stetit, insuper addes

[163] Cf. 1. 4. 54: *non satis est puris versum perscribere verbis* ('It is not enough to compose verse out of plain words').

[164] Cf. 2. 8. 42 (*putet*).

pressaque Venafranae quod baca remisit olivae. (63–9)

You ought to know the recipe for double sauce. The simple version consists of sweet olive oil, which should be mixed with rich neat wine and fish brine, the kind that stinks in a Byzantine jar. Stir these together and boil with chopped herbs and Cilician saffron: let it stand, then add extra virgin oil from the Venafran olive.

Sadly, many of the subtleties of this recipe are lost on us. For a start, it is not clear whether the simple sauce (*ius simplex*) is being opposed parenthetically to the double sauce, or whether it is the first stage in its composition.[165] In any case, duplicity is significant here. The pun *ius*, gravy or the law, is often used, as we have seen, to suggest the messiness of the law or the quasi-legal formulas of gastronomy or entertaining. *Ius* was used earlier in *Satires* 2 in a legal context (1. 82, 2. 99, 2. 131, 3. 72), but in 17 the two senses were merged, when the word was surrounded by culinary metaphors for speech: pus and poison, bitter speech, sour contestants, and salty insults. The two-faced nature of contemporary law and the need for skill in negotiating its pitfalls are hints that bubble from this recipe.[166] The recipe also steals from Varro's 'culinary' definition of satire, 'sprinkled (*conspersimus*) with hilarity, mixed (*admixta*) with philosophy'. Horace echoes this with *miscere* (65) and *sparsum* (68).[167] There are shades, too, of Lucilius' farcical *court-bouillon* of Lupus. And finally, this recipe looks back to the comic cooking-lists that epitomize the messy or spicy mixture of the works in which they appear.

The sauce certainly seems to be made from the best ingredients available: Byzantine *muria*, Cilician saffron, Venafran olive oil. It contains a peculiarly Alexandrian combination of bad and sweet smells: sweet olive (*dulci olivo*, 64), stinking *muria* (*putuit*, 66) and aromatic herbs.[168] There

[165] Gray (1987: 103) mentions an 'unrecorded' Italian double sauce; cf. also Apic. 8. 1. 5, where *ius simplex* is the juice from meat, to which herbs and other ingredients are added.

[166] The remark about ignorance of sauces at 2. 4. 38 (*ignarum quibus est ius aptius*) must remind us of Ofellus' legal quandaries at 2. 2. 131: *vafri inscitia iuris*.

[167] Ap. Cic. *Top. Acad.* 1. 2. 8.

[168] Compare the 'piquant' contrast between rotting animals at the end of *Georg.* 3 and honey and herbs at the beginning of *Georg.* 4; or the honey later produced from the putrescent carcase. Herbs are significant in Virgil's poetry, with their sweet

are no sharp flavours, nothing acrid or harsh, no vinegar. Nasidienus' sauce recipe at 2. 8. 45–53, which contains white pepper, vinegar, and bitter elecampanes, and the ingredients of the legal pickle in 1. 7 are obviously designed as contrasts. This is a prescription for satire in a literary atmosphere full of restrictions. The satirist must mellow (cf. *nil nisi lene*, 26), and not embitter his critics. In 2. 1 a satirist who composed aggressive poems was threatened with legal reprisal (*ius iudiciumque*, 82–3). Now he must acquaint himself with the law and learn not to be *nimis acer* (2. 1. 1). The ingredients of Horatian satire are mixed (*confusum*, 67), but within the bounds of an appropriate recipe (*miscere . . . decebit*, 65; cf. *miscebat*, 24, *decet*, 26; *miscet*, 55, eliminating *aliena*, 57).

The outlying passages of dialogue can now be read differently too. *Nova praecepta* (2) are prescriptions for new satire (where 'new' contrasts with *Satires* 1 and Lucilian satire, signifies 'neoteric' and draws an analogy with New Comedy). *Res tenuis tenui sermone peractas* is not only 'a subtle science', but also 'a fine Callimachean subject expressed in fine language'. *Adire ad fontes remotos* is as much a Callimachean metaphor as a Lucretian one. And the answer to the enigma *celabitur auctor* (11, 'the author is a secret') is, ironically enough, Horace himself.[169] Both Rudd and Classen come unwittingly close to this conclusion: Rudd draws an elaborate parallel between writers and cooks, literary and convivial decorum, but forgets the analogy when interpreting 2. 4; Classen's article is entitled 'Horace—a Cook?', but deals only with the pseudo-Epicurean element.[170] This poem can only be deciphered by recognizing Horace's blend of legal, literary, and culinary themes elsewhere in the satires.

At least four levels of meaning can be found, then, in this poem: superficially, the culinary meaning; by extension, the

smells, thin (*tenuis*) juices, and concentrated flavour: cf. *Ecl.* 2. 11: *herbas olentis*; 2. 48, 49.

[169] Other suggestions have included Maecenas, Epicurus, Ennius, Curtillus, Catius (see Classen 1978: 343–4). The only other supporter of Horace is Wieland (*Horazens Satiren* (Leipzig, 1786), 134)—for the very different reason that Horace was the member of Maecenas' circle most likely to need to combine elegant living with a low budget.

[170] Rudd (1966: 202–3); Classen (1978).

social meaning; underneath, a philosophical meaning; and an aesthetic meaning. This new aesthetic meaning has not been unearthed to displace any of the others. In satire, literary issues are inseparable from social, moral, and political ones anyway, and to couch a literary programme in gastronomic ones is a way of pointing out that modern satire is a social skill: entertaining readers is no longer carefree.[171]

Even so, the aesthetic meaning clashes with the philosophical one in an uncomfortable way. Although the vocabulary common to philosophy, cooking, and literature makes these systems look parallel, it also draws attention to the fact that they are ultimately divergent. In Horace's other works, as we have seen, writing, diet, and morals are all governed by the same framework of rules. The *Ars Poetica*, a system of decorum similar to Catius' culinary precepts, justifies writing as a moral activity. Horace teaches the proper way to nurture a poet:

> unde parentur opes, quid alat formetque poetam,
> quid deceat, quid non, quid virtus, quo ferat error.
> (*AP* 307–8)

The sources of creativity, what should nurture and shape the poet, what is appropriate, what not, and where right and wrong will lead him.

Poetic decorum and morality are so fused[172] that the meaning of line 309 of the poem (*scribendi recte sapere est principium et fons*: 'judgement is the primary source of good writing') is unclear. *Sapere* either looks ahead to a section on moral subject-matter, or back to lines on stylistic etiquette. Horace then resorts to a culinary analogy for poetic decorum which has much more in common with Catius' rules than with Ofellus', without any apparent difficulty. A poem with

[171] Contrast Lucilius and his victims capering about informally while the cabbage cooks: *discincti ludere donec / decoqueretur holus soliti* (2. 1. 73–4).

[172] Cf. Cic. *Orat.* 70: *ut enim in vita sic in oratione nihil est difficilius quam quid deceat videre* ('In literature as well as life, nothing is harder than to see what is appropriate'); Plato, *Phaed.* 274b: *Τὸ δ' εὐπρεπείας δὴ γραφῆς περὶ καὶ ἀπρεπείας, πῇ γιγνόμενον καλῶς ἂν ἔχοι καὶ ὅπῃ ἀπρεπῶς, λοιπόν* ('It remains for us to discuss the matter of propriety and impropriety in writing, how it should be done and how not').

faults, he says, is like a sophisticated banquet where the slightest fault results in a fiasco:

ut gratas inter mensas symphonia discors
et crassum unguentum et Sardo cum melle papaver
offendunt, poterat duci quia cena sine istis. (*AP* 374–6)

Like a pleasant meal, where a tuneless orchestra or coarse ointment or poppyseeds with Sardinian honey give offence, because the dinner could have been served without them.

In the context of morals, however, Horace uses misbehaviour at the dinner-table as an example of a trivial error that does not offend him (*Sat.* 1. 3. 90–4). It becomes clear that while sophisticated cookery and poetics are compatible, or poetics and morals, sophisticated cookery and morals are anathema to each other.

In satire, however, inconsistencies come to the surface, and *Satire* 2. 4 gives its poetic/culinary instructions a negative moral framework. However internally correct, they are symptoms of an obsession that is morally awry; they are no less tedious for having a double meaning. Horace's satirical response turns out to be a commentary on his own poetic principles. At the end of the poem, when Horace says *mihi cura / non mediocris inest* (93–4), he seems to be saying ironically that he shares Catius' immoderate, unepicurean desire to learn culinary precepts. Horace is the satirist laughing at a fool. But one important concession of Book 2 is that the satirist is willing to expose himself to Saturnalian abuse (2. 3. 5, 2. 7. 4). As Horace says, using the imagery of Christmas games, the man who laughs at others finds a tail tied to his own back (2. 3. 53). If these Cordon Bleu lessons are also a creative writing course, Horace is the victim of his own *cura non mediocris*, inconsistent with the other principles of Epicurean ἀταραξία that govern his life and philosophy: this *cura* is his obsessive perfectionism in writing.[173] At *Ep.* 2.1. 224–5 Horace complains that no one appreciates the subtle work that goes into his poems:

cum lamentamur non apparere labores

[173] Cf. 1. 10. 67 ff. Horace's moral catchphrase *iam satis est* is often replaced by *non satis est* in an aesthetic context (e.g. *AP* 99. *Sat.* 1. 4. 40–1, 54).

nostros et tenui deducta poemata filo.

when we moan that the labours involved in our finely-spun poems go unnoticed.

In fact, he has been at pains throughout the satires to acquaint us with it. To discuss poetic principles in this new guise of a tract on *nouvelle cuisine* is to observe one kind of decorum: the metaphor is appropriate for the culinary associations of *satura* and the comic tradition of self-important cooks. At the same time, it is extremely indecorous (which is, again, apt enough for such a self-effacing, consciously sullied genre). Horace downgrades his poetic principles and makes them comic by identifying them with a newfangled craze, part of a corrupt society and morally suspect. *Satire* 2. 4 is a Saturnalian skit on the dignity and authority of Horatian poetics.

Unexpected support for this reading of the poem can be found in an eighteenth-century parody, William King's *Art of Cookery*.[174] This is a translation of the *Ars Poetica* which substitutes culinary principles for poetic ones at every opportunity. Its wit lies in the way it exploits literary words in the original Latin that are metaphors from the senses. For example, the line *non satis est pulchra esse poemata; dulcia sunto* (*AP* 99, 'It is not enough for poetry to be fine: it must be sweet too') is applied to patisserie:

Unless some Sweetness at the Bottom lye,
Who cares for all the crinkling of the pye?

Horace's supreme dictum on taste, *scribendi recte sapere est et principium et fons* (*AP* 309) is made to recall his own puns on the culinary and aesthetic meanings of *sapere*:

The fundamental Principle of all,
What ingenious cooks the Relish call.

And Horace's plea for realism (*AP* 340 ff.) becomes a caution against denatured dishes:

Meat forc'd too much, untouch'd at Table lies,

[174] The poem is discussed by Fuller (1976). See also Owen (1977) on its implications for 18th-c. culinary aesthetics.

Few care for carving Trifles in Disguise,
Or that fantastick Dish, some call *Surprise*.

By carving trifles in disguise in *Sat.* 2. 4, Horace, it seems, had already written his own parody.

3 *Horace,* Satire *2. 8*

Satire 2. 8 is much easier to stomach than 2. 4: here we have a drama, not a 'disembodied voice'.[175] The dramatic quality of the poem seems to encourage us to be satisfied with the poem's surface meaning alone. A dinner party is described, at which the guests, a mixed collection of literary figures and shady parasites, have the time of their lives making fun of the pedantic gourmet who is their host. At the end they bolt unexpectedly without finishing the meal. Faced with this incompleteness, critics have always resorted to external 'reality' to explain the poem: Palmer (1925) argues that Nasidienus, the host, is a pseudonym for Salvidienus Rufus, who was executed by Octavian; the scanty remains of contemporary gastronomy are used to assess the extravagance of the menu; Rudd (1966) plays down the cultural strangeness of the poem by comparing it with twentieth-century satires on vulgarians. A less literal approach to the poem will show that this dinner is very specifically staged as the consummation of all the connected themes of Book 2: the social and legal position of the writer, the codification of cookery and the law, the proper or enforced limits on consumption, writing, and speech. The books which began with the image of a contented feaster (*uti conviva satur*) end with an appropriate solution to the satirist's dilemma.

As we have already seen, a literary *convivium* is often simultaneously 'lifelike' or self-contained, and the metaphorical focus for larger concerns. Here, the tight, almost abrupt narrative of Horace's poem invites us to see this meal as a figurative event, especially as it is the finale to a book that has dwelt on the themes of social compatibility and the

[175] Rudd (1966: 221): 'Catius is a mouthpiece and his teacher a disembodied voice.'

perversion of communal eating. Not only is the dinner party a 'natural' setting for satire, as a miniature of social organization and the unstable civilizing of animal instincts: its hybrid qualities alone make it an attractive theme. The dinner party is often summed up as a blend of physical and cerebral pleasures, jokes, and philosophy (an image inspired by the central feature of the stewpot or mixing-bowl). Rambling sympotic works, like Athenaeus' *Deipnosophistae* or Macrobius' *Saturnalia*, where the food on the table provides philological fodder, play on the shared vocabulary of physical and mental satisfaction.[176] Above all, the ideal dinner is a mixture of hilarity and seriousness. At a meal described by Pliny, pleasures are 'seasoned' with serious pursuits (*Ep.* 3. 1. 9); at Xenophon's *symposium* a jester is introduced because the guests, 'though well fed on seriousness, are hungry for laughter' (*Symp.* 1. 13). This of course gives it a structural affinity with satire. Varro's recipe for satire is a mixture of hilarity, philosophy, and dialectic; Horace himself prescribes equal quantities of seriousness and humour (*et sermone opus est modo tristi, saepe iocoso, Sat.* 1. 10. 11).

This balance between fun and seriousness may of course shift. Roman literary *convivia* are often parodies which spoil the Greek *symposia* idealized by Plato.[177] The Roman *cena*, which, unlike the *symposium*, necessarily involved eating, was, as we have seen, thrust into literary prominence because it helped to embody the differences between an intellectual and a materialist culture. In *Sat.* 2. 8 Horace uses food to measure the distance from Plato's *Symposium*.

Satire 2. 4, we saw, restored Platonic 'feasts of words' to their material form. A similar form of parody occurs at the beginning of 2. 8. Lines 4–5:

da, si grave non est,
quae prima iratum ventrem placaverit esca

Tell me, if it is not too tedious, what food first appeased the wrathful stomach . . .

[176] See above pp. 29–30.

[177] Cf. Cameron (1969) on the *Cena Trimalchionis*: it is 'a play upon a serious and respectable literary genre, a hybrid blend of dialogue and meal description'.

parody Plato, *Timaeus* 17b:[178] εἰ μή τί σοι χαλεπόν, ἐξ ἀρχῆς διὰ βραχέων πάλιν ἐπάνελθε αὐτά. ('If it is not a bore, give us a brief summary from the beginning'). The hidden point of the allusion is that *Timaeus* also opens with feast metaphors: a group of philosophers have taken turns to be hosts (ἑστιατόρες) and guests (δαιτυμόνες) at philosophical 'feasts' of words (τὴν τῶν λόγων ἑστίασιν 27b).[179]

Plato's *Symposium* is the perfect compromise between metaphor and reality. An actual feast becomes the setting for philosophical discussion on an erotic (therefore suitably sympotic) theme. This partial descent into reality is complemented by the 'mixed' nature of the text. Agathon's party, like a satyr-play, takes place after the performance of his own tragedy (173a); the guests include Aristophanes the comedian and the Silenus-like Socrates (215a–b); Socrates calls Alcibiades' farcical seduction story a σατυρικὸν δρᾶμα . . . καὶ σιληνικόν (222d, 'a Silenic satyr-play').[180] The speeches are a mixture of different styles; internally Agathon's speech combines seriousness and playfulness (197e, τὰ μὲν παιδίας, τὰ δὲ σπουδῆς μετρίας); Socrates ends the evening discussing the virtues needed to write tragic and comic poetry (223d).

Food itself, however, plays little part in the dialogue, except as metaphor or simile. Socrates 'thirsts' after wisdom; Aristophanes compares the split human soul to an egg, a sorb-apple, and a turbot.[181] At Horace's debased *symposium*, apples and turbot become real (*ilia rhombi*, 30; *melimela*, 31). Nasidienus' praise of his own food would be foreign to the spirit of Plato (at 177b Eryximachus appeals for love to be considered as a theme, reminding his audience that someone once wrote a panegyric on salt). The weight of food at Horace's feast marks its stylistic descent into full satire. The Platonic parody with which it begins alerts us to a series of

[178] The allusion is pointed out by Fraenkel (1957: 137), who also notes that *Sat.* 2. 2. 2 is based on the beginning of Eryximachus' speech in the *Symposium* (177a), and that 2. 4 begins with an allusion to the beginning of *Phaedrus* (228b). It cannot be an accident that these satires are the ones which deal most with food: the weight of material represents the lapse of modern culture from the Platonic ideal.

[179] See above p. 41 n. 185 and p. 140 on the image of the philosophical feast of words.

[180] I owe this point to John Henderson.

[181] 175e, 190e, 191d.

fallings-off from ideal Platonic feasts. To begin with, Horace, who was not present himself, asks his friend Fundanius about the food, not the conversation, trivializing philosophical vocabulary in the process: *bonus* is used of 'a good time': *numquam / in vita fuerit melius* (3–4, 'I had the time of my life'); *beatus* (1) means 'rich'; *vita* (4) is simply a collection of social experiences; the search for knowledge (*nosse laboro*, 19) is restricted to the names of guests at the dinner.[182] This opening suggests some of the ambiguities of the *convivium*, which can both epitomize social relations and parody them.

The feast is recalled by the comic poet Fundanius, and his representation of it vitally determines the way in which we experience the event. His indecorous mixture of tragic and comic styles,[183] hyperbole and bathos, precipitates the descent of the philosophical 'feast of words' begun in the *Symposium*. Two of Nasidienus' guests, the tragedian Varius and the comedian Fundanius, complement this mixed style, just as Agathon and Aristophanes complemented the mixture of solemnity and fun at the *Symposium*. Fundanius writes a tragicomedy (or is it a satyr-play?) to corrupt the perfect propriety of Nasidienus' gastronomic design.

Many incidental details can be read as reminders of this. In the description of the page-boys, *alte cinctus* (10, 'in formal dress') and *purpureo* (11, 'crimson') suggest preparation for a lofty epic, programmatic girding-up for a majestic ὕψος.[184] The most visible catastrophe of the piece, a collapsing tapestry (*suspensa gravis aulaea*, 54), is a figure for satirical irreverence, the bathetic treatment of the tragic style. *Gravis* suggests high seriousness; *aulaeum* is a stage-curtain as well as a tapestry, which would open on a tragedy (or comedy).[185] Falling from the heights, tragedy is reduced

[182] Proper philosophical discussion takes place at Horace's meal in the country (2. 6. 70–6).

[183] *AP* 231–3: to confuse tragic and comic style is like making a prudish matron dance with satyrs; ibid. 90–1: a tragic feast should not be described in low language (*indignatur item privatis ac prope socco / dignis carminibus narrari cena Thyestae*).

[184] Cf. *praecincti* (70, 'well-groomed'); cp. *discincti* (2. 1. 73, 'in casuals', at Lucilius' informal dinner).

[185] For *aulaeum* used figuratively of lofty or tragic language, cf. Apul. *Met.* 1. 8: *aulaeum tragicum dimoveto . . . et cedo verbis communibus* ('Remove the tragic curtain, and I shall return to plain words').

to the sphere of the dinner plate: *ruinas / in patinam fecere* (54–5, 'it collapsed on to the plate'), trailing an epic simile behind it:

trahentia pulveris atri
quantum non Aquilo Campanis excitat agris.

dragging with it more black dust than the North wind stirs up from the Campanian plains.

Wine-jars upturned into simple earthenware cups (*invertunt Allifanis vinaria tota*, 39) also suggest the bathetic irreverence of the narrative, the decanting of high language into a humble form.[186] Horace's request, *da, si non grave est* (4), is not only a parody (of epic as well as tragedy: this chronicle of the *iratum ventrem* is also a bogus wrath-poem), but a prescription too: 'if it is not serious', as well as 'if it is not a bore'. While Nasidienus weeps, the tragedian Varius can scarcely stifle his laughter in his napkin (*mappa compescere risum / vix poterat*, 63–4).[187] The host's fiasco, caused by the mocking goddess Fortune (*gaudes illudere rebus / humanis*, 62–3), unintendedly provides entertainment: *ludos spectasse* (79, 'watching the games'). Nasidienus tries to salvage his performance by varying the theatrical display of dishes;[188] Nomentanus tries to raise his spirits (*ni amicum / tolleret* at 60–1 suggests the raising of a stage curtain[189]); his host returns after a scene-change: *mutatae frontis* (84, 'with a new front').[190] Finally, the whole occasion dissolves into farce.

Nasidienus' feast is not only the antithesis of Plato's *symposium*: it also reverses Lucilius' informal Republican supper (2. 1. 71–4), the hospitality of the country mouse and Horace's relaxed philosophical dinners in the countryside in 2. 6. The host puts a brave face on things by using words that suggest the ideal unity of the *symposium*: *commoda*, comfortable, *conviva*, companion, *comis*, congenial (76); at the same

[186] See Commager (1962) on wine in *Od.* 1. 20.

[187] This comic mask contrasts with the appropriately tragic one he wears in *Sat.* 1. 5: *flentibus hinc Varius discedit maestus amicis* (93, 'a sad Varius left his weeping friends').

[188] On the theatricality of Roman convivial *apparatus*, see Rosati (1983).

[189] The Roman curtain was lowered to reveal the stage: Phaedr. 5. 7. 23, *aulaeo misso*; Apul. *Met.* 10. 29, *aulaeo subducto*.

[190] For *frons* as stage-scenery, backdrop, etc., see Virg. *Georg.* 3. 24, Vitr. 5. 6. 8.

time, words of separation mark the dissolution of true conviviality: *dissimilem* (28, dissimilar), *districtum* (68, distraught), *divisos* (78, divided), *discerpta* (86, disjointed), *avulsos* (89, torn apart). The easy, high-minded conversation and *laissez-faire* spirit of Horace's country dinner (*solutus / legibus insanis*, 68–9) contrast with the restricted and hypocritical behaviour of Nasidienus' guests, which the theatrical imagery reinforces. The gastronome Nomentanus,[191] like the informers who lurked behind *Satires* 1 (*indice digito*, 26), discloses the true identity of dishes that have something to hide: *si quid forte lateret* (25, 'whatever lay underneath'); *longe dissimilem noto celantia sucum* (28, 'vastly different from their normal flavour' 28).[192]

If we are to see this poem as a cultural product, it is not enough, with Rudd, to regard it simply as a satire on food-snobs: it must be seen as a representation of Roman culture itself, as the adulteration of conversation with secrecy, freedom with rules and manners. We are reminded once again of the central theme of *Satires* 2: the predicament of the modern satirist faced with hostile critics and the risk of prosecution (a double *ius iudiciumque*), whose writing can no longer be acrimonious or easy. *Satire* 2. 4 was Horace's new recipe for satire in gastronomic terms. *Satire* 2. 8 is best read as a final critical trial: a tragicomic *convivium* which sums up all the difficulties facing the satirist, a poet who has to live in society (*convivere*) while attacking it. The guests leave in disgust, but they have created their own entertainment out of the disaster: in some sense, even a satirical fiasco can be successful.

We are told specifically that Horace was absent from this meal, looking unsuccessfully for a *conviva* (2): in the narrowest sense, a dining-companion; in the widest, someone socially compatible. As usual this evasion should be taken with a pinch of salt.[193] For a start, Fundanius' mixed and at

[191] *Sapiens* (60) only in culinary matters.

[192] Informers: e.g. 1. 3. 60–1, 1. 4. 81 ff.

[193] Cf. 2. 2. 2 *nec meus hic sermo* ('these are not my words'); 2. 4. 11 *celabitur auctor* ('the author will be a secret').

times poisonous description is typical of a satirist. The miscellaneous composition of the guests can be explained if we see them as representatives of the people Horace writes satire for and against. Maecenas, Varius, and Viscus are among the poets Horace is happy to please (1. 10. 81–3); Fundanius is praised at 1. 10. 40–2. The other guests are typical satirical victims: parasites, *scurrae*, and spendthrifts (Vibidius, Balatro, Porcius, and Nomentanus).[194] According to Trebatius, these are the people it is more dangerous to satirize:

> quanto rectius hoc quam tristi laedere versu
> Pantolabum scurram Nomentanumque nepotem.
> cum sibi quisque timet, quamquam est intactus, et odit.
> (2. 1. 21–3)

How much better this [writing epic] is than using miserable verse to wound the party-goer Pantolabus and the playboy Nomentanus. The more fearful your victims, the more they hate you, even if you leave them alone.

Nasidienus, the host, can also be seen as a transformation of Horace. At *Epistle* 2. 2. 58 ff., where Horace compares himself to a host trying to please guests with different tastes: he uses words that are both gastronomic and aesthetic (*sal, palatum,* and *acidum*) to reinforce the analogy. Nasidienus' meal is also a critical sampling; gastronomic flavours are also literary flavours, which excite unexpected responses in the guests/audience. 'Convivial' puns on words with gastronomic, moral, and aesthetic meanings run riot in this satire. There must, for example, be a point to the name Nasidienus ('the man with the nose').[195] It looks like a well-chosen pun

[194] Buried in paupers' graves at 1. 8. 11. Nomentanus is a type name for a prodigal at 1. 1. 102, 2. 3. 175, 224. The name Panto-labus ('snatching everything') seems to be paraphrased in the description of the greedy Porcius: *totas simul absorbere* (24).

[195] Rudd (1966: 146) rejects this: 'But when we are asked to note the significance of . . . Nasidienus ("The nose") . . . then it is time to call a halt.' Roos (1958) argues from inscriptions that Nasidienus was a real name, and that Horace might therefore be referring to a real person.

on *nasus*, the organ of smell and aesthetic taste, critical sneering and satirical wit.[196] Like a nervous satirist, Nasidienus has stomachs and palates to appease (*iratum ventrem* 5, *stomachum* 9, *subtile palatum* 38).[197] At every stage of the meal he flaunts his good taste: like Horace he is a perfectionist, separating good from bad (*deterior*, 44; *melius*, 53), despising the commonplace (*longe dissimilem noto celantia sucum*, 28), selecting (*delecta*, 32), eliminating superfluity (*sublegit quodcumque iaceret inutile quodque / posset cenantis offendere*: 'he removed any waste that might offend the guests', 12–13). In all this he follows Catius' master and is the antithesis of the slapdash Lucilius.[198] The precariousness of the host's endeavours is similar to that of the poet's:[199] he has to ensure that the food is properly cooked and seasoned, and that the pages are well-groomed.[200] Few appreciate the labours that go into a successful meal: *eoque / responsura tuo numquam est par fama labori* (65–6, 'Your fame is never equal to your labours'). Horace, too, laments that his poems are

[196] Sneering: *Sat.* 1. 3. 29–30; *Ep.* 1. 5. 23; Mart. 1. 3. 6, 5. 19. 17; Plin. *NH* 11. 158. Gastronomic taste: *Sat.* 2. 2. 89–90 *non quia nasus / illis nullus erat* ('not because they had no taste'); discrimination: Mart. 1. 41. 18; wit: Plin. *NH* pref. 7 (of Lucilius).

[197] *Stomachus* = taste: Mart. 13. 3. 8; Cic. *Fam.* 7. 1. 2; = anger: Cic. *Att.* 16. 2. 3; Hor. *Od.* 1. 6. 6. *Palatum* = critical sense: Hor. *Ep.* 2. 2. 62; Quint. 6. 3. 19. For angry stomachs cf. 2. 2. 18: *latrantem stomachum bene leniet* ('appeases the growling stomach'). *Latrare* is also used of satirical aggression at 2. 1. 84; and poets are described as an irritable race at *Ep.* 2. 2. 102.

[198] See Shero (1923) on the links between 2. 8 and Lucilius' dinner-party satires. Two possible allusions suggest that Horace is challenging Lucilius' supremacy as a dinner-party writer. Anderson (1963) notes that Horace always portrays Lucilius in the act of dining. *Sic ut mihi numquam / in vita fuerit melius* (3–4, 'I had the time of my life') recalls fr. 204W: *cenasti in vita numquam bene. Gausape purpureo mensam pertersit* (11, 'he wiped the table with a crimson cloth') recalls fr. 598W: *purpureo tersit tunc latas gausape mensas.* This malicious choice of quotation implies that Lucilius was neither a good nor a neat writer.

[199] The parallel between host and general at 73 supports the idea that Nasidienus is based on Salvidienus Rufus (for his career see Vell. Pat. 2. 76. 4 and Suet. *Aug.* 66). Perhaps the equation suggested the similarly vulnerable position of both politician and writer, both answerable to a supreme *iudex* (cf. *iudice Caesare*, 2. 1. 84).

[200] The pun in English has its equivalent in the Latin: *ut omnes / praecincti recte pueri comptique ministrent* (69–70, 'that all the page-boys are smartly dressed at table'). See *AP* 446–7 for verses that are *incompti*. Another disaster, a footman who slips (*pede lapsus*, 72) and smashes a plate, may also have literary significance: see *AP* 80 for a pun on metrical and bodily feet; Cic. *Brut.* 185 for *labi* of literary lapses.

taken for granted.[201] Failure, however, is always glaring.[202] The guests, too, correspond to different kinds of readers. Maecenas is treated as one of the initiated and offered four extra wines.[203] Porcius, the comic pig-man, eats cakes whole, like a reader who appreciates only the sops, not the subtleties.[204] It is left to Nomentanus to explain the underlying truth: *si quid forte lateret* (25).

Despite his good taste, however, Nasidienus is also a *distorted* representation of Horace. His 'nose' is false, or disproportionate; he is a comic Pulcinello. What he cannot judge, unlike Horace, are the proper ends or limits of speech. Just as Catius' gastronomic lecture bored Horace in 2. 4, so Nasidienus' commentary spoils the guests' appetite for his dishes: *suavis res, si non causas narraret earum et / naturas dominus* (92–3, 'delicious things, if only our host hadn't insisted on telling us about their origins and properties').[205] His fears that they will numb their appreciation by drinking (*fervida quod subtile exsurdant vina palatum*, 38) come true: they start to decant his wine into common cups, an image of their satirical disrespect. Nasidienus himself becomes the object of the critical nose: the joker Balatro sneers at him: *suspendens omnia naso* (64, 'dangling it all from his nose'[206]) and offers him ironic consolation: *haec est condicio vivendi* ('That's life'). The host of the satire and the satirical victim change roles.[207] In one sense, to sum up the human condition

[201] *Ep.* 2. 1. 224–5.

[202] At *AP* 374 ff. Horace uses images of mistakes at banquets (an out-of-tune orchestra, over-heavy scent, and a mixture of poppyseeds and bitter Sardinian honey) as analogies for catastrophic mistakes in poetry.

[203] Maecenas himself wrote a *Symposium*, where Horace, Virgil, and Messalla were the guests (Serv. ad *Aen.* 8. 310).

[204] At 2. 4. 47–8 the cook is told not to spend all his energy making cakes (*crustula*), which I argued stood for jokes, from Horace's analogy between bribing children with cakes and softening satire with jokes. Porcius (described as *ridiculus*) only appreciates humour.

[205] Ps. Acron sees the paradox: 'Nasidienus, a Roman knight, is elegant in every respect except for his rotten taste (*putidus*) in listing all the delights that are on display.'

[206] At 1. 6. 6 Horace thanks Maecenas for not despising him: [*nec*] *naso suspendis adunco / ignotos* ('you do not dangle me from that aquiline nose').

[207] Persius saw Horace in Balatro when he described the poet's critical acumen: *callidus excusso populum suspendere naso* (1. 118). Other features link Balatro with Horace: he has servile origins, implied in the name Servilius (Horace is *libertino*

in terms of a cook's problems is a banal alternative to true philosophy.[208] Yet in another it is an appropriate analogy for the life of a satirist. Like the other Saturnalian poems, this satire is self-parody: Nasidienus' commentary caricatures Horace's own programmatic writings and brings about the end of Book 2 with the premature departure of the guests.

The food served at the dinner deserves careful scrutiny, as it must somehow be representing the satirical product that is being sampled. So far, however, it has only been discussed as testimony for Roman eating-habits, and assessed for extravagance and bad taste.[209] As a whole, its connections with 'reality' are important: airy, sympotic philosophy is being abased into solid substance. Yet each course is also composed of words that bring metaphorical meanings into play, a mixture of literary-critical puns and words that suggest the hidden corruption in Roman culture. It may be Nasidienus' commentary that causes the guests' disgust, but the food on their plates provides a gauge of their irritation. The style of cooking is very different from the recipe for blandness dictated by Catius. It is not too far-fetched, perhaps, to say that these dishes live up to the two criticisms levelled against Horace's work at the start of 2. 1: that he was either *nimis acer* (too sharp) or *sine nervis* (enervated).

This would for a start explain the curious lack of a *promulsis*, the standard introductory course with honeyed wine (from *mulceo*, to soothe).[210] Catius (2. 4. 26) had recommended serving only *lene mulsum* on a empty stomach. Here, the first dish, Lucanian boar surrounded by sharply flavoured vegetables and pickles, seems designed to excite hostility:

> in primis Lucanus aper; leni fuit Austro
> captus, ut aiebat cenae pater; acria circum
> rapula, lactucae, radices, qualia lassum
> pervellunt stomachum, siser, allec, faecula Coa. (6–9)

patre natum, 1. 6. 6); and he is a parasite (cf. 2. 7. 29–35, and Augustus' letter to Horace in the *Vita Horati* urging him to leave *ista parasitica mensa*).

[208] One sense of *condicio* is 'the acceptance of a dinner invitation' (*OLD* s.v.); *condicio vivendi* also shadows *convivendi*.

[209] See Palmer (1925), Rudd (1966), Ricotti (1983).

[210] Palmer (1925) ad loc. assumes that Horace has just forgotten to mention it.

The first course was Lucanian boar; it had been caught during a mild South wind, according to our host; it was surrounded with bitter turnips, lettuces, radishes, all sorts of things that pep up a jaded appetite—rampion, fish-sauce, dregs of Coan wine.

The dish exemplifies the ancient belief that animals and those who eat them absorb the nature of their environment.[211] Boars are traditionally ferocious.[212] The surrounding garnish (*acria circum / rapula*, etc.) is an appropriate recreation of the boar's natural vegetable habitat and a 'tasteful' aetiology of its savage temperament. At the same time, it cannot fail to remind us of the dangers of satirical acerbity. However proper this acrid first course may be in culinary terms,[213] it is destined to have an acrimonious reception. It does not placate the angry stomach so much as exacerbate it. The guests' jaded appetites (*lassum stomachum*) are violently assailed by the pungent vegetables, roots (*radices*) by some strange inversion doing the uprooting (*pervellunt*).[214] Vindictiveness and hostility are the emotions nourished: the guests later become *acris potores* (36–7), bent on revenge (*ulti*, 93).[215] The choice of a maple-wood table (apparently only a poor second to citrus in the hierarchy of elegance, and therefore not an immediately obvious choice) can be also be explained along these lines: *acernam* (10) sounds like *acer* or *acerbus*.

By contrast the page-boys are described as though they were *sine nervis*, effete or emasculated. The simile *ut Attica virgo / cum sacris Cereris* (13–14, 'like an Athenian maiden with the sacred objects of Ceres') not only distances this page-boy from the ideal world of Athens: it also suggests that he is effeminate. *Alcon Chium [ferens] maris expers* (15, 'Alcon bearing Chian wine untouched by salt') sets up another ambiguity. Does *maris expers* mean unsalted wine, or wine produced in Italy rather than overseas? Or does the

[211] See Gourévitch (1974).

[212] Cf. *Epod.* 2. 31–2: *acris . . . apros*; Virg. *Ecl.* 10. 56: *acris . . . apros*.

[213] Cf. Celsus 1. 2. 8 recommends first courses of pickles and vegetables.

[214] Cp. Cato, *Agr.* 161. 2: *vellito ab radice*; Virg. *Aen.* 3. 28: *radicibus vellitur*. *Pervellere* can also be used of carping criticism: e.g. at Cic. *de Or.* 1. 265.

[215] See Ch. 5 below on *Epod.* 3, where indigestible garlic causes Horace to belch out savage invective.

phrase mean 'castrated' (from *mas, maris*—male, male parts)? Grammatically it could be applied to Alcon as well as to the wine.[216]

The mood of the whole piece is summed up in the murky depths of Nasidienus' sauce recipe: *his mixtum ius est* (45, 'this is how the sauce is mixed'). This one obviously contains many more bitter ingredients than Catius' sauce (2. 4. 63–9): white pepper (*pipere albo*), vinegar (*aceto*), green rocket (*erucas viridis*), bitter elecampanes (*inulas amaras*)—all of which, again, feed the guests' mood of acrimony.[217] *Illutos echinos*, unwashed sea-urchins, another ingredient, also suggests prickliness.[218] At the same time, Curtillus, the cook mentioned as the source of this ingredient (52), could be translated 'little gelding'.[219] Finally, the motley limbs served at the end of the meal lack those parts associated with virility: they are 'loin-less' (*suavius . . . quam si cum lumbis quis edit*, 90).[220] Horace appears to be reviving the criticisms of 2. 1. 1 ff., turning a culinary expert into the victim of the chief critics and the satirical victims themselves. By contrast, Catius' prescriptions were designed to tone down the satirical recipe and make it bland: *nil nisi lene decet* (2. 4. 26).

Ironically Nasidienus' placatory comments are full of words associated with softening, but their effect is even more exacerbating than the acrid flavours of the dishes; these, we are told, would otherwise be *suavis res* (92, 'sweet' or 'tasty'). The boar, he says, was seduced into being captured by a gentle wind (*leni fuit Austro / captus*, 6–7). Nomentanus mentions *melimela* (31, honey-apples), the quintessence of sweetness.[221] The joints of meat served at the end are chosen

[216] See Palmer (1925) ad loc; Housman (1913) on Persius 6. 39 (*nostrum hoc maris expers*) argues that the phrase *maris expers* means 'emasculated'.

[217] Cf. the metaphors for bitter speech in *Sat.* 1. 7: *sermonis amari*; *pus atque venenum*; *salso*; *aceto*.

[218] At *Epod.* 5. 27–8 Horace describes the witch Sagana with her bristling hair (*capillis . . . asperis*), like a sea-urchin or a running boar.

[219] Cf. 1. 6. 104: *curto mulo*, a gelded mule, the satirist's means of transport.

[220] Cf. Pers. 1. 104: *delumbis*, 'emasculated, enervated'; Tac. *Dial.* 18. 5: *elumbis* (coupled with *fractus*).

[221] Cf. Mart. 13. 24. 2, 7. 25. 7, where *melimela* are used as a parallel for *dulcia epigrammata; mel*, honey, is often used as a type for sweetness, charm, or blandness: see Bramble (1974: 52), Rocca (1979).

to be *suavius* (89). *Esca* (5) is commonly used of flatterers' bait.[222] There are shades here of the legacy-hunters (*captatores*) of 2. 5, with their gifts of food to the childless rich, whom they themselves metaphorically 'consume' (44, 82–3). One of these happens to be called Nasica, and he and Nasidienus have different ways of courting favour. Nasidienus' lamprey, we are told, was caught while spawning: *haec gravida / capta est, deterior post partum carne futura* (2. 8. 43–4, 'This was caught pregnant: the meat is worse after eggs are laid'). This smacks of Tiresias' advice to Ulysses to avoid a man who has a child or fertile wife: *sperne, domi si gnatus erit fecundave coniunx* (2. 5. 31). The connection is reinforced by the imagery of fishing for legacies in 2. 5 (25, 44) and the appearance of the ubiquitous word *ius* at 2. 5. 29: it is better to support a childless man, however corrupt, in the courts, even if he has the cheek to indict (*vocet in ius*) a better man than himself.[223] Nasidienus, of course, devises *ius* in the shape of a sauce-recipe; his only child is the dinner itself; he is called *cenae pater* (7, 'father of the dinner'), and, when it goes wrong, weeps as though his son had died young (*si filius / immaturus obisset*, 58–9).

There is another hidden ingredient in these dishes. They seem, on the surface, to be the perfect products of *haute cuisine*, carefully selected, presented in a witty, artificially 'natural' state: the fierce boar grazing among pungent roots and vegetables, the lamprey afloat amid swimming prawns.[224] Even so, there is something insidiously rotten in the way Fundanius recalls them. His descriptions paradoxically 'taint' nice food by conjuring up the margins of inedibility at the heart of civilization: poison, rottenness, cannibalism, and omophagy.[225]

First, the boar: the details of its capture (*leni fuit Austro / captus*) 'unintentionally' recall 2. 2, where Ofellus talks of

[222] *OLD* s.v.

[223] Cf. Varro's pun at *RR* 3. 17. 4: *hos pisces nemo cocus in ius vocare audet*.

[224] *Squillas inter murena natantis*, 42 (the word-order suggesting the arrangement on the plate). Cf *SHA Elag*. 24. 1: fish served in blue sauce as though swimming in the sea.

[225] Cf. the title of Murray's (1963) translation of 2. 8: 'A Taint of Elegance'.

the honest rancid boars of the past (89), and calls down corruption on morally unwholesome boars of today:

> praesentes Austri, coquite horum obsonia, quamquam
> putet aper . . . recens (41–2)

Come, south winds, and bake this food: however fresh, the boar is already rotten . . .

In fact, the phrasing is so suggestive that Rudd concludes that the boar Nasidienus serves is slightly high.[226] This would not correspond, however, with the change in taste that 2. 2 records: rancid boars are now gastronomically unacceptable.[227] Even so, Nasidienus' remark succeeds in putting the guests off their food. Another inconsequential remark, about reddened honey-apples being picked by the light of a waning moon (*melimela rubere minorem / ad lunam delecta*, 31–2), recalls the moon that blushes over Canidia's rites in 1. 8 (*Lunam . . . rubentem*, 35), anticipating the reference to her poison at the end of the poem.[228]

Other details highlight the paradoxical connections between elegant food and the margins of eating, putrefaction, and poison. Even the sauce recipe contains staleness in the form of the finest (sour) vinegar (*aceto / quod Methymnaeam vitio mutaverit uvam*, 49–50) and filthy sea-urchins (*illutos echinos*, 52).[229] If we read this recipe (*his mixtum ius*) in the light of the double meaning of *ius* (which the satires encourage us to do throughout), it appears on one level to be an indictment of the Roman legal system which exposes its corrupt and adulterated nature. The recipe also debases some of the concerns of Platonic word-feasts: the search for a definition of justice (δίκη—Latin *ius*) in the Republic; and the myth of the creation in *Timaeus*, where the universe is blended in a giant mixing-bowl (41d). Nasidienus' reception by the guests recreates the critical trial (*ius iudiciumque*) which threatens the aggressive satirist in 2. 1. The mixture of acrid and decadent ingredients in the sauce (see above) is a

226 Rudd (1966: 216).

227 Cicero (*In Pis.* 67) despises Piso's old-fashioned dinners, with their piles of half-rancid meat: *exstructa mensa . . . multa carne subrancida*.

228 *Rhombus*, a fish mentioned at 30, is also a magician's instrument.

229 Cf. 2. 4. 66: on the best (rotten) fish-sauce.

culinary transformation of the kind of satire that excites critical hostility.[230] In his efforts to escape an adverse verdict by constantly changing his dishes (from boar to birds), Nasidienus recalls the Protean defendant described at 2. 3. 72:

> cum rapies in ius malis ridentem alienis,
> fiet aper, modo avis, modo saxum et, cum volet, arbor.

When you haul him into court for laughing at other people's misfortunes, he turns into a boar, then a bird, then a rock, then even a tree.

Occupying the central position in the poem, the recipe sums up this satire too: *mixtum ius*, a critical trial described in mixed language which exposes the messiness of social *mores*.[231] Depending on the perspective, then, the sauce recipe can be read as a culinary hint, a description of satire, or a parody of contemporary civilization.

Nasidienus' *pièce de résistance* (literally) is a choice selection of special joints from animals of specific sex,[232] brought in on a huge trencher:[233]

> deinde secuti
> mazonomo pueri magno discerpta ferentes
> membra gruis sparsi sale multo, non sine farre,
> pinguibus et ficis pastum iecur anseris albae,
> et leporum avulsos, ut multo suavius, armos,
> quam si cum lumbis quis edit; tum pectore adusto
> vidimus et merulas poni et sine clune palumbes.
> (85–91)

[230] This may explain some descriptive details close to Horace's own autobiography: *Lucanus* (6) recalls 2. 1. 34: *Lucanus an Apulus anceps* ('either a Lucanian or an Apulian'); *citra mare nato* (47, 'born overseas') recalls 1. 10. 31: *natus mare citra*.

[231] Cf. *mixto* (2. 4. 19), *miscet* (2. 4. 55), *miscere* (2. 4. 64). The trial (*ius*) of 1. 7 is more obviously a blend of bitter flavours (cf. *compositum* 1. 7. 20). Nasidienus' stuffed dishes recall the etymology of 'satire': *gravida*, of the lamprey (43) and *pastum*, of the liver (88), could both be replaced by *satur*.

[232] *Grus* (87, 'crane') is usually feminine; *anser* (88, 'goose') is usually masculine (see Palmer 1925). This sex-change should perhaps be seen as contributing to the 'unnatural' nature of the dish.

[233] *Mazonomo*: cf. Varro, *RR* 3. 4. 3; Nemesianus, *de Aucupio* 1. 17.

Then came slaves bearing a huge platter, on which lay the dismembered limbs of a crane sprinkled with salt and wrapped in pastry, the liver of a white goose fed on rich figs, and the torn-off shoulders of hares, much more delicious, we were told, than if they had been eaten with their loins; then we saw blackbirds served with breasts burned off, and pigeons without their rumps.

This is partly a mockery of gastronomic fussiness,[234] but it also has other shades of meaning. While the effect is meant to be pleasant (*suavius*, 89, *suavis res*, 92), Nasidienus' pseudo-aetiological commentary and Fundanius' description make this the most corrupted dish of all.[235] The elegant dish is transformed into a tragic *sparagmos*: culinary preparation becomes violent butchery. *Discerpta / membra gruis sparsi sale multo* (85–6) is reminiscent of Virgil's description of Aristaeus' dismembered body: *discerptum latos iuvenem sparsere per agros* (*Georg*. 4. 527: 'The young man's dismembered body was scattered across the fields'). The juxtaposition of *leporum, avulsos*, and *suavius* highlights the ambivalent nature of civilized cooking, in which barbaric acts are performed to produce something *lepidum* or *suave*.[236] *Membra gruis sparsi*, 'the limbs of a scattered crane', are narrowly restored (with seasoning) to the culinary sphere: they become 'crane joints sprinkled with salt' (*membra gruis sparsi sale multo*). *Foie gras*, even fed on rich figs (*pinguibus ficis*), still reminds us of the liver Canidia cut from her boy victim at *Epod*. 5. 37. *Avulsos* (89), another violent verb, recalls Canidia's style of butchery, with her bare teeth: *pullam divellere mordicus agnam* (*Sat*. 1. 8. 27).[237] The blackbirds are like wild Amazons (the sound of *mazonomo* (86) adds to this impression): *pectore adusto* should mean 'with breasts

[234] See Rudd (1966: 220). Cf. Pliny *NH* 10. 140 on cooks who insist on particular cuts of poultry, the haunch or breast.

[235] *Si non causas narraret earum et / naturas dominus* (92–3, 'if only the host hadn't insisted on telling us about their origins and properties') is a parody of Virg. *Georg*. 2. 490 on Lucretius: *felix qui potuit rerum cognoscere causas* ('Happy the man who can discover the secrets of the universe'). Cf. *Sat*. 2. 4. 45: *natura*.

[236] Cf. Eubulus *Auge* (K2. 170 = Ath. 14. 622e) where dinner guests commit acts of savagery on goslings' limbs: διεσπάρακται θερμὰ χηνίσκων μέλη.

[237] We could also compare the dead Priam at *Aen*. 2. 558: *avolsumque umeris caput et sine nomine corpus* ('a head torn from its shoulders and a body without a name').

burned off' rather than 'with burnt or crisped breasts', as usually suggested.[238]

This description of a perfect dish in the end violates tragic decorum by bringing acts of violence or metamorphosis on to the stage:

> ne pueros coram populo Medea trucidet,
> aut humana palam coquat exta nefarius Atreus,
> aut in avem Procne vertatur, Cadmus in anguem.
> (*AP* 185–7)

Medea should not butcher her children in public, or wicked Atreus cook human entrails openly, or Procne turn into a bird, or Cadmus into a snake.

It becomes still more sinister when we remember that Nasidienus was described as *cenae pater* (7), and that when his dinner went wrong he wept 'as though his son had died young' (58–9). These well-garnished but mutilated limbs reek of a Thyestean feast.[239] Fundanius casts the final blight on it with his parting simile: *velut illis / Canidia adflasset peior serpentibus Afris* (94–5, 'as though Canidia had blasted it with breath worse than African serpents'). The guests will not touch the last dish, as though it had been poisoned by the Canidia: as a finale, the witch who has infused so much of the dinner with her lingering poison is brought on stage.[240]

The broken limbs are the culmination of a series of fiascos or breakages: the *ruina* of the curtain (54), a broken plate (mentioned as a possibility at 72: *si patinam . . . frangat*) and a broken wine-jar (similarly, at 81: *fracta lagoena*).[241] The split

[238] Rudd (1966: 219) argues that they were only crisped: if they were burnt, *suavis res* would make no sense. But cf. 2. 8. 68, *panis adustus*, which unambiguously means 'burnt, so it can no longer be used'. Compare Curtius 6. 5. 28 using *adustus* to describe mutilated Amazons: *aduritur dextera [papilla]*.

[239] Cf. Pers. 5. 17–18: *mensasque relinque Mycenis / cum capite et pedibus* ('leave meals of heads and feet at Mycenae'). One of the guests, Varius, was author of a *Thyestes* (29 BC).

[240] See below Ch. 5 on *Epod.* 3 (esp. 6–8), where Horace suspects Canidia of poisoning his food.

[241] See Tanner (1979: 54–5) on the part played by breakages at the Cena Trimalchionis: they show 'the close connection between over-abundance and rubbish in the society of Trimalchio. There is a similar degeneration of words and music and gestures in the entertainment offered at and around the table, just as the feast offers a degeneration into perverted sexual activities. They all collapse together.'

between host and guests has grown wider; the guests exchange private whispers (*secreta divisos aure susurros*, 78). So the final dish also represents the dismemberment of the corporate spirit, what Cicero calls the *vitae coniunctio* that characterizes the true *convivium* (*de Sen.* 13. 45), or the spirit of love informing the *Symposium*: οὗτος δὲ ἡμᾶς ἀλλοτριότητος μὲν κενοῖ, οἰκειότητος δὲ πληροῖ (197d, 'This is what empties us of alien feelings, fills us with intimate warmth'). It is also a parallel for Nasidienus, tortured by anxiety before the dinner (*torquerier omni / sollicitudine districtum*, 67–8), and savaged in a civilized way by his critics. If we inspect the damage, we find that the cook or host, for so long a parallel for the poet, fulfils Horace's joking prophecy in another context: *invenias etiam disiecti membra poetae* (1. 4. 62, 'you will find the dismembered limbs of a poet').

The ending of the poem has often seemed disappointing. Rudd, for example, remarks:

> Structurally it is the weakest ending in the book, and (however fictional in content) it puts the guests in an exceedingly poor light—to our way of thinking. But this cannot have been Horace's intention, for the piece, like 1. 9, was written for the entertainment of these very people. He must have seen their departure as a dramatic gesture which paid the host back for his absurd and vulgar display. (1966: 222)

This explanation assumes that Horace had either a faulty sense of structure or an inept conception of manners, which is unlikely. Instead, we should see this as a deliberately premature ending, complementing the premature ending of the dinner and that unexpected shortness of the book as a whole (only eight poems, though 2. 3 is long enough for three poems) which Fraenkel (1957) attributed to a waning of ideas.

If we follow the stage-directions offered in the text (where the scene-changes are marked by theatrical images), we can deduce that the poem is an unfinished play by Fundanius, not five acts long, as Horace dictates in the *Ars Poetica*, but only four.[242] The ending of 2. 8 is in quite a different sense a

[242] M. Schmidt (1937) needlessly strains the poem, which she sees as 'eine wahre Komödie', to fit Horace's five-act rule: viz. lines 6–17, 20–41, 42–59, 60–78, 80–95.

dramatic ending. It has something in common with Cicero's well-known description of a mime: *Mimi ergo iam exitus, non fabulae; in quo cum clausula non invenit, fugit aliquis e manibus, dein scabilla concrepant, aulaeum tollitur* (*pro Cael.* 27. 65, 'It ends like a mime, not a play: there is no finale, just an escape-scene, then the clapping begins and the curtain falls'). In 2. 8 the guests escape (*fugimus*, 93) inconsequentially, and the tragic language and scenery of the dinner are dissolved *sine clausula*.

At the same time, allusions to earlier poems do give 2. 8 a structural rounding-off, even in the process of stressing its comparative incompleteness. The image of departing guests brings *Satires* 2 full-circle back to the end of *Sat.* 1. 1, where the man who is satisfied with his life leaves it like a contented feaster: *uti conviva satur*. Nasidienus' guests, by contrast, leave without tasting the food: *ut nihil omnino gustaremus* (94). The unfinished meal is a model or gauge of incomplete writing.[243] Once again, the etymology of 'satire' must be in play: Horace's *satura* is not *satura* (full); his *sermo* has no ending. The final image summarizes or solves the inhibitions of the satirist in Book 2: a complete representation of society would be too free, too susceptible to criticism. In 2. 1 the satirist's works are likened to Canidia's poison (48), and there the poem ends with a satirist's lucky escape at the hands of his *iudices*: *solventur risu tabulae, tu missus abibis* ('The indictment is dissolved in laughter, and you get off scot-free'). 2. 8 has the same ending, all malice dissolved into farce. Horace finally takes Trebatius' advice, to be silent (*quiescas*, 2. 1. 5), seemingly not before time. But the final line is a backlash, a sting in the tail, which reminds us of the satirist's power to poison what he breathes on:

velut illis
Canidia adflasset peior serpentibus Afris. (94–5)

The poem fits more easily into four scenes: viz. 6–33, 33–53, 54–77, 84–95 (with an interval from 77 to 82). The theatrical images that mark the scene-changes are: *vertere pallor / tum parochi faciem* (35–6, 'then pallor transformed the host's mask'); *aulaea ruinas / . . . fecere* (54–5, 'the curtain collapsed'); *redis mutatae frontis* (84, 'you return with a new face/back-drop').

[243] Contrast the end of the *Eclogues*: *ite saturae . . . capellae* (10. 77: 'go home, contented goats').

Persius

After Horace's refusal to contemplate even a satisfying finish, any new experiment in satire was bound to seem superfluous. Persius' response is to compress and intensify the Horatian recipe still further, coming up with *aliquid decoctius*, 'something more boiled-down', an astringent, flavoursome concentrate.[244] Mashed-up quotations from his predecessors, Lucilius and Horace, float on the surface of a violently boiled brew.[245] Compression, of course, creates difficulty, and difficulty is the essence of Persius' writing. Satire that was once easy and convivial is condensed into indigestible language, poetry that expects no guests, a recipe whose instructions are inscrutable.[246]

If we look at the origins of the metaphor *decoquere*, we find that the word was normally used of boiling down juices, for example olive-water, meat-juice, and in particular grape-juice, *sapa*.[247] All our sources make it clear that the process reduced liquid to half, or more usually a third of its original volume[248] Persius wrote only six satires, exactly a third of Horace's eighteen. *Sapa*, concentrated grape-juice, has the same roots as *sapere*, which took its double meaning from the ancient notion that intelligence flowed in the bodily juices. So Persius' concentrated juice is not only *decoctius*: it is also *sapientius*, 'more flavoursome' or 'inspired by a higher wisdom'. The flavour of satire has changed again. The language of Persius' philosophy, Stoicism, is no longer inane verbiage, as Horace represented it, but a tightly knotted, unparaphraseable riddle, which we must work to decipher. Unlike Horace's sunny Epicureanism, Stoicism is an uphill

[244] 1. 125.

[245] e.g. 1. 1: *o curas hominum! o quantum in rebus inane!* ('Oh, the woes of mankind, the emptiness of life!'), borrowed, according to the scholiast, from Lucilius' first book; 1. 118 (of Horace): *callidus excusso populum suspendere naso* ('clever at dangling the public from his well-blown nose'), a conflation of Hor. *Sat.* 2. 8. 64: *suspendens omnia naso* ('dangling everything from his nose'; cf. 1. 6. 5) and *Sat.* 1. 4. 8: *emunctae naris* ('cleaned-out nose').

[246] See Anderson (1982: 169).

[247] e.g. Virg. *Georg.* 1. 294; Col. 11. 2. 71; Plin. *NH* 17. 264; 23. 62; 31. 40.

[248] e.g. Plin. *NH* 23. 62: *vino cognata res sapa est musto decocto, donec tertia pars supersit* ('*sapa* is related to wine, and consists of grape-juice boiled down to one third of its original volume'); cf. 14. 86; 32. 39; 21. 119.

struggle, a goal with a strict regimen attached: sleepless nights, a morbid pallor, and stodgy meals. Horace's *securum holus* (safe vegetables) is now *durum holus* (hard vegetables), the Stoic's diet of unappetizing beets and heavy lentils.[249] Horace sweetened his readers with jokes; but Persius, who shares his name with the hybrid duellist from Horace's bitter satirical pickle (1. 7), gives satire back a barely muzzled bite.[250] His poetry, he says, smacks of bitten nails (*demorsos sapit unguis* 1. 106);[251] instead of soothing his readers, he stings them with a biting solution of vinegary truth or purges them with hellebore.[252]

Within his small compass, Persius digs deeper than Horace and distorts his easy balance. The truth is spluttered and muffled, delivered in fits and starts. Instead of a domesticated scarecrow-satyr, guardian of a restored society, Persius is a naughty boy who urinates in sacred places, and exposes the emperor's nakedness.[253] Gone are Horatian *mediocritas* and men in their prime. Instead, Rome becomes centrifugal, polarized into youth and old age.[254] Cooking still provides a model for society, but man in Persius is either raw or overcooked. The uncooked soul of the ephebe (*non cocta*, 3. 22) waits to be baked in the right mould, and take on the flavour of maturity (*sapimus patruos*, 1. 11).[255] Meanwhile, the

[249] Hor. *Sat.* 2. 7. 30; Pers. 3. 112; cf. 3. 114, *plebeia beta* ('plebeian beet'), 3. 55, *siliquis et grandi pasta polenta* ('fed on pods and hearty polenta').

[250] Hor. *Sat.* 1. 7. 32: Persius was splashed with Italian vinegar (*postquam est Italo perfusus aceto*). 1. 109–10: *canina littera*: the dog-like snarl of satire (see Anderson 1958).

[251] *Demorsos* conflates *decoctus* with *mordere*: biting poetry is produced by a writer whose own nails are gnawed by strain and effort. Cf. Hor. *Sat.* 1. 10. 71: *vivos et roderet unguis*.

[252] Vinegar: 1. 107, *sed quid opus est teneras mordaci radere vero*; 1. 126, *inde vaporata lector mihi ferveat aure*; 5. 15, syringed ear of receptive Stoic philosopher (*Stoicus hic aurem mordaci lotus aceto*); cf. 5. 63: purged ears (*purgatas aures*). Hellebore: 4. 16: *Anticyras . . . sorbere meracas*.

[253] Hor. *Sat.* 1. 8; Pers. 1. 112–14.

[254] Even the ancient response to Persius was sharply divided. In some quarters, his slim volume acquired the reputation of an elixir: the public ripped it to pieces in their enthusiasm, according to the *Vita*. Martial, predictably, valued it more highly than any epic saga: 4. 29. 7–8, *saepius in libro numeratur Persius uno / quam levis in tota Marsus Amazonide*. St Jerome, however, burned the book because it was so obscure (Simon Verepaeus, *de Epistolis Latine conscribendis libri V* 1. 43. para. 9).

[255] 3. 21 ff. (a metaphor of unbaked clay): *maligne / respondet viridi non cocta fidelia limo. / udum et molle lutum es*. Vidal-Naquet (1981b) uses Lévi-Straussian

inveterate old fool is already frying in the sun: *adsiduo curata cuticula sole* (4. 18); *aprici senes* (5. 179); *figas in cute solem* (4. 33); or fat with corruption: *sed stupet hic vitio et fibris increvit opimum / pingue* (3. 32–3). In a kaleidoscopic world (*rerum discolor usus*, 5. 52), men are distinguished by almost culinary consistencies: one bloated with sleep (*satur turgescere*, 56), one thinned by gambling (*decoquit*, 57), another putrid with lust (*putri*, 58); while cargoes of spices themselves take on the different textures of discoloured human skin: wrinkled peppercorns and pallid cumin.[256] In *Satire* 2, muddled Rome (*turbida Roma*, 1. 5) becomes an abattoir: wading through dripping lard, chitterlings, lights and tripe, huge stews, and fatty sausages, Persius exposes man's own *scelerata pulpa*, spoiled flesh (63), in their midst.[257] The satires' set-meals are tainted by corruption: putrefying groceries in the pantry, greasy festival food on the windowsill, the tattered dregs of dying vinegar.[258] The sluggard on the morning after, with his bile bubbling, needs to be skimmed like a jar of wine: *stertimus, indomitum quod despumare Falernum / sufficiat* (3. 3–4). And even Persius' recipe for salvation uses unsettlingly culinary metaphors:[259]

conpositum ius fasque animo sanctosque recessus
mentis et incoctum generoso pectus honore. (2. 73–4)

A blend of justice and integrity, the recesses of the mind untainted, and a heart stewed in noble honour.

Sidestepping the issue of lost political liberty, Persius has turned inwards to the concerns of individual freedom and philosophical wholeness.[260] As John Bramble has taught us, the body was never far from even the most abstract ancient

cooking imagery to describe the rituals that prepared the Athenian ephebe for adult life. The article in French was originally called 'Les Jeunes, le cru, l'enfant grec et le cuit'.

256 5. 55–7.

257 Cf. 2. 30, greasy lights and chitterlings; 42, stews and sausages; 47, heifers' tripe; 48, intestines and rich cakes. See Flintoff (1982: 352): 'Meat himself, he turns everything around him into meat.'

258 3. 73–6; 5. 180–4; 4. 32.

259 See Flintoff (1982: 352–3).

260 I have greatly benefited from reading John Henderson's article (forthcoming in *Ramus*), which sees the *Satires* as a Neronian document for the 'care of the self'.

writing, and aesthetics took its language from wider cultural debates about bodily regimen: sex, diet, age, and disease.[261] In Persius' satires, these metaphors double up to turn literary decadence into a symptom of a wider moral malaise. Taste and wisdom (conflated in the verb *sapere*) become parallel concepts.[262] Both the philosophical subject and the literary text are exposed as bloated or emaciated, virile or emasculated, rugged or silky-smooth, while Persius' bizarre, conflated anatomizing suggests a world out of joint.

For Horace, as we saw, Lucilius' greatest gift was his ability to strip off the skin of society and reveal the corruption underneath. Persius not only skins the body: he digs right down underneath the bones to the gristle, pulp, offal, fibres, veins, and secretions that make up human flesh. Under the handsome outer skin, the white toga (1. 16, 1. 110, 5. 33), lurk putrescent ulcers, swollen stomachs, blocked ears, layers of adipose tissue, varicose veins, and foul juices: bile, saliva, sulphurous fumes, and bad breath.[263] Literature and morals are both fat with corruption. The heir stuffing himself on goose-offal, the priest trembling with lard, and the sick man clutching at greased hors-d'œuvres,[264] these all share their symptoms with the dyspeptic literati, who snort out rancid titbits, spout Virgilian sponge-cake (*spumosum et cortice pingui / ut ramale vetus vegrandi subere coctum*), ingest rich meaty chunks of heroic song (*robusti carminis offas*), or gnaw on the dismembered limbs of tragedy (*mensasque relinqui Mycenis / cum capite et pedibus*).[265] Persius' first satire, set at a stomach-turning literary orgy, is smeared with literary pastiche: mincing Neoteric epyllia and varicose,

[261] Bramble (1974) esp. 26–66: 'The nature and sources of Persius' imagery'.

[262] e.g. 1. 11, 1. 105, 6. 38.

[263] Ulcer: 3. 113; swollen stomachs: 1. 57 (*pinguis aqualiculus*), 3. 98 (*turgidus hic epulis*), 5. 74 (*ast illi tremat omento popa venter*), 5. 187 (*deos inflantis corpora*), 5. 56 (*turgescere*); blocked ears: 1. 126 (*inde vaporata aure*), 5. 86 (*Stoicus hic aurem mordaci lotus aceto*); fat: 3. 32 (*fibris increvit opimum pingue*), 6. 77 (*Cappadocas pinguis*); varicose veins: 1. 76 (*venosus liber Acci*), 5. 189 (*varicosos centuriones*); bile: 2. 13 (*acri / bile tumet*), 3. 8 (*turgescit vitrea bilis*), 4. 6 (*commota fervet plebecula bile*), 5. 144 (*mascula bilis*); saliva: 1. 104 (*delumbe saliva*), 2. 33 (*lustralibus salivis*); sulphur: 3. 99 (*sulpureas mefites*); bad breath: 3. 89 (*gravis halitus*).

[264] 6. 71, *tuus iste nepos olim satur anseris extis*; 6. 74, *ast illi tremat omento popa venter*; 3. 102, *uncta cadunt laxis tunc pulmentaria labris*.

[265] *Rancidulum quiddam*: 1. 33; Virgil: 1. 96–7; epic: 5. 5; tragedy: 5. 5. 17–18.

warty Roman archaisms.[266] Isolated in all this is his own alternative recipe, *plebeia prandia* (5. 18). 'Plebeian lunches' stand simultaneously for simple diction and for the wholesome food that nourishes Stoic redemption: astringent beetroot, hearty pods, and stinging vinegar.

Or so it seems. It would be a mistake, however, to imagine that Persius' satire is rude with health or even curative. The self-criticism that began to erupt through the surface of Horace's second book of satires here becomes an obsession: the author's voice splits into schizophrenic dialogues between sage and novice, doctor and patient, and it is no longer certain where to divide their speech.[267] The poet's ulcers, hangovers, false starts, vacillations, and eruptions of spleen must alter any impression that Persius 'is the steady incarnation of *sapientia*'.[268] Diagnosis of one's own sickness, however, may be the first critical stage in achieving health. But Persius even so displays the symptoms of the imbalanced world he observes around him.

The same applies to his analysis of literary imperfection. Persius may use bodily or culinary metaphors to flesh out his satire on contemporary style, but these grotesque shapes and discordant flavours are brewed up within a genre, *satura*, that is itself a bodily or culinary metaphor (a crowded dish or a stuffed gut) complete with its own suspect associations. A strange contradiction arises.[269] Many of Persius' most distasteful images are the very ones that most recall the origins of *satura*. In *Satire* 2, heavy stews and greasy sausages (*grandes patinae tuccetaque crassa*, 2. 42)[270] represent the offerings of a materialistic society to its gods, and there must be a link between these and the bloated dishes that embody

[266] 1. 99–106; 1. 76–7. Persius' vision of decadence recalls his contemporary Petronius' post-mortem on the *corpus orationis*, the corpse of rhetoric (*Sat.* 1): decayed, effeminate, swollen with cloying dainties, flatulent, spotty, and plague-ridden.

[267] See Housman (1913: 16–21), Rudd (1970: 286–8) on the problem of speakers in *Satire* 3; Henderson (forthcoming) on the ambiguity of Persius' role as student and teacher.

[268] Anderson (1982: 179).

[269] Bramble (1974) never mentions the etymology of satire in the course of his rich discussion of Persius' imagery.

[270] The scholiast defines *tucceta* as pieces of spiced beef or pork (perhaps like modern salami): *condimentis quibusdam crassis oblita ac macerata*.

inflated style in the first satire. At the same time, by an odd coincidence, these stews and salamis remind us of the culinary origins of *satura*, the full dish and the stuffed sausage.[271] The fat stomachs, distended with tripe, that fill Persius' pages not only embody moral laxity and crass sensibilities: they are also walking figures of the text itself, the *farcimen* or stuffed gut.[272] If this were Virgil and the object described were a slender reed pipe (*tenuis avena*), we would seize on it as an internal image for the creation of bucolic poetry. The mixed and bloated forms of satire, however, represent the excesses the poet abhors in the act of exemplifying the poem itself. In *Satire* 1, there is an even clearer reminder of the sort of stuffed text we are reading: *Romulidae saturi*, the bloated Romans on their dinner-couches (1. 31), are the physical embodiments of 'Roman satire'.[273] Their moral and literary connotations have been recognized, but no one has yet mentioned their uneasy relationship with the frame, or tackled the apparent contradictions involved when a writer of *satura* pits himself against bodily and stylistic tumidity.

It could be argued, of course, that Persius, like Horace, is writing a deliberately contradictory kind of satire: *satura decocta*, the bloated dish boiled down.[274] He is at least sharply aware of his own faults, not stupefied by layers of insensible fat.[275] His satires do sting like vinegar. Even so, it is clear that compressing Horace's satires produces not a pure distillation, but a raucous cacophony. Callimachean refinement is stained with pastiche of faulty styles; and, by relentlessly dissecting the diseased bodies around him, Persius himself comes to seem morbid and sick. It is no surprise, then, that the metaphors his readers have lighted on to describe his style are inspired by his own negative subject-matter: 'scabrous and hobbling'; 'grossly distorted, as it were

[271] Both *grandes* and *crassa* paraphrase the idea of amplitude in *satur*.

[272] *Pinguis aqualiculus propenso sesquipede extet* (1. 57, 'his podgy pork-belly sticks out a full two yards'); *tremat omento popa venter* (6. 74, 'the priest's stomach wobbles with tripe').

[273] Cf. the heir stuffed with goose innards at 6. 71: *satur anseris extis*.

[274] *Satur* and *decoquere* are opposite or alternative conditions at 5. 56 f: *hic satur inriguo mavult turgescere somno . . . hunc alea decoquit*.

[275] See the description of the irredeemable Natta: 3. 31–4.

urinary Latin'.[276] In other words, *plebeia prandia*, the simple, stodgy meals of the Stoic novice that are always taken as the basis of Persius' poetics, could not be further from this contorted, arcane language; and *verba togae* is a strange idiom for a poet who so disregards his audience.[277] A much better image for Persius' *satura*, ironically enough, is the sizzling frying-pan of speech (*sartago loquendi*, 1. 80), which he uses to describe a style of speaking foreign or extraneous to Rome.[278]

Plebeia prandia may not be what Persius is providing so much as a theoretical cure for his illness, like the tough vegetables the sick man spews out at the end of *Satire* 3. It fits better with Persius' mortified confessions of his own failings to assume that he is only an imperfect student of good writing. It has always been assumed, for example, that the following stylistic prescription is a summary by Cornutus of Persius' own style, his gift for sharp juxtapositions, his scathing moralizing and ability to pinpoint other people's faults:[279]

> verba togae sequeris iunctura callidus acri,
> ore teres modico, pallentis radere mores
> doctus et ingenuo culpam defigere ludo. (5. 14–16)

Clever at piquant juxtapositions, you pursue the language of the toga, pursing your lips smoothly, knowing how to scrape away at pallid morals and impale each fault with your well-bred humour.

Such flattery would not, however, be consistent with Persius' obsession with self-criticism. In a conversation where, as usual, the parts are never definitely assigned, these lines could just as well be an admiring description of his teacher Cornutus' style.[280] Meanwhile, the list of negative qualities

[276] Dryden, *Discourse of Satire*; Richlin (1981: 187).

[277] See Bramble (1974: 154–5) on Persius' contempt for the crowd.

[278] Bramble (1974: 122) thinks that the metaphor suggests 'exotic variety of diction' rather than a mixture of hissing sounds, as Isidore (*Etym.* 20. 8) and Augustine (*Conf.* 3. 1) imply. He compares Eubulus fr. 109 (2K): λοπὰς παφλάζει βαρβάρῳ λαλήματι (Jahn) (where sizzling sounds are likened to unintelligible speech: 'the pan burbles in double-dutch').

[279] See e.g. Bramble (1974: 2–3).

[280] The *Vita* identifies Cornutus as tragedian as well as philosopher.

that precedes it ('you do not squeeze the wind from puffing bellows, nor, hoarse with pent-up murmuring, caw something solemn fatuously to yourself, nor stretch inflated cheeks to burst them with a pop')[281] in fact sums up many of Persius' own idiosyncrasies. He has just, for instance, deflated his own epic appeal for a hundred mouths by talking about 'rich meaty chunks of poetry'. 'Pent-up murmuring' repeats his many intimations about the stifled atmosphere of the satires. The cawing crow may refer back to the parrots and magpies of the prologue, but St Augustine writes that crows always sound as though they are squawking '*cras, cras*', 'tomorrow, tomorrow', and that happens to be the message of this poem: *cras hesternum consumpsimus* (68, 'we have eaten up yesterday's tomorrow').[282] It is also clear that Persius has no place in his infected hotchpotch for literary decorum. His use of the phrase *olla Thyestae* (5. 8, 'the stewpot of Thyestes'), for instance, seems to be specifically flouting Horace's dictum in the *Ars Poetica* that tragic feasts should never be described in common language.[283]

Persius sacrifices purity and consistency in satire, then, for the sake of burrowing into the filthy entrails of a diseased society. Does he go too far for his own good? His view of his own place in that society has often been misunderstood: 'Such diversity is bad. In the ideal world that Persius seems to imagine, everybody would be without personality or individuality, all living according to that single aim of Stoicism which he has adopted for himself.'[284] This vision of perfect automata ignores the fact that satire is a representation of an imperfect society, not a hygienic model for an ideal one. Persius' satires are both precocious, prematurely cooked (*praecox*), and *aliquid decoctius*, overripe.[285] His discoloured, disjointed outlook is a permanent state of mind,

[281] 5. 11–13; adapted from Guy Lee's translation in Lee and Barr (1987).

[282] Aug. *Serm.* 82. 14: *cras, cras corvi vox est*; cf. 224. 4.

[283] *AP* 90–1: *indignatur item privatis ac prope socco / dignis carminibus narrari cena Thyestae*. See Coffey (1989: 115).

[284] Anderson (1982: 180).

[285] Cf. *Inc. Pall.* 95: *odi puerulos praecoqui sapientia* ('I hate precocious little boys'); Gellius on ripe fruit: *quae neque cruda et inmitia sunt neque caduca et decocta* (10. 11. 3, 'Those which are neither raw and bitter, nor overripe and pulpy').

never perfectly cured. His writing shares the symptoms of the malaise he records, a recipe for adolescence that is more of a specimen than a tonic.

Juvenal

As we have seen, the attempts of Horace and Persius to depict an engorged society in slender or compressed language result in some hybrid products: *tenuis satura* and *decocta satura*. Among their different food-images, simple and plebeian or mixed and sophisticated, it is difficult to choose which best represent their writing. Even the images of grease and mess which they conspicuously reject spoil any appearance of stylistic purity. Persius' satire, in particular, which could have been a pure 'essence', presents the most grotesque images and a concentration of jarring styles.

With Juvenal, all this changes. He abandons the contradictions, and re-expands satire to make it worthy of its name, to saturate it to its limits. He openly adopts the high style, claiming that he is returning to Lucilius, who, as he conceives it, tried fearlessly to emulate world-scale vice;[286] he completely bypasses the attempts of Horace and Persius to pare satire down to Callimachean principles. Only the grandest style is adequate for vice which exceeds even epic proportions; only tragic diction will do justice to Rome, which, as he says in *Satire* 2, now equals the whole world. For the first time, satire represents Rome *en masse*, with all its sprawling rubble: monuments, statues, crowds, and traffic. The picture of a crowded inn at 8. 174 ff.—a general lying among sailors, thieves and fugitives, hangmen, coffin-makers, and a castrated priest of Cybele—vastly expands the clientele of Horace's inn in *Satire* 1. 5. We also go beyond Rome, to Egypt (15), to the Alps (10), and (hypothetically at least, 2) to the ends of the earth. Geographical extremes extend satire in line with its new uninhibitedly full and all-embracing character, breaking the bounds of possibility.

[286] See Bramble (1974: 170): this is a parodic reinterpretation of Horace's now stereotyped literary picture of Lucilius.

Satire is fulfilled at last. But political freedom is still even more curtailed: instead of named individuals, we are shown a panorama of rhetorical types and historical shades.

Juvenal's literary claims are part of his satirical pose: as someone excluded from his rights to property, status, and power, as a genuine Roman reclaiming his inheritance. One of his basic techniques is polar opposition, contrasts between large and small, paucity and excess. In particular, there is the rhetorical contrast between *tot* and *unus*, or 'all or nothing', which suggests an extreme and disproportionate state of affairs, another way of appealing to the reader's yearning for melodramatic demonstrations of injustice.[287] He aspires to heroic sublimity with the very metaphors that Horace and other Callimachean poets used to reject it.[288] To describe women poisoners, he adopts the tragic buskin, breaking the bounds imposed by his predecessors:

> fingimus haec altum satura sumente cothurnum
> scilicet, et finem egressi legemque priorum
> grande Sophocleo carmen bacchamur hiatu. (6. 634–6)

I must be inventing it all, mincing in tragic boots and breaking all the bounds set by my satirical ancestors, yelping grand poetry complete with Sophoclean sound effects.

A modern incident of cannibalism becomes more monstrous than any tragedy: *cunctis graviora coturnis* (15. 29). In other words, the satirist means to expose excess in a style which aspires to that excess: as Quintilian says, hyperbole or exaggeration is justified when the subject-matter is something that has exceeded all limits.[289]

Yet even this epic exaggeration has its own in-built deflating devices. Kenney calls the metaphors of grandeur

[287] e.g. 1. 137–8; 1. 24: *patricios omnis opibus cum provocet unus* ('when one man challenges the whole aristocracy with his wealth'); 3. 251: *tot vasa ingentia, tot res / inpositas capiti . . . servulus infelix* ('poor little slave, with all those huge urns, all that paraphernalia balanced on his head'); 3. 208: *et tamen illud / perdidit infelix totum nihil* ('and yet the poor man has lost all the nothing in the world that he owned'); 3. 10: *dum tota domus raeda componitur una* ('while his whole house is packed into one cart').

[288] See Bramble (1974: 156 ff.).

[289] Quint. 8. 7. 76.

'rhetorical claptrap', and says: 'the assault is in the grand style, so much so as to verge on parody'.[290] Not only does this draw attention to the difference between tragic or epic grandeur and modern behaviour (men of today may be more monstrous, but they are also less heroic): it also makes Juvenal's satire very uneven. Like Rome, the city jerry-built on unsure foundations (*tenui tibicine fultam*, 3. 193), his mountains of rhetoric can be undermined at will. A private literary joke, for example, ends a catalogue of the horrors of living in Rome at 3. 7–9:

> incendia, lapsus
> tectorum adsiduos ac mille pericula saevae
> urbis et Augusto recitantes mense poetas.

Conflagrations, sudden slumps in housing, and all the other thousand hazards of life in the urban jungle, not to mention poetry recitals in mid-August.

A lady's hairstyle is built up into an edifice of multiple layers: *tot premit ordinibus, tot adhuc conpagibus altum / aedificat caput* (6. 502–3), 'Wave upon wave, lock upon lock, the steep juggernaut rises . . .' But, seen from behind, without stilettos, she is suddenly exposed as a pygmy: *post minor est, credas aliam* (6. 504), 'From behind she's so small you wouldn't recognize her'. The effect is always achieved with a combination of heroic and ignominious words. At 10. 64 Sejanus' heroic statue is melted into pitchers, basins, frying-pans, and chamber-pots: *urceoli, pelves, sartago, matellae*. Later, Hannibal climbs the Alps: 'Nature throws in his path high Alpine passes, blizzards of snow: but he splits the very rocks asunder, moves mountains—with vinegar' (10. 152–3). And Juvenal's most tragic satire ends with an image of pork sausages: *exta et candiduli divina tomacula porci* (10. 355), 'the snow-white pig's intestines and holy sausages'. As Bramble says: 'vernacular elements: obscenities, diminutives, Grecisms, words from the lower literary

[290] Kenney (1962: 33, 32). Cf. Anderson (1957: 35): 'He exaggerates his passion almost into bathos.'

strata—these have the function of insisting on reality, and when they are placed next to grandiloquent language what often greets us is a sense of dislocation, of feeling that there is a gap between life as it might be and life as it is.'[291] The theme of the satire, the futility of human endeavour, is perfectly suggested by this self-defeating style.

In the light of all this, another remark made by John Bramble needs to be reconsidered. He has written: 'Juvenal is less obedient than Martial to the rules of his genre and sometimes even anarchic, his language a mixture of high and low.'[292] This of course begs the question whether a genre like satire has any rules, whether it is not really an anti-genre, an all-containing genre, whose potential earlier satirists had inhibited. Juvenal is not so much breaking rules as exploiting the full capacity of the anarchic genre; widening gaps between opposites and stretching language to shocking extremes.

Food, though it is only part of the mixed stuffing of the satires, once again plays a unique role as subject and metaphor. It shows us the links that Juvenal is making between his verse-form (the full dish), his style, and his material. He tends to polarize food into two extremes: the bloating luxurious food eaten by the rich and the meagre scraps eaten by the poor. While Horace and Persius expressed satisfaction with simple food, for Juvenal these scraps are cause for resentment. Can we also see these two extremes as two opposing and coexisting aspects of the word *satura*: the stuffed body and the collection of scraps? Juvenal may aspire to a 'full' style, just as he claims a right to be included in the meals of the rich. But he also deflates fullness with incongruous words and details: his style becomes scrappy and miscellaneous, and the final impression is of heterogeneity as much as fullness. Despite the rhetoric of protest, Juvenal's accumulation of vices is no less a series of fantasies on the themes of anomaly and abnormality which stretch mixed satire to unprecedented limits.

[291] Bramble (1982: 616–17).
[292] Ibid. 559.

The first satire, a programme for the book, overtopples everything that Horace and Persius stood for.[293] Juvenal appropriates all the metaphors of the high style, swelling sails (*utere velis, / totos pande sinus*, 149–50) and war-trumpets (*tecum prius voluta / haec animo ante tubas*, 168–9).[294] At the same time, he uses the old culinary etymology to slot his work into the satirical tradition. We recognize an allusion when Juvenal talks of the 'fodder' which fills his book, a mixed 'mash' or *farrago* of human activity.[295]

> quidquid agunt homines, votum, timor, ira, voluptas,
> gaudia, discursus, nostri farrago libelli est (1. 85–6)

All the different doings of mankind, prayers, fears, anger, pleasures, delights, hustling and bustling: this is the fodder of my little book.

Farrago is not, as it has been suggested, 'a plebeian, culinary metaphor', but strictly something even lower, a mash of cereals used to feed animals.[296] Again, Juvenal is doing nothing by halves: the traditional etymology is reduced to its most ignominious form. This image of a book being fed comes ultimately, as we have seen, from Callimachus' description of Antimachus' *Lyde* as 'fat writing'.[297] It is an image consistent, then, with the impression we are receiving of Juvenal as an unusually greedy, undiscriminating poet.

In fact, Callimachean 'fat book' imagery crops up even earlier in the poem, though this seems to have escaped notice. Juvenal launches into satire as a resentful, silent reader at a poetry recitation of stale epics and tragedies. These books can be recognized as direct descendants of 'fat

[293] For *Satire* 1 as a programme, see Kenney (1962: 29–40).

[294] Sails: Prop. 3. 9. 4; Virg. *Georg*. 2. 4. 1; Hor *Od*. 4. 153–4; Plin. *Ep*. 6. 33. 10; see also Kenney (1958); trumpets: Mart. 8. 3. 21–2.

[295] Coffey (1976: 15) argues that *farrago* means 'the fodder that goes to feed my book' rather than 'the hotchpotch that my book is'. He adds (211 n. 25) that the idea of feeding a book with its material comes from Hor. *Ep*. 1. 20. 5, where the book is personified as a slave (*liber nutritus*).

[296] Martyn (1970). *Farrago*: cf. Virg. *Georg*. 3. 205; Plin. *NH* 18. 142; Col. 11. 2. 75; as slave food, Pers. 5. 77.

[297] Callim. fr. 398 Pf.; see Bramble (1974: 56–8) for discussion of the literary metaphor *pinguis*; and see Wimmel (1960); Clausen (1964).

Lyde': they are said to 'eat up' the whole day (*consumpserit*) and be 'full' even in the margins (*plena iam margine*):[298]

> impune diem consumpserit ingens
> Telephus aut summi plena iam margine libri
> scriptus et in tergo necdum finitus Orestes? (1. 4–6)

Will lumbering *Telephus* munch up the whole day unstopped, or will it be *Orestes*, cramming the margins to the edge, sprawling on to the back, and still no end in sight?

These books are all too familiar, and we duly expect Juvenal to offer some alternative. However, such obvious eating imagery, at the start of a book of satires, must remind us of the origins of the word *satura* as well. Here at last, ironically enough, are the kind of books that are fully capable of holding satire. And it turns out that this is exactly what Juvenal has in mind. He plans to emulate the consumption of these stale tragedies by 'filling' his own hungry tablets: *nonne libet . . . ceras inplere capaces* (1. 63, 'Can't I fill my own greedy notebooks?') Two capacious forms compete for space, then, in *Sat.* 1: the inflated tragedy and the full *satura*. Books that seem antipathetic set the mould for Juvenal's own style: sensational, shattering, full right up to the margins, running over to the other side of the paper, and still not doing justice to the subject.

Another familiar Callimachean opposition is between untrodden paths (untapped subject-matter) and the well-worn highway. Juvenal again defeats our expectations by choosing the busy highway—to be precise, a crossroads in the heart of Rome (*medio . . . quadrivio*, 63–4). Paradoxically, *this* is the material with scope, not the tired themes of epic. Juvenal's little book (*libellus*) will be *satura* with all the heterogeneous ingredients of society, walking contradictions and perversions.[299] Along his highroad parades a freak-show

[298] For the perpetually 'unfinished' epic (*necdum finitus*), cf. Petr. *Sat.* 118. 6: *nondum recepit ultimam manum*. Juvenal himself characterizes histories as mammoth works at 7. 100–1: *nullo quippe modo millensima pagina surgit / omnibus et crescit multa damnosa papyro* ('all too soon, the thousandth page looms up: they think paper grows on trees').

[299] Martyn (1970) notes that Juvenal uses the word *monstrum* more than any other satirist, and is the first to use it of outrageous human beings.

of Roman monsters and anomalies: a married eunuch, a bare-breasted matron turned gladiator, a millionaire barber, a schoolboy adulterer.[300] A man in his litter (*plena ipso*, 33), full with himself alone (like the fat books, *plena iam margine*), is contrasted with a professional informer who nibbles away at what is left of the Roman aristocracy (*nobilitate comesa*, 34). This is the first hint of a parallel between Juvenal, alone and excluded from his proper share of literary space, and the rightful owners of Rome, deprived of their share in Roman capital by greedy upstarts. The imagery of eating (*consumpserit diem* of the fat books and *nobilitate comesa* of the excluded aristocrats) unites the two themes of resentment. The book's subject-matter, this iniquitous battle for a share of the Roman pie, helps to explain the largely fantastical discursus on the dole-queue. A throng (*turbae togatae*, 96, *densissima lectica*, 120–1) of impostors and genuine Romans squabble for the *parva sportula*. The master himself, in a stunning example of hyperbole, dines alone in an empty room on all the products of the world:[301]

nam tot de pulchris et latis orbibus et tam
antiquis una comedunt patrimonia mensa. (137–8)

From those fine, spacious, antique tables, they eat whole legacies at one sitting.

But this bloated figure is soon destined for death by dyspeptic peacock (*crudum pavonem*, 143).

The 'fodder' for Juvenal's book will be the stream of human activity and the abundant supply of vice that he sees at the crossroads:

quando uberior vitiorum copia? quando
maior avaritiae patuit sinus? (87–8)

There's been a boom in vice: greed has never had so much spending power.

The pocket (*sinus*) of avarice has never gaped wider. Society has never been so vicious:

nil erit ulterius quod nostris moribus addat

[300] Ibid. for analysis of these.
[301] *Orbibus* could mean 'round tables' or 'the world' itself.

posteritas, eadem facient cupientque minores,
omne in praecipiti vitium stetit. (147–9)

Future generations can go no further. Our children's children will go through the same performance; our own vices are teetering on the edge.

In the end, Juvenal rises to his own challenge (*unde / ingenium par materiae?*: 'where is the genius equal to such a theme?', 150–1), by launching into epic: *utere velis*, *totos pande sinus* ('Spread your sails, open them to their full swell.'). *Sinus*, once a purse, now a billowing sail, is unfurled to engulf the vast expanse of greed. *Satura* is poised on the edge, ready to swallow up the unprecedented evils of the whole world.

Food remains the most graspable reminder of iniquity in Rome in all the satires. Juvenal portrays himself above all as a *hungry* poet: his notebooks are *capaces*, waiting to be stuffed. In *Satire* 7, he extends this into an appeal for sustenance for the leonine appetites of all modern poets,[302] who are no longer 'satisfied' by their patrons, as Horace was:[303] *satur est cum dicit Horatius 'euhoe'* (7. 62), 'Horace sang "Alleluia" on a full stomach'). Yet the poor client's greatest hope is for a half-eaten hare, a morsel of boar's haunch, and a mini-capon (5. 167–8, *semesum leporem . . . aliquid de clunibus apri / . . . minor altilis*). *Minor altilis* is the ultimate oxymoron: a bird that has been made first bigger, then smaller than its original size. The beggar, meanwhile, can only expect scraps.[304] A street bully who hurls insults at a poor Roman spits on his meagre diet:

cuius aceto,
cuius conche tumes? quis tecum sectile porrum
sutor et elixi vervecis labra comedit? (3. 292–4)

Whose vinegar, whose beans have you just farted? What cobbler has shared your meal of chives and the lips of a castrated ram?

[302] 7. 78; cf. 7. 7: Clio (i.e. the historian) is *esuriens*; 7. 29: the poet's bust is *macra*.

[303] For the contentment of the past compare the peasants in *Satire* 14, with their saturated clods and steaming cauldrons of porridge: 166 (*saturabat glebula*); 171 (*grandes fumabant pultibus ollae*).

[304] 3. 210: *frusta rogantem*.

This food exudes powerlessness and inferiority: *acetum* is sour wine; *tumes* (lit. swell) suggests unjustified pride (caused by eating the flatulent bean); *sectile* and *vervecis* (a castrated ram) hint at impotence.[305]

Stews are a persistent image for the portion of the dregs of society, reminding us at the same time of the negative associations of mixed verse-forms like *satura*. A decomposing stew is the first item in Juvenal's miser's larder.[306]

> neque enim omnia sustinet umquam
> mucida caerulei panis consumere frusta,
> hesternum solitus medio servare minutal
> Septembri nec non differre in tempora cenae
> alterius conchem aestivam cum parte lacerti
> signatam vel dimidio putrique siluro,
> filaque sectivi numerata includere porri. (14. 27–33)

Even *he* cannot bear to eat mouldy hunks of blueish bread, though he is used to recycling yesterday's stew in mid-September, and setting aside summer beans and a labelled tail-end of mackerel or half a stinking catfish, counting out each strand of chives and locking it away.

This collection of rotten (*hesternum, putri*) and broken left-overs (*frusta, minutal, parte, dimidio, sectivi*) is a string of variations on the theme of inedibility. 'Rehashed cabbage' (*crambe repetita*, 7. 154), the metaphorical equivalent of the poor client's meagre diet (compare *caulis*, 1. 134), is, of course, Juvenal's name for stale, monotonous recitations, the kind he cursed at the beginning of *Satires* 1. It could be argued, then, that rehashed food is not a good metaphor for his work, since that was always presented as an alternative to these clichéd repetitions. At the same time, we have seen how satire identified itself with its own suspect and rancid subject-matter: the irony of *Satire* 1 is that 'fresh' subject-

[305] *Secto* can mean 'to castrate': Mart. 5. 41. 3; 9. 5. 4; Tib. 1. 4. 70; Ov. *Fast*. 4. 221. For the significance of *vervex*, a castrated ram, see Mart. 14. 211: *caput vervecinum*, with its unspoken play on *colla*, neck, and *coleos*, testicles; Plaut. *Capt*. 820, *verveci sectario*; as a term of abuse: Juv. 10. 50; Plaut. *Cas*. 535; Dio 65. 3. 3.

[306] From *minutus*, 'consisting of fine pieces'. For scorn poured on mincemeat, cf. *penthiacum*, a stew described as 'rubbish' (*nenias*) by Trimalchio (Petr. *Sat*. 47. 10); and *catillum concacatum* (ibid. 66. 7).

matter can be ransacked from the rotten, unpalatable material that surrounds us every day.

Satire is primarily a miscellany, and the gastronomic origins of our word 'miscellany' appear in *Satire* 11, where Juvenal charts the precipitous descent of a gourmet from his world-wide search for dainties (*elementa per omnia*, 14) to the mixed scraps of the gladiators' school: *sic veniunt ad miscellanea ludi* (11. 20). The link between the humble stew and the dregs of society is one that lingers on in European literature.[307] Juvenal's stale casserole is the ancestor of the *olla podrida*. For Eugène Sue, in *Les Mystères de Paris*, the *arlequin*, a mixed stew of scraps from the tables of the rich, becomes 'a metonym of the ghetto, of the heterogeneity obtained when people are thrown together by misfortune only to form a "ramassis", an admixture of human refuse'.[308] Juvenal's stews, ragouts, and mashes exploit the same kinds of associations, linking miscellaneous *satura* with the wretched mixed dishes in the bowels of the Roman empire.[309]

At the other end of the spectrum, the rich are characterized by their unending capacity for ingestion. A priest of Cybele is a rare specimen of an outsize gut (*rarum ac memorabile magni / gutturis exemplum*, 2. 113–14). The courtier Montanus is absorbed by his own stomach, which heaves into view, dragging its load of tripes: (*Montani quoque venter adest abdomine tardus*, 4. 107). *Abdomen*, a word more commonly used of pigs' stomachs, especially as food,[310]

[307] 'Dregs' is a metaphor in Latin too: see Cicero *Att.* 1. 19. 4: the cesspool of the city (*sentinam urbis*); *Pis.* 9: 'the dregs of the city' (*faece urbis*).

[308] J. Brown (1984: 97). This is Sue's definition of the *arlequin*: 'un ramassis de viande, de poisson et de toutes sortes de restes provenant de la desserte de la table des domestiques des grandes maisons.' Cf. J. Brown (1984: 96) on the ragout: 'an aggregate of disparate elements which lacks the consistency, coherence and planning of bourgeois cuisine; whereas the latter signals a kind of cultural uniformity, the ragoût, because of its varied combinations, connotes dissemination and discrimination, hence is the metonym of the ghetto as cauldron, as hell'. See also Aron (1979) on Parisian leftovers: 'analysis of the inedible is one of the better ways to study a society'.

[309] Dupont (1977: 133–5) mentions the *arlequin* in her study of the *Satyrica*, arguing that the mixed dishes of the *cena* embody the medley form of the Menippean satire.

[310] Pig's stomach: e.g. Plaut. *Cur.* 323, Val. Max. 3. 5. 3, Plin *NH* 8. 209, Juv. 2. 86.

suggests the man's bestial qualities, blurring the distinction between his paunch and the sort of bloated food that goes into it. Describing human organs as meat is a deflating device Juvenal shares with Persius: *ventriculum* (pig's stomach) is used of an actor's stomach at 3. 97; *sinciput* (pig's head) is used of a boy at 13. 85. *Vulva*, a succulent paunch at 11. 81, is used earlier of a human womb, that of Domitian's niece Julia, seduced by him, then forced to have an abortion:

> cum tot abortivis fecundam Iulia vulvam
> solveret et patruo similes effunderet offas. (2. 32)

When Julia flushed out her fertile womb with all those abortions, discharging chunks of flesh that were dead ringers for her uncle.

Offas, the word for foetuses here, inspires disgust because it is normally used of lumps of food.[311] Julia's womb becomes an edible stuffed pig's womb disgorged of its contents: the pregnant woman, *satura*, dissolved into misshapen bits.

Juvenal's epic description of an incident of cannibalism in modern Egypt, *Satire* 15, seems at the furthest extreme from the Roman world: the boundaries of satire are being extended on an Odyssean scale. Cannibalism, which is now commonly seen by anthropologists as a cultural label designed to 'normalize' one civilization and marginalize another,[312] is clearly one of Juvenal the xenophobe's most melodramatic stage properties. His prejudice against Egyptians can be cited from a number of passages.[313] With Greeks, Jews, Asians, and Africans, they 'adulterated' or mixed up the pure stock of Rome. His description of these savages tearing up their human victims perverts the epic formula for describing animal sacrifice, the uniting ritual of civilized life.[314]

[311] e.g. Plaut. *Mil.* 760: *offam porcinam.*

[312] Vidal-Naquet (1981a) sees cannibalism in the *Odyssey* as one of several styles of eating opposed to Ithacan agriculture that separate the concepts of wildness and civilization. Braund (1986) discusses Juvenal 15 from an anthropological perspective (she compares Strabo on cannibalism among the Irish).

[313] 1. 26–7; 1. 130; 3. 83; 6. 82 ff.

[314] See e.g. Hom. *Il.* 1. 159 ff.; Virgil *Aen.* 1. 210 ff. For further examples which misapply the same epic formula in the context of cannibalism, compare Ovid on Lycaon (*Met.* 1. 228 ff.) and Seneca's description of Atreus cooking the children of Thyestes (*Thy.* 1059 ff.).

At the same time, this account of 'otherness'[315] in eating contains sinister reminiscences of society nearer home:

> ast illum in plurima sectum
> frusta et particulas, ut multis mortuus unus
> sufficeret, totum corrosis ossibus edit
> victrix turba, nec ardenti decoxit aeno
> aut veribus, longum usque adeo tardumque putavit
> expectare focos, contenta cadavere crudo. (15. 78–83)

The exultant rabble chopped him into bite-sized pieces, so there would be enough to go round, and crunched the lot, right down to the bones: they could not be bothered to boil it in a hot saucepan, or grill it; it was too much of a grind to build a fire, so they gladly ate the corpse as *crudités*.

Here the phrase *in plurima sectum / frusta et particulas* smacks of the *minutal*, the *miscellanea*, and the plain *frusta / frustula* used by Juvenal as images of the beggar's portion.[316] This sense of *déjà vu* is justified. Although Juvenal expresses tongue-in-cheek indignation that the Egyptians are too uncivilized even to cook their victim, he uses similar images elsewhere of cooked human flesh to describe aspects of his own society, at home in Rome. At 13. 84, for example, a lying father swears by his son's head (a conventional form of oath) that he is telling the truth. But there is something disquieting about his language. He swears that he is prepared to eat his son's head boiled with vinegar:

> 'comedam', inquit, 'flebile nati
> sinciput elixi Pharioque madentis aceto.'

He says, 'I'll eat my son's head boiled and weeping with Egyptian vinegar.'

The word *sinciput* (usually an animal's head as food[317]) signals the breaking of a taboo. So does the too-detailed description of the dressing of the head: lightly boiled and weeping with vinegar. The father can envisage and even

[315] The word 'adulterate' is from Latin *ad-altero*, 'to make other': *OLD* s.v.

[316] Cf. 16. 11, where a man is beaten up by a soldier: *nigram in facie tumidis livoribus offam* ('a black bruised dumpling on his face').

[317] *Sinciput*: Nov. com. 13; Petr. 135. 4, 136. 1; Pers. 6. 70; Plin. *NH* 8. 209.

relish a hypothetical event.[318] And the fact that the vinegar is Egyptian (*Phario*) looks ahead to the undisguised cannibalism of 15. Elsewhere, Roman governors suck the bones of the provinces (*ossa vides rerum vacuis exucta medullis*, 8. 90). And cannibalism is the accusation hurled at upstart informers in *Satire* 1, who nibble at the leftovers of the old nobility (*rapturus de comesa / nobilitate quod superest*, 1. 34–5).[319] Juvenal's pious conclusion in *Satire* 15 is that pity is what separates men from the beasts. But even this is phrased in language that, in the context of cannibalism, suggests a macabre gastronomic pun. Our *mollissima corda* (131, 'succulent hearts'), he says, are the best part of us: *haec nostri pars optima sensus* (133).[320]

Juvenal's *œuvre*, then, stretches the definition of satire in different directions. It contains both fat stomachs (deflated by gastronomic language) and meagre collections of scraps, which stand both for the unequal portions belonging to different sections of society and for the people themselves. Juvenal aspires to the 'fatness' of more privileged books, yet admits that his subject-matter is a rich supply of corruption (*uberior vitiorum copia*, 1. 87), a humble mash (*farrago*, 1. 86) of human misbehaviour.

In his own invitation poem, *Satire* 11, however, Juvenal does not offer *minutal* or *miscellanea* (though a slave does steal scraps, *exiguae ofellae*, from the table). His menu is

[318] *Flebile*, 'tearful', becomes conflated with *madentis*, 'dripping': the boy's tears are provoked by the acidity of the vinegar. Cf. Ennius *Sat.* 12–13W: mournful mustard and sad onion (*triste sinapi . . . caepe maestum*); Lucilius 194M = 216W: the tearful onion (*flebile cepe*).

[319] For cannibalism as a social metaphor, cf. Hor. *Sat.* 2. 5, Petr. *Sat.* 116 f., both descriptions of voracious legacy-hunters. Cf. 14. 79–81: Juvenal uses an image of vultures teaching their children to bring carrion back to the nest to describe parents who pass on bad morals to their children. Other forms of greed, avaricious and sexual, are often expressed in terms of gluttony, reduced immediately to a more basic level, e.g. at 2. 53, *coloephia*, mutton chops, is given an obscene complexion by the scholiast: *pulmentum sive membrum virile*. At 9. 3–5 a link is made between different kinds of licking: *nos colaphum incutimus lambenti crustula servo?* ('Would we take a cudgel to a slave who licks the tarts?'). See Adams (1982: 134–6) on the ambiguities of *lambeo*. For avarice as eating cf. 7. 218: *[aera] praemordet* ('he nibbles at the cash'); and at 1. 138: *una comedunt patrimonia mensa* ('they eat legacies at one sitting').

[320] Puns on the meaning of *cor*, anatomical and culinary: cf. Mart. 14. 219. 2; 7. 78. 4.

purified of urban vice: a succulent kid (*pinguissimus haedulus*, 65–6), with more milk in its veins than blood; mountain asparagus (*montani asparagi*, 68–9); champion eggs with their mothers (*ova grandia ipsis cum matribus*, 70–1). Only the cornucopia that ends the meal:

> Signinum Syriumque pirum de corbibus isdem
> aemula Picenis et odoris mala recentis (11. 73–4)

Signian and Syrian pears in pear-twig baskets, and apples that challenge Picenians with their freshness and scent . . .

might be an allusion to the purest form of *lanx satura*. The menu aspires to grandeur and perfection—*pinguissimus, grandia, aemula Picenis*—so much so that it has been read as a parody of simple meal-description.[321] Juvenal has something in common with those mendacious innkeepers described by Macrobius, who display food in glass bowls of water to magnify its true size.[322]

However, the very fact that the menu is hedged in on either side by invective against urban gluttony suggests how vulnerable purity is to corruption. This meal is unnaturally free from blood, rot, and poison.[323] But the monstrosity that *satura* must really emulate is the all-consuming stomach that girdles this tiny set piece, Juvenal's hyperbolic metaphor for Rome's consumption of the world:[324]

> quis enim te deficiente crumina
> et crescente gula manet exitus, aere paterno
> ac rebus mersis in ventrem faenoris, atque
> argenti gravis et pecorum agrorum capacem. (38–41)

What way out is there now your funds are exhausted and your gullet ever-swelling, your debts engulfing your whole inheritance

[321] By Colton (1965).

[322] Macr. 7. 14. 1.

[323] 68: the kid with more milk in its veins than blood; 74–6: fresh apples, past all risk of poisonous raw juice.

[324] See Weisinger (1972: 232) on 38–41: 'As the man's desires increase, his resources dwindle, with the result that the man is reduced to a mere stomach digesting everything around it'; and ibid. 238 on 127–8 (*hinc surgit orexis, / hinc stomacho vires*): 'the *orexis* and the *stomacho vires* have become the entire man.' The Greek word for asparagus, ἀσφάραγος, is the same as a word for the gullet: cf. Hom. *Il.* 22. 328. Juvenal's menu paradoxically contains the very purest form of mountainous gullet in *montani / asparagi* (69–8). Cf. 4. 107: *Montani . . . venter*.

in a stomach capacious enough for mounds of silver, flocks and estates?

Satires 4 and 5 also deal in different ways with mammoth consumption. We need to ask whether they can also be read as formal experiments on different aspects of the *satura* theme, bringing to life some of the eating metaphors of the first poem.

1 *Juvenal*, Satire *4*

Satire 4 is an epic parody. The main body describes a council called by Domitian to discuss how to serve a monster fish (*rhombus*) presented to him by a poor fisherman;[325] the introduction dwells on the cruelty of the Egyptian courtier Crispinus. The sense of unevenness and dislocation Juvenal provokes is to be found at its most extreme in this poem, and it is achieved through the distortion of different scales. We have seen how Juvenal builds up a histrionic style only to deflate it with bathos. How is he going to describe the unlaughable enormity of Domitian's regime? The solution lies in the art of understatement. Juvenal takes a marginal or trivial event from Domitian's reign, magnifies it out of all proportion with an exaggeratedly heroic style, and suggests the full magnitude of the reign of terror obliquely: it is somewhere off the edge of the map, or the edge of the plate, as we shall see.

Like a courtier, Juvenal flatters the trivial event, and this choice of style may be in part political: it is an attack on epic poets like Statius who misapplied their art to fawn on the emperor.[326] Another influence is Lucilius' satire, the *Council of the Gods*, which gives us a fictional trial of the career of Lentulus Lupus (a parody of Ennius' description of the gods meeting to deify Romulus). Juvenal reworks Lucilius' fish

[325] Presenting a monster fish to a king: cf. Hdt. 3. 42; Suet. *Tib.* 60; Mart. 13. 91. The best modern analysis of the poem is by Sweet (1979).

[326] Specific parody of Statius is thought to be lurking in parts of the poem. See Sweet (1979). The only fragment to survive from Statius' epic on Domitian's German campaign, preserved in Valla, describes a cabinet of war of AD 83, and names three courtiers in common with Juvenal.

puns on an imperial scale for his own evaluation of Domitian's reign. He takes the parody of epic a stage further by realizing the culinary metaphor. As the fish is amplified, so Domitian's reign is trivialized. Juvenal's magniloquent language, too large for the *nugae* described, reproduces the confusion of priorities under Domitian's regime, where a monster fish is more of an emergency than even an imminent invasion. But the poem also indirectly satirizes Domitian's more serious crimes: even epic satire, it is implied, is inadequate to describe the real monstrosities.

The poem begins with an attack on Crispinus, an Egyptian courtier whom we saw in *Satire* 1, twirling his summer ring from a carriage window. Here we learn how he raped a Vestal Virgin and bought a 6-pound mullet for 6,000 sesterces. This introduction has proved controversial in the past, because it is linked so tangentially to the main part of the poem. Scholars either use it as an example of Juvenal's most inept writing or labour to find geometrical principles of organization.[327] The fact that both sections are about monster fish and monstrous men is apparently not good enough. It is extremely difficult, as we have already seen, to make sense of questions about coherence and incoherence in satire. There is a looseness about the medley form that makes it especially hard to separate mess from design.[328] In fact there are an enormous number of minute connections between this section and the main one, if the obvious ones are not enough. But it is the differences between the two that matter most, and the tenuousness of the connection is itself significant.

The point of the introduction, above all, seems to be to establish a scale of monstrosity against which to measure Domitian.[329] Crispinus is a monster (*monstrum*, 2), an outsized freak of nature with no redeeming feature and unquantifiable possessions: *quid refert . . . quantis . . . quanta . . . quot* (5–7). Seemingly, then, no one could surpass this: there

[327] Attacking Juvenal: Highet (1954: 257); Coffey (1963: 206); defending him: Helmbold and O'Neil (1956); Heilman (1967); Kilpatrick (1973).

[328] Fronto 1. 42 (212N) sees messiness as the main characteristic of a hotchpotch (*ut ea quae per saturam feruntur*).

[329] See Helmbold and O'Neil (1956: 70); Crispinus is 'a tiny reflection of the larger, more savage and more ridiculous Domitian'.

could be *nil ulterius*, nothing more extreme, as Juvenal says in the first satire.

Everything Crispinus does, Domitian magnifies. Crispinus ordered the burial of a Vestal Virgin while her blood was still quick in her veins (*sanguine vivo*, 10); Domitian tore apart the half-dead world (*semianimum orbem*, 37) and dripped with the blood of slaughtered aristocrats (*Lamiarum caede madenti*, 154). The line-ending *senis abstulit orbi* (19, 'of a childless old man') is almost repeated at line 15: *claras quibus abstulit urbi*. Crispinus might have robbed an old man, but Domitian robbed an entire city of its citizens. That word *orbi*, 'childless', looks ahead to the doubled use of the word *orbis*, sphere, later in the poem: it means both the round platter made for the fish and the whole world which the dish symbolizes. Domitian's reign is seen through a fish-eye lens: the focus at the centre on the round dish (*spatiosum orbem*, 132) takes in on the peripheries the whole world itself (*semianimum orbem*, 37). Ring-composition locks the two halves together, with trivial things, Crispinus' *facta leviora* (11) and Domitian's *nugae* (150) in the middle, and serious crimes at both margins, spilling over the edge of the paper.

From the melodrama of a Vestal Virgin interred alive, Juvenal turns quickly to a lighter subject, Crispinus' fish:

> mullum sex milibus emit
> aequantem sane paribus sestertia libris,
> ut perhibent qui *de magnis maiora* locuntur. (4. 15–17)

He bought a 6-pound mullet for 6,000 sesterces, and, if you are the sort of person who likes making big things seem bigger, Juvenal says, that means 1,000 sesterces for each pound of its weight. Already, we can see the fisherman's hands stretching wider and wider. The authority on ancient fishes tells us that no mullet could possibly be this big, so Juvenal is telling a tall story.[330] In a poem where the word *magnus* occurs disproportionately often, this aside looks ahead to the epic magnification of the central event.[331] It also warns us about Domitian's fulsome flatterers. Juvenal draws

[330] See Thompson (1957),
[331] As Sweet (1979) argues.

attention to his own magnifying glass and damns it in the same breath.

Crispinus, too, does not escape this arbitrary telescoping process. In the space of one line, he turns from a hero (*succinctus patria* suggests an epic girding of the loins or a Homeric genealogy) into a parcel of take-away food: *succinctus patria papyro* (24, 'wrapped in his national papyrus'). *Papyro*, usually understood as a 'fishmonger's apron', is also the substance used for baking fish in or wrapping the ancient equivalent of fish and chips, and this is surely Juvenal's joke here.[332] Another distortion of scales follows: the fish, too, is reduced by synecdoche (*squamas*, 25).

The unsettling mixture of magnification and deflation continues:

> qualis tunc epulas ipsum gluttisse putamus
> induperatorem, cum tot sestertia, partem
> exiguam et modicae sumptam de margine cenae,
> purpureus magni ructarit scurra Palati
> iam princeps equitem, magna qui voce solebat
> vendere municipes fracta de merce siluros? (28–34)

What sort of feasts do we suppose the emperor himself guzzled, when the court jester in his robes belched out all that money on just a tiny fraction of some side-dish for a modest dinner—the chief of the Knights, who once hyped provincial catfish that fell off the back of a boat?

Induperatorem, the emperor, archaic and dignified, collapses into the colloquial *gluttisse*, to guzzle; *purpureus magni* builds up to an ignoble belch: *ructarit scurra*; *princeps equitum*, the chief of Roman knights, betrays his origins as a provincial fishmonger: *magna qui voce solebat / vendere municipes fracta de merce siluros*.

This transition into the Domitian story may seem far-fetched, but that can be explained: the size of feasts is a deliberate 'evasion' of the main issue, comparative degrees of monstrosity. The fish story is itself an oblique and partial

[332] Catullus (95. 8) and Martial (3. 2. 4, 4. 86. 8, 3. 50. 9, 13. 1. 1) speak of this as a fate worse than death for their own manuscript paper: see Thomson (1964); Pasoli (1970–2).

way of looking at Domitian's reign. Crispinus' mullet may have been only a side-dish (30).[333] Crispinus himself is small and modest compared with Domitian, his behaviour a mere hors-d'œuvre to Domitian's cruelty. In the same way, the first part of the poem is 'small' compared with the centre (even though it seems hyperbolic), and the centre is small compared with the full truth of the reign of terror. Crispinus, only a part of Domitian's court (108), is a monster, so Domitian is left to the imagination, too horrific even for the most magniloquent language. Juvenal is once again emulating the 'fat' books of *Satire* 1, *plena iam margine* (1. 5): if even his margins are full, that is a device for measuring the inexpressible.

Juvenal moves from margin to centre by upgrading his diction into epic, epic that is constantly being toppled. He calls on the Muses, only to insist that they sit, not stand: *non est / cantandum: res vera agitur* (34–5). This true story does not need artificial exaggeration. Juvenal flatters the Muses by calling them virgins, a joke in bad taste after the rape of the Vestal, which also prepares us for the unctuous flattery of the main story. Truth is brought on as the rhetorical prop for Juvenal's hyperbole. Domitian is *Flavius ultimus* (37–8), the last Flavian, or, the worst, most extreme Flavian, which reminds us again of the phrase *nil ulterius*, the motto or motivating force behind *Satire* 1. In the next line he is exposed as a bald Nero: the last *Flavius*, which suggests blondness, is an emperor with no hair.

The fish swims into view with its astonishing bulk (*spatium admirabile rhombi*, 39), prize specimen of the world that is Domitian's personal safari-park (*vivaria*, 51). Crispinus, from the margins of the empire, cannibalistic Egypt, once sold broken, inferior fish (*municipes fracta de merce siluros*, 33). Now even his monster mullet is small fry. The ambiguous words *implevit sinus* (41), literally 'it filled the fold', begin the fishy distortion of scales in the poem. The fish either filled the fisherman's net, or it filled an entire gulf: we

[333] Unlike Calliodorus' mullet (Mart. 10. 31. 4), which was the *pièce de résistance* of a whole dinner (*cenae pompa caputque*).

can magnify or telescope it at will, tell our own tall story (*qui de magnis maiora locuntur*).[334]

Sweet (1979) argues that the fish is symbolic of Domitian's regime, but this is not the only analogy that can be made. It is also a microcosm of the world, of Domitian, of his courtiers, and even of the barbarians. Many words in this satire come from the vocabulary of prodigies and portents (*admirabile, miratrix, monstrum, prodigio, attonitus, dirus*), turning the fish into a perverted miracle; it is misinterpreted as a good omen by the courtier Veiento (124–8).[335] At the same time, it is like the freaks and monsters emperors exhibited in the circus as a sign of their total control over the world.[336] The ostentatious dish was the circus in miniature: Vitellius' favourite dish, the Shield of Minerva, which contained all the products of the world in miniature, summed up and trivialized supreme power.[337] Granted an audience with the emperor at 64, the fish is given central importance (in the shape of the line as well), while the senators are thrust to the margins: *exclusi spectant admissa obsonia patres* ('Shut out, the fathers watched the entrance of the entrée'). The fisherman magnifies his emperor with flattering words:[338]

propera stomachum laxare sagina
et tua servatum consume in saecula rhombum. (4. 67–8)

Quickly clear your stomach of small fry, and then consume a turbot preserved for your reign alone.

Like the fish with its raised fins (*erectas in terga sudes*, 128), the emperor's coxcomb rises to the bait:

'ipse capi voluit.' quid apertius? et tamen illi
surgebant cristae. (4. 69–70)

[334] See Ferguson (1979) ad loc.

[335] Deroux (1983) argues that it is specifically an omen sent by the gods to signal the disruption of the *pax deorum*.

[336] Veiento praises other sorts of *spectacula*, 121–2. See Vegetti (1981) on Pliny's portrayal of the world as macrocosm of the circus where *mirabilia* are displayed.

[337] Suet. *Vit.* 13. 2. Vitellius, like Domitian, had a huge platter specially made to contain the dish (Plin. *NH* 35. 163–4; Dio 64. 3. 3), and was nicknamed *patinarius* ('trencherman': Suet. *Vit.* 17. 2).

[338] The word *sagina* means the small fry that feed larger fish, as well as 'stuffing'. See Courtney (1980) ad loc.

'He wanted to be caught.' What could be more explicit? And yet the emperor's head swelled.

The final compliment is double-edged: both emperor and fish are willing to be hooked.[339]

If Domitian is the biggest fish in the biggest pond, his courtiers are smaller sea-monsters, like fat giant tunnies borne down to the Black Sea (mentioned to give an idea of the scale of the monster, *rhombus*), sluggish and fat after a long winter (*desidia tardos et longo frigore pingues*, 44). The courtiers, too, are fat and sluggish from the long metaphorical winters of Domitian's reign (*deformis hiems*, 58; *tempora saevitiae*, 151). One of them, Montanus, is a smaller version of Domitian, consumed by his own stomach (*Montani quoque venter adest abdomine tardus*, 107). Courtiers only survive the winter by not resisting the current (89–90); if they do survive, they are, like the *rhombus*, prodigies. To live to old age is a miracle among the nobility (*prodigio par est in nobilitate senectus*, 97). Some, like the fish, survive only to be killed (*mors tam saeva maneret*, 95); or eaten, like the courtier Fuscus, whose guts were destined for Romanian vultures (*vulturibus servabat viscera Dacis*, 111).

These geographical details extend the range of the satire, and also chart the lapse of heroic voyages into gastronomic odysseys. The experience of the elder statesman Montanus is entirely that of a gourmet, a parody of Homer's account of Odysseus' experience of many cities and peoples (with *Circeis*, 140, as a Homeric allusion):

> nulli maior fuit usus edendi
> tempestate mea: Circeis nata forent an
> Lucrinum ad saxum Rutopinove edita funda
> ostrea callebat primo deprendere morsu,
> et semel aspecti litus dicebat echini. (4. 139–43)

He was a gastronomic pundit for our times: at first bite he could identify oysters born in Circeii or near the Lucrine rocks or washed up in the nets at Richborough, and pinpoint a sea-urchin at a glance.

[339] Cf. Hor. *Sat.* 2. 5. 25–6: *si vafer unus et alter / insidiatorem praeroso fugerit hamo* ('if one or two are cunning enough to nibble the bait and escape your trap'); Sen. *Ben.* 4. 20. 3: *captator est et hamum iacit* ('he is a con-man and casts his hook').

The blind flatterer Catullus is a massive, eye-catching monster (*grande et conspicuum . . . monstrum*, 115), a shadow of Virgil's hideous blinded Cyclops (*monstrum horrendum, informe, ingens*). And the setting of the main story, Alba, birthplace of heroic Rome, is symbolically in ruins (*diruta*, 60). Domitian may have other fish to fry, the marauding tribes of Germany (147–9), but these are pushed, again, to the margins of the poem.

The central agenda of the meeting is to find a dish large enough to hold the fish: *sed derat pisci patinae mensura* (72, 'There was a shortage of plates to fit the fish'). Epic periphrases like *patinae mensura* ('the dimensions of a dish') are not only parodies but evidence of the dislocation of values, when the part or the measurement is more important than the whole.[340] Domitian's only words in response: '*quidnam igitur censes? conciditur?*' (130, ' "So, what do you suggest? Shall we cut it up?" ') are sinister and pregnant. His instincts have always been for cutting up (*cum iam semianimum* laceraret *ultimus Flavius orbem*, 37; *Lamiarum* caede *madenti*, 154). Only the fish escapes this indignity.[341] Montanus suggests that a huge dish be made specially to contain it, a deep casserole with thin walls: *testa alta paretur / quae tenui muro spatiosum colligat orbem* (4. 131–2). This dish is also a clue to the dangers of Domitian's reign. *Orbis*, like *sinus*, covers a vast range of sizes, from a dish to the entire world.[342] *Tenui muro* usually means 'a fine distinction'.[343] There is also a thin whisper between belonging to Domitian's world and leaving it:[344] *tenui iugulos aperire susurro* (4. 110, 'Opening jugular veins at the slightest whisper').

The huge fish and the dish built to contain it are also, in the way we have come to expect, metaphors for epic satire

[340] See Sweet (1979). Cf. *Montani venter* (107); *spatium admirabile rhombi* (39).

[341] '*Absit ab illo / dedecus hoc*' (130–1); unlike Domitian's noble citizens at 151–2: *abstulit urbi / inlustresque animas et vindice nullo*.

[342] See above p. 47 on the confusion of style (cosmogonic) and scale (culinary) that surrounds *orbis* in the *Moretum*.

[343] e.g. at Cicero *Rep*. 4. 4.

[344] Domitian is immured in his hilltop villa (*arx* 145), like Crispinus' rich widow, on a smaller scale, in her sedan-chair with closed shutters (*quae vehitur cluso latis specularibus antro*, 4. 21). Like hers, his domain excludes everyone but flatterers (cf. *admissa obsonia*, 64).

and its material. Once again, the plate in satire is a clue to the dimensions and ingredients of the verse-form in which it is contained. Horace, as we saw, offers a friend vegetables off a *modica patella* (*Ep*. 1. 5. 2), an emblem of modest writing, and uses *puris catillis* to suggest refined, Callimachean *satura* (2. 4. 75). In the same satire, his advice not to squeeze large fish on to too narrow a plate is also a warning against overstraining a limited form:

> immane est vitium dare milia terna macello
> angustoque vagos piscis urgere catino. (77)

It is a cardinal sin to shell out a fortune at the market, and then squeeze floundering fish into a narrow dish.

At the other end of the scale from Juvenal, Martial's epigram on a *rhombus* also suggests that a hyperbolic subject is inappropriate for epigram, the most minuscule kind of verse:

> quamvis lata gerat patella rhombum
> rhombus est latior tamen patella. (13. 81)

However wide the plate is that bears the turbot, the turbot is still wider than the plate.

Juvenal's capacity, on the other hand, will stretch to contain its subject. The search for a container big enough for the fish, *sed derat pisci patinae mensura* (72), rephrases Juvenal's rhetorical question in *Satire* 1: *unde / ingenium par materiae?* (150–1, 'Where is the talent equal to such material?'). If the geometrical shape, the rhombus, that gives the fish its name is superimposed on a circle (compare *orbem*, 4. 132), it will either flap over the sides or leave vast areas of the plate unfilled: the fit will not be perfect. Juvenal needs to rise to the occasion. But he also wants to suggest that Domitian's monstrosity is almost out of his range. The fat stomachs and full dishes in this poem help to give us a kind of scale for measuring Juvenal's huge *satura*. However, like the degrading word *farrago* in *Satire* 1, these pompous stomachs have all their stuffing taken out of them: *abdomen* (107) is usually used of a pig's stomach, and *sagina* is usually animal food.[345]

Juvenal 'fills out' this satire with epic bombast. That ambiguous phrase *implevit sinus* (41, 'filled the net or filled

[345] *Sagina* as food: Plaut. *Mil*. 845, *Mos*. 236; Cic. *Flac*. 17. 7. 6. 1.

the gulf') recalls the programming of *Satire* 1, where Juvenal punned on the word *sinus* (a sail, or fold pocket, for holding money) to make his launch into epic (*utere velis / totos pande sinus*, 149–50) equal to his monstrous subject-matter, the gaping pocket of avarice (*avaritiae sinus*, 87). *Surgebant* (70) and *surgitur* (144) suggest epic sublimity.[346] Crispinus has a loud voice (*magna voce*, 32); winter screeches (*stridebat*, 58), doors stretch wide (*patuerunt*, 63).

Even so, the discrepancy between the inflated style and despicable subject-matter has its deflating effect. Juvenal's *satura* shatters itself into motley parts like Crispinus' scraps of broken fish: the expression *diversis partibus orbis* (148), the scattered parts of the circular world, appears on the edges of the poem. Juvenal's ultimate model is, of course, the first ring-composition: Homer's shield of Achilles. But this miniature epic, or huge satire, on Domitian has more in common with Vitellius' gastronomic 'Shield of Minerva'. In the end, however, the poem is not trivial, because its subject stands for something far more momentous. Juvenal seems to be saying that he wants to extend the *lanx satura*, to serve up the whole world on a plate; at the same time some subjects, in this case Domitian's reign of terror, are out of its sphere.

2 *Juvenal*, Satire 5

Satire 5 takes us to yet another iniquitous dinner party, which gives Juvenal another chance to experiment on the hyperbolic gulf between the host's magnificent food and the disgusting scraps left over for his clients. *Satire* 4 polarized Crispinus' pathetic fish scraps and the *spatium admirabile rhombi* (39). Here, Juvenal develops the theme of the corrupted client–patron relationship first explored in *Satire* 1, with the dole-queue outside the house and the solitary patron dining alone off seven courses indoors. The text here is the oxymoron *luxuriae sordes*, which Juvenal coined in *Satire* 1: the filth of luxury (1. 140).

The divisive dinner party seems to have been a charged theme of the period. It is condemned several times by

[346] *Surgere*: cf. Juv. 7. 100 on histories: *millensima pagina surgit*: 'the thousandth page looms'. *Surgit opus*: Prop. 4. 1. 67; Man. 1. 113; jokingly, Ov. *Am.* 1. 1. 27.

Martial, who relishes the contrast between plumped oysters and clogged-up mussels, doves with huge rumps and dead magpies. The question is also addressed, more philosophically, in one of Pliny's letters.[347] What seems to lie behind all these outbursts is a debate between two cultural traditions. The communal meal, especially a public one, was above all a symbolic re-enactment of the social hierarchy. At Roman sacrifices, for example, the meat was traditionally distributed according to the rank of those taking part.[348] At one open-air festival, the emperor Domitian is said to have distributed baskets of food to everyone in the crowd: even so, he gave the senators larger baskets than the plebs.[349] According to another concept of civilized life, however, the meal was the one occasion where social distinctions ought to be ignored, and, theoretically at least, equality reigned.[350] Now that Republican freedom was long since lost, the equal meal was a precarious reminder of social liberty. Juvenal uses this unequal meal as another focus for *indignatio*; the inequalities of the dinner stand for the whole *iniqua urbs*, the unequal city, which Juvenal could not tolerate in *Satire* 1. The central plea of the poem is *ut cenes civiliter* (112), that we should dine in a civilized way, like proper citizens.

The chief function of the client is, it seems, like Thyrsites at the banquet of the gods, to provide light entertainment for the sadistic host:[351]

> nam quae comoedia, mimus
> quis melior plorante gula? (157–8)

No comedy-sketch or pantomime is better entertainment than a grumbling stomach.

His agony gives rise to a number of suitably 'convivial' jokes, though here they have their own biting edge. For example,

[347] Martial 3. 60; also 2. 43, 3. 49, 4. 85, 6. 11; Plin. *Ep.* 2. 6. See Adamietz (1972) for a survey of the models for this poem. For other niggardly meals, cf. *Anth. Pal.* 11. 313, 314 (punning on πίναξ, a dish, and πεῖνα, hunger).

[348] See Lincoln (1985); Santini (1985).

[349] Suet. *Dom.* 4. 5.

[350] D'Arms (1984; 1990) deals specifically with the problem of whether Roman meals were ever 'equal'.

[351] See Fehr (1990) on the *akletoi*, beggars who earned their meals by providing entertainment for the guests at their own expense.

there is a pun on the two functions of bile and gritted teeth, to digest food, or, in its absence, to express anger:

> ergo omnia fiunt,
> si nescis, ut per lacrimas effundere bilem
> cogaris pressoque diu stridere molari. (158–60)

If you hadn't realized, everything's arranged so that you'll be forced to weep tears of bile and gnash your back-teeth.

At the beginning of the poem, Juvenal may also be playing on the double meaning of *ius*: *tantine iniuria cenae* (9), 'Is the dinner worth the kick in the teeth?'). The pun inverts Trimalchio's punning sentiment at the start of his dinner: *hoc est ius cenae* (Petr. *Sat.* 35. 7). Now the right to a good dinner becomes inextricably linked with social rights in general. Finally, the verb *sapere*, used of Virro at 170 (*ille sapit qui te sic utitur*, 'he must have good taste, if he treats you like this'), is another sick gastronomic pun.

The food itself spans the extremes of Roman culture: for the host, the affluence of Rome, for the client its effluence. The entire resources of the Republic are dismissed as *frivola*, little crumbs (59). But Juvenal, like Fundanius at Nasidienus' dinner, has ways of subtly tainting the fruits of Roman imperialism just as much as its debris. Virro's bejewelled cups may resemble Aeneas' scabbard (43–5); his boar is like the mythological Calydonian boar (115–16); his apples are those of Homer's Phaeacians or the Hesperides (151–7); the amber on his cups is like the tears of the daughters of the Sun: *Heliadum crustas* (38). The whole world is ransacked for this one table (at 1. 137, *tot pulchris et latis orbibus* meant either round table-tops or the globe itself).

However, it is also clear that Virro's food does not deserve these heroic comparisons. When Juvenal speaks of *iniquas mensas* (3–4) and *inaequales / phialas* (38–9), he is not just referring to the inequalities between patron and client. He is also suggesting the inequalities between the patron's food and the heroic past that inspires it. As the host of the dinner, Virro is technically its king, *rex*, a secularized title from the sacred term *rex sacrorum*, the official who presided over sacrifices. Kings were, of course, also the old threat to Roman equality, and it is a contradiction in terms for a *rex* to drink Republican wine, bottled in the days of the bearded consuls (30). Even more pointed is the passing allusion to the Republican heroes Thrasea and Helvidius Priscus, Stoic martyrs who opposed Nero and Vespasian and drank this wine on the birthdays of the freedom-fighters Brutus and Cassius (36–7). *Albanis aliquid de montibus* (33), wine from the Alban hills, marks the contrast between the origins of Rome, at Alba Longa, and modern decadence, a suggestion made in *Satire* 4 as well. The Social wars between Rome and her allies (31) are reduced to dinner-party brawls (26), where the guests fight with bread-rolls. Religious vocabulary also points to the hubris of the secularized dinner: *salva sit artoptae reverentia* (72, 'holy reverence be to the bread-basket'); *excelsi minister* (83, 'the lofty minister').[352]

The first course of the meal marks the social differences between patron and client: a king-prawn or lobster (the exact difference in scale is difficult to guess) versus a pathetic shrimp. The lobster (80–2), which adds distinction to the plate (*distinguat*) and is fenced round with asparagus (*saepta*), behaves like the swaggering street bully in *Satire* 3. Puffing out its broad chest, *longo pectore*, it reminds us that Virro is not large-hearted like the heroes he aspires to. The lobster looks down its nose at the frightened guests, or rather down its tail: *despiciat cauda* (82), mooning at the assembled company. A contemptuous look doubles as an obscene gesture.

Meanwhile the guest's portion, the tiny shrimp, is hemmed in (*constrictus*, 84) by half an egg, intimidated by this hostility. And half an egg is not much of an improvement on the half-piece of begging mat that was the only other resource of the poor client at 8–9: *tegetis pars dimidia brevior*. The tiny dish is practically a funeral feast, *feralis cena*: one would have to be disembodied to eat it. The exiguousness of funeral food recalls another sadistic feast, the famous joke-meal staged by Domitian for his senators, where they were whisked away in the middle of the night to a

[352] Cf. Macrob. 3. 16. 8: a fish brought into dinner and treated more like a *numen* than *deliciae*.

room hung in black with funeral food on the table and each man's tombstone by his place.[353]

The rest of Trebius' food is broken, stale, rotten, or poisonous, or in some way associated with the grotesque parasites of the past: wine too harsh to be absorbed by sweaty sheep's wool (24); wine drunk from a cracked mug named after Nero's *scurra*, the long-nosed cobbler Vatinius (46);[354] gum-grinding bread (67–8); oil scraped from lamps or used by Africans in the baths (86).[355] Instead of luxurious fish from over-trawled seas and ransacked markets, the client is offered an eel (*anguilla*), first cousin of the viper, or a fish that crams itself on the sewage gushing from the market district of Rome, the Subura (a favourite seedy haunt of Roman satire), into the River Tiber:

> vos anguilla manet longae cognata colubrae
> aut †glacie aspersus† maculis Tiberinus et ipse
> vernula riparum, pinguis torrente cloaca
> et solitus mediae cryptam penetrare Suburae (5. 103–6)

All that's left for you is an eel, long-lost cousin of the viper, or else the slum-child from the Tiber's banks, spattered with ice-spots, stuffed on sewer-spillage, lurking under inner-city manholes.

The identity of this second fish is in some doubt, though various clues suggest that it is probably the *lupus* (wolf-fish or sea-bass). Macrobius tells us that Lucilius called this fish, caught between the two bridges of the Tiber, *catillo* (licker of plates).[356] And it has even been suggested that *glacie*, ice, is a corruption of *glutto*, glutton, which would suggest a more obvious allusion to Lucilius' scavenger.[357] If there is any reference to the *catillo* here, it would fill out the list of despicable *scurrae* who lick the plates at dinner.

353 Funeral food: Aug. *Conf.* 6. 2; Plut. *Crass.* 19; *Qu. Rom.* 95. Domitian's meal: Dio 67. 9. Elagabalus is said to have tortured his parasites with food made of glass or painted on canvas (*SHA* 25. 9).

354 Vatinius: cf. Mart. 14. 96. 1, 10. 3. 4; Tac *Ann.* 15. 34.

355 *Alveolis* (88) means platters, stomachs, and pig's troughs.

356 Macr. *Sat.* 3. 16. 18.

357 See Campbell (1945: 46). He also reorders 103–6, reading 103, 106, 104, 105, as the *anguilla* is more likely to go up sewers than any other fish. See Courtney (1980) ad loc. for a discussion: *glacie aspersus maculis* has otherwise been explained as a reference to a fungus that only survived in icy water. *Catillo*: Paul. Fest. 144M

The host's dishes meanwhile become more and more tainted under the surface. At 114–18 Juvenal builds up to a swelling crescendo of bloated food: a huge goose-liver;[358] a capon as big as the goose itself;[359] a boar, inflated by an epic simile; and truffles, gathered after grandiose thunderstorms, which make the dinner even grander (*cenas maiores*):

> anseris ante ipsum magni iecur, anseribus par
> altilis, et flavi dignus ferro Meleagri
> spumat aper. post hunc tradentur tubera, si ver
> tunc erit et facient optata tonitrua cenas
> maiores. (5. 114–18)

The host is served a liver from a golden goose, a capon the size of the goose itself, and a frothing boar worthy of blond Meleager's sword. And then, if it's spring, truffles may be sprung on him: the right bolt from the blue can make these dinners mushroom.

The choice of words links the food with stylistic tumidity: *Tubera* are the essence of swollen protuberance; *tonitrua*, thunder, is also a metaphor for the high style;[360] *cenas maiores* are not just 'longer dinners', but 'dinners of epic proportions'. Once again, the exaggeration is self-deflating. *Tuber* reminds us of the Latin proverb quoted by Apuleius (*Fl.* 18): *ubi uber, ibi tuber*; 'where there is abundance there is also malignancy'.

The client may be eating dubious toadstools, but Juvenal's simile for Virro's mushroom implies a death-wish for the host: it was like the last one eaten by the emperor Claudius (147–8).[361] The final course is apples (150–5), for Virro one of mythological perfection, whose smell alone you could feed off (small comfort for the client, who can only feed off smells anyway); for Trebius an eczematous or maggoty fruit: *tu scabie frueris mali*. Here, *frueris*, in the context of fruit, takes us back to line 14, where food is classed as 'the rotten fruit of

defines it as an archaic name for a glutton; cf. Plaut. *Cas.* 553. *Glutto*: cf. Juv. 4. 28 *gluttisse*.

358 For the stuffed goose-liver, cf. Pliny *NH* 10. 27. 52; Hor. *Sat.* 2. 8. 88.

359 Cf. Mart 13. 58: liver bigger than a goose; 3. 60. 3: gorged oysters (*ostrea saturata*); 7: a pigeon with outsized rump (*inmodicis turtur . . . clunibus*).

360 Cf. Prop. 3. 17. 40; Plin. *Ep.* 1. 20. 19.

361 Cf. the name of Trebius' wife—Mycale—from Gk. μύκης, mushroom. Cf. Mart. 1. 20. 4, *boletum qualem Claudius edit, edas*: 'I hope you eat a mushroom like the one Claudius ate.'

friendship with the great': *fructus amicitiae magnae cibus*. There may also be a pun on *malum*, evil, and *malum*, an apple, thrown in.[362] The client enjoys the rotten fruits of evil, which are no better than the apples given to a performing monkey dressed as a soldier. This mock-heroic costume recalls the dinner-party brawl between guests (26–8) and the buffeted clown (171–2). Once again, Trebius descends to the level of a cheap and subservient entertainer, whose weapons give him no powers of resistance.[363]

So the client's food links him with monkeys, dogs, snakes, sewers, rot, death, half-shares, tiny crumbs, poison, and immigrants: the dregs and debris of Roman society. Is it possible that Juvenal chose Virro's name because of its similarity to the word for snake-poison, *virus*? We have already seen how Virro feeds the client on snake-like eels, and he also serves him oil only good for warding off *serpentibus atris* (91). This line may well be interpolated, but in any case there is another allusion, in *sororibus Afris* (152), to the last words of Horace's dinner-party satire: *serpentibus Afris* or *atris*, black or African snakes (2. 8. 95).[364] There, the guests left the dinner without touching any of the food, as though the witch Canidia had breathed her snake-poison all over it. This final invocation of Canidia, Horace's witch-muse, resuscitates memories of the satirist's power to poison what he breathes on.

Juvenal's satire on the divisive dinner contains one irony that was not available to Martial or Pliny. The verse form is *satura*, yet the poem is a complaint from someone who is *ieiunus*, 'hungry'. The two words are opposites, as we can see from a statement about rhetorical decorum made by Cicero.[365] *ut quidque erit dicendum ita dicet, nec satura ieiune*

[362] For which there is a precedent in Plautus, despite the difference in vowel-length: *Amph*. 723.

[363] Lucian (*Merc. Cond*. 24) compares a parasite to an ape, and also to a hooked pike (λάβραξ, = Lat. *lupus*, the fish that Juvenal uses), which again connects the host's control over his clients with his control over their food.

[364] A further source is Virgil *Georg*. 1. 129: *ille malum virus serpentibus addidit atris* ('Jupiter added deadly poison from black snakes').

[365] Pliny's description of unequal food uses a similar contrast: *sibi et paucis opima quaedam, ceteris vilia et minuta ponebat* (*Ep*. 2. 6, 'He served himself and a few friends with rich food; the rest got cheap scraps').

nec grandia minute (*Orat.* 123, 'He will use the correct style for every subject, not jejune language for full subjects, nor scrappy language for grand themes'). So *Satire* 5 has an extra dimension: it brings to life the eating metaphors that informed Juvenal's programme for his own writing. He began *Satire* 1 by condemning the bloated tragedies that consumed whole days and burst at the seams, and aspired to make his own work a gargantuan consumer by filling note-books with a farrago of human corruption (*uberior vitiorum copia*). In *Satire* 5, his sense of *indignatio* is again inflamed by the hunger of society's victims. By the end of the poem, Juvenal has turned on the client and accused him of deserving his own fate: he is worthy of this kind of patron (*tali dignus amico*) if he can tolerate such humiliation. Juvenal injects his cause with his own *indignatio*; the satirical impulse is alive even at a dinner that commemorates the loss of political liberty. And Trebius' role as spectator at the feast (*spectes*, 121)[366] is not only presented as a social disgrace. It also brings us back full circle to Juvenal's groan at the beginning of Book 1: *semper ego auditor tantum*? ('Must I always stay in the audience?').

Both menus at Virro's dinner provide fodder for Juvenal's poison pen. *Cenas maiores* are deformed by their own excesses. Trebius' tiny plate (*exigua patella*, 85) may recall Horace's *modica patella* and *angustus catinus*, but this old image of modest satisfaction has now become an outrage. Juvenal is no longer satisifed with limited consumption. Trebius' nasty fish, fat on sewage, *pinguis torrente cloaca*, is a much better metaphor for a greedy kind of satire, which absorbs the filth it feeds on. The same could be said of the ibis, mentioned in passing at 15. 3 (the ibis, gorged on snakes: *saturam serpentibus ibin*), a bird that fed on poison and its own excrement, and personified satirical venom.[367]

[366] Morford (1977: 237) quotes Highet's reply to Ribbeck, who assumed a lacuna at this point: 'There is indeed a lacuna, but it is in the clients' bellies.' Cf. Mart. 1. 43. 11: we were only spectators (*tantum spectavimus omnes*).

[367] Aelian (*NA* 10. 29) records that the ibis fed insatiably on poison; see Thompson (1936: 107–14). The bird's name was used by Callimachus and after him Ovid as a pseudonym for a hated opponent. The *catillo*, similarly, becomes fat on waste from the sewers. For sewers as the metaphorical excretory ducts of the city of Rome, see pp. 14–15 above. For a connection between *satura*, the stuffed gut

Juvenal takes revenge on his society by emulating or simulating its poisonous nature in writing, and seeking out a richer supply of corruption (*uberior vitiorum copia*).[368] Foraging in the sewers and filth of Rome, he makes his *saturae* grow fat on vice.

(*farcimen*) and these gut-like conduits of the city's waste, cf. Cic. *Sest.* 77: sewers stuffed with citizens' bodies (*corporibus civium . . . cloacas refarciri*).

[368] For poisoning in the satires as another form of perverting and corrupting food see 1, 70–2; 6. 631 ff.

4
A Taste of Things to Come
INVITATION POEMS

Introduction

If a Roman writer were a dinner party, what sort of dinner party would he be? This sounds like a trivial game, but it is a question we can also ask more seriously about the Latin invitation poem. The handful of invitations that survive are still being read literally: such windfall lists of food seem far too good to waste, when they could be used to tell us what the Romans really ate.[1] By now, however, it should go without saying that the style of the party that a writer offers his guests is also some sort of projection of the style in which he entertains his readers. And these proffered poetic meals are the first that Latin scholars have been ready to see as 'declarations of taste as well as means'.[2] The loaded table is no longer what it seems. Lurking under the heavy weight of material are more abstract notions of style: the modest dish represents small-scale poetry, the clean napkin pure diction, the wine-jar Greek lyric, and so on.[3]

[1] One ancient historian recently mentioned his plan to assess the nutritional value of Martial's menus. These poems are also used as evidence for Roman eating in most general surveys of the subject: e.g. Ricotti (1983), Dosi and Schnell (1984).

[2] Trimpi (1962), quoted by Race (1978: 188).

[3] See Commager (1957: 326) on wine in Hor. *Od.* 1. 20: 'Roman content, Greek container, modest cups, the emphatic *ego ipse*—the phrasing suggests that Horace's real gift to his patron is not so much the promised wine as the poem itself . . . By the time that Maecenas finishes reading the poem he has, in effect, already imbibed Horace's promised tribute.' Contra, Nisbet and Hubbard (1975: 248) on *Od.* 1. 20. See Mette (1961) on the parallels between Horace's *mensa tenuis* in the *Odes* and the *genus tenue* of his style. Race (1978) sees Hor. *Od.* 1. 20 and other invitation poems as examples of *recusatio*, the poet's tongue-in-cheek apology for limited resources: so Horace's offer of a small plate (*modica patella: Ep.* 1. 5. 2), cheap wine (*vile*

This new kind of interpretation does at last give Roman writers due credit for being complex and allusive. But does it over-compensate for the very literal approach it rejects? To conceive of the invitation poem as exclusively real or exclusively metaphorical is to misunderstand the blend of ingredients it contains. As we have seen, food is only half-suitable for representing poetic style, and could be regarded as its direct opposite. In fact, all the poems discussed here make a very clear distinction between the food on the table and the entertainments or conversation, and the different kinds of pleasure these provide. The singer or reader at a Roman dinner party is a more obvious parallel for the writer than the cook. Although lists of food can also be read as tokens for something more abstract, like the mood of the dinner or the writer's relationship with his readers, that must always be weighed against their purely material function. The first-person address adds another dimension: how easy is it to go through a list of one's own food without embarrassment? Cultural attitudes inevitably come into play in deciding the weight and treatment food receives in these poems. And that in itself can tell us something about the poet's own work. In the poems discussed here, we shall find that the balance between food on the one hand, and conversation or entertainment on the other, is a very good gauge of the ratio of tasteful and vulgar, ephemeral and lasting, and bodily and insubstantial elements in the author's work as a whole.

The structure of the invitation poem usually separates these two groups of elements. This is a kind of poem that pretends to be a real letter; it is one of those representations that try to close the gap between literary and functional writing.[4] It manages this with such ease that one critic, Edmunds (1982), has assumed that literary invitations are a good guide to the structure of invitations in real life. Catullus 13, Horace *Epistle* 1. 5, and all Martial's invitations, he argues, have the same tripartite structure: first, the invitation

Sabinum: Od. 1. 20. 1) and tender calf (*tener vitulus: Od.* 4. 2. 54) can be tokens of slender writing.

[4] Cf. Cat. 16: abusive 'graffiti'; Prop. 3. 23, where love-elegy is represented as love messages scratched on writing-tablets.

proper, specifying time and place; then the menu; and finally the entertainment. Poems that do not have this structure, for example Horace, *Odes* 1. 20, 3. 29, 4. 12, have to be excluded from this scheme, despite the fact that they are clearly invitations too. Similarly, he admits that some simple meal-descriptions, for example in Ovid's 'Baucis and Philemon', have some overlap with invitation menus, as they exploit the same cultural opposition between simplicity and luxury.[5] The Roman invitation poem is distinctive because it is based on a specifically Roman cultural form. We might well expect a link with Hellenistic epigrams.[6] However, Philodemus' letter to Piso (*Anth. Pal.* 11. 44) is the only surviving example of the Greek invitation in epigram form, and it has a number of specifically Roman elements: the ninth hour as the time for the dinner, the rejection of *οὔθατα* (sows' udders, a specifically Roman delicacy) and the Roman patron, Piso.[7]

Edmunds supports his argument with two 'real-life' invitations, Pliny, *Epistle* 1. 15 and the freedman's invitation at Petronius, *Sat.* 46. 2. However, neither of these is a very good control. Pliny's letter certainly has the proper tripartite structure: the opening address (*heus tu promittis ad cenam nec venis*: 'you accepted my dinner-invitation but didn't turn up'[8]), a list of modest dishes, and finally the entertainments. But Pliny is very suspect as a source for natural, uncontrived invitations, and 1. 15 will be regarded here as a fiction. And

[5] Ov. *Met.* 8. 638–78f. The topos of the modest meal has Greek symposiastic precedents: Bacchylides fr. 21 (= Ath. 11. 500b), Xenophanes fr. 1 (comparing modest *θαλίη*, feast, and *μολπή*, songs, at a festival) and Anacreon 96D. Race (1978: 195 n. 19) also compares some other contrasts between modest and luxurious food (though these are not strictly invitations): Parmenion (*AP* 9. 43), Sen. *Ep.* 114. 11, 115; Pers. *Sat.* 5, contrasting Mycenean feasts (17) with his own plebeian lunches (18) and modest table (44) (see Bramble (1974: 102–18, 143–6, 156–64)); Diog. Laert. 2. 139–40 (Menedemus' modest dinners and temperate conversation); Plut. *Mor.* 150C–D. See also Race's useful appendix on other contrasts between modesty and grandeur in Horace outside the context of food (1978: 191–3).

[6] Nisbet and Hubbard (1975: 244) categorize *Odes* 1 in this way; Cairns (1972: 74) classes the *vocatio ad cenam* as a type of epideictic *kleticon*.

[7] Edmunds acknowledges that there is some overlap with other Greek epigrams: themes of modest resources (Lucill. *AP* 11. 10. 1 *δειπνάριον*; Nicaenetus 2703 ff. GP; cf. Alc. fr. 71 LP); inappropriateness of dancers (Lucill. *AP* 11. 11); poetry recitations (Lucill. *AP* 11. 10, 137, 140, 394); menu (Rhianus 3246 ff. GP).

[8] *Venire*: cf. Martial 10. 48. 5, 11. 52. 2.

the second example is hardly formal enough to give us any idea of the correct wording:

aliqua die te persuadeam, ut ad villam venias et videas casulas nostras? inveniemus quod manducemus, pullum, ova: belle erit . . . inveniemus ergo unde saturi fiamus. (46. 2)

Can I persuade you to come to the country one day and visit my cottage? We'll find something to chew on—chicken, eggs: everything will be fine and dandy . . . we're bound to find something to fill up on.

The two-clause phrase *pullum, ova* seems deliberately incomplete: this scheme for casual browsing is welcome relief from the contrived and satiating dishes of the *cena* itself.[9] Edmunds concludes that Philodemus is reproducing a cultural form specific to Roman life: 'The freedman is not imitating the tone of the invitation poem nor is Pliny imitating its structure. Rather, we see in these two places the conventions of Roman social life upon which the invitation-poem rests.'[10]

This conception of the real invitation appears, then, to be based entirely on its fictional versions. We get a very different impression if we put literary invitations next to some more obviously real or unencumbered invitations, those that survive in the Oxyrhynchus Papyri. These are of course by no means ideal controls either, as they belong to a different time and place (second- and third-century AD Egypt), but they are likely to bear more relation to the Roman social invitation than literary excerpts from Petronius and Pliny. A real invitation would in fact have looked more like this:[11]

Dioscorides invites you to dine at the wedding of his son on the 14th of Mesore in the temple of Sabazius from the ninth hour. Farewell.

[9] The phrase *unde saturi fiamus* contrasts with Encolpius' sensation of mental satiety even before Trimalchio's dinner: *his repleti voluptatibus*, 30. 5; *admiratione iam saturi*, 28. 6.

[10] Edmunds (1980: 188).

[11] 2678. Cf. 110, 111, 523, 524, 747, 926, 927, 1484–7, 1486, 1487, 2791, 2792, 3501. For a comprehensive list, see Skeat (1975: 253 n. 2).

Religious feasts, weddings, children's birthdays, coming-of-age parties are all represented, but in none of these perfunctory invitations is there any mention of food or entertainment.[12] So we can infer that, far from being an accurate reproduction of a social form, the literary invitation actually flouts social convention, or at least adds extra information.

There are two possible reasons for this. First, the menu and list of entertainments are there to tell us something about the author's tastes and values; whether he sets store by display, whether he is confident that a congenial atmosphere will more than make up for a modest table. Secondly, the menu and entertainments are there to provide a contrast to each other. Our attention is drawn to the difference between the material elements of the dinner, which may seem lumpish or coarse, and its abstract but more essential ingredients. Since the emphasis does always seem to be in favour of abstract pleasures—wit, laughter, friendship, poetry—and since the food offered always does seem to be simple and cheap, it is easy to see how this distinction between the two sections can be blurred, and the food, too, can be seen as the embodiment of modest stylistic principles. In particular, of course, these are Callimachean principles. Thus Philodemus (*AP* 11. 44) advertises his own humble cottage (*λιτὴν καλιάδα*, 1), honest companions (*ἑτάρους παναληθέας*, 5) and sweet conversation (6),[13] and banishes *οὔθατα* (luxurious sows' udders) and Chian wine (4).[14] Similarly, Horace's *modica patella* (*Ep.* 1. 5. 2) and Martial's *parva cenula* (5. 78. 22) are tokens of their allegiance to slender writing. Martial's rejection of the thick book (*crassum volumen*, 5. 78. 25), another Callimachean *παχὺ γράμμα*, in favour of the flute-player *parvi Condyli* (30), can be read in the context of his more obviously programmatic writing, most clearly in a poem where he contrasts his slim flute (*angusta avena*) with epic

[12] Verbal invitations in Plautus are similarly informal: e.g. *Curc.* 728: *tu, miles, apud me cenabis* ('soldier, you'll dine with me today'); *Rud.* 1417: *hic hodie cenato, leno* ('dine with me today, pimp'). Cf. Lucian, *Gallus* 9.

[13] 'Sweet' (*μελιχρός*) bridges the gap between the the sense of taste and more abstract feeling.

[14] Race (1978) suggests that the adjective *Χιογενῆ*, alluding to Homer's birthplace and his use of compound words, also signals the rejection of Homeric epic. (cf. Phaeacia, mentioned at 6).

trumpets.[15] Pliny follows suit, contrasting simple vegetables (*olivae betacei cucurbitae bulbi*) and tasteful entertainments with luxurious dishes (*ostrea vulvas echinos*) and coarse Spanish dancing girls. As Race sums up: 'Suffice it to say that Pliny is talking more about the style of his letters than he is about some real dinner.'[16]

Literary polemic is undoubtedly central to these poems. They are provocative but sympathetic statements which link literary modesty with more general resistance to materialism and hierarchy in Rome. And the modest menu undoubtedly adds to this effect. Even so, to read poetic meanings into the food has its drawbacks. One drawback is that this disregards the opposition between the two halves of the meal, the food and the entertainment, an opposition that is important in marking the contrast between substance and words or bulk and essence. And this contrast between food and conversation or food and poetry mirrors another contrast, between the poet's paper letter and the patron's material support. In crude terms, the poems are the immaterial half of an economic exchange between poet and patron or friend; they are payment for real dinners.[17] This notion of poetry as a currency exchangeable for food goes back to the primeval banquets of the *Odyssey*. And comparisons in early lyric poetry between food and music at banquets or *symposia*[18] suggest that the poems that contain these comparisons are themselves imitations of convivial entertainments;[19] their authors at least pretend to be singing for their supper. For

[15] 8. 3. 21–2. Virg. *Ecl.* 1. 2 *tenui . . . avena*; see Bramble (1974: 164–72) on metaphors for grand and humble styles.

[16] Race (1978: 186).

[17] Martial 11. 52 contrasts the grand dinners of his patron Stella (invited to dinner in 10. 48) with his own meagre (and, of course, only fictional) meal. Cf. 1. 44: two epigrams on the subject of hares are equivalent to two dinners of hare with Stella. There may be a pun here with insubstantial *lepos* (charm), despite the difference in vowel length; cf. 4. 23. 6: *qui si Cecropio* satur *lepore* / *Romanae* sale *luserit Minervae* ('if he, sated with Attic charm, toys with the salt of Roman Minerva'); Plaut. *Cas.* 217.

[18] E.g. Xenophanes fr. 1, where food and songs are equally pure (καθαρόν, 1, 8; εὐφήμοις μύθοις, 14; καθαροῖσι λόγοις, 14); Anacreon 96D, where the full mixing-bowl (κρητῆρι . . . πλέῳ) is juxtaposed with martial epic. Anacreon uses a convivial image to describe his own subject-matter: 'mixing (συμμίσγων) the radiant gifts of Aphrodite and the Muses' (cf. the etymology of κρατῆρ from κεράννυμι, to mix).

[19] See Giangrande (1968).

the Roman writers discussed here, however, the bardic function of the poet is an anachronism to be parodied; the author is now the host as well as the performer. Yet conviviality remains not only the context of these poems, but their very essence. In them, we see the author at his most playful or most genial.

There is another important element in these exchanges, which the 'poetic' interpretation ignores: *urbanitas*, urbane wit, the essential salt sprinkled on a Roman dinner party, which infuses the invitation poem too. The first-person address carries with it all the paraphernalia of social banter that other contrasts between simple and luxurious meals lack; it is not enough simply to extract their common elements. *Urbanitas* relies on the tacit and shared recognition of social codes, yet it usually works by breaking those codes: the understatement and hyperbole it uses are deliberate distortions which indirectly signal the author's appreciation of proper scale. The point is to adjust the face-value of the speaker's words. Quintilian, for example, points to the word *satagere* as an example of *urbanitas*: 'to do enough' is a witty and understated way of saying 'to overdo it'.[20] This elusive but all-important quality of *urbanitas* tends to register itself through the ironic presentation of its exact opposites.[21] Add to this the fact that to go through with a straightforward and literal description of one's own food was socially embarrassing; this sort of inventory was not part of the normal invitation-form or polite behaviour in general. We already have enough to explain the ironic and uneven tone of these inventories. To alternate between cheek and deference, boasting and apology, disavowal and solicitude is one solution to the problems of straight exposition.

This means that we must assume that lists of food are often trying to lead us astray. Any possibility of a simple one-to-one correspondence between items of food and, say, literary qualities is out of the question. And where the correspondence seems to work especially well, it may be that the fit is meant to be a paradoxical one. Inventories of

[20] Quint 6. 3. 54.

[21] See in general on this subject Ramage (1973).

ingredients parody the idea of correspondence as much as they exploit it. In a roundabout way, they show the impossibility of summing up either a poetic style or an ideal dinner party in material terms.

What the menus do often display is the kind of bantering wit associated with Roman convivial behaviour. Certain kinds of wit, we are told, were not thought compatible with gentlemanly behaviour, though the limits of propriety were by no means well defined. To complicate matters, wit in general treads a thin line between good and bad taste. Cicero sees it as the ability to observe grossness in a manner which is not itself gross, Quintilian as the ability to misrepresent the true state of things: *ut aliter quam est rectum verumque dicatur.*[22] Quintilian's example of this is a convivial joke attributed to the emperor Galba, who, confronted with some leftover fish placed uneaten side uppermost, remarked: *Festinemus, alii subcenant* ('We'd better hurry up. They're eating it from below').[23] Too much distortion or exaggeration, however, can turn wit into *stultitia.*[24] Not surprisingly, the 'spice' of wit is thought to be ambiguity,[25] which casts responsibility for bad taste on to the listener. To complicate matters further, remarks in bad taste can display the innate good taste of the joker, if the listener is open to irony. Irony and innuendo are the hallmarks of convivial wit, since they remove the need for openly defying the limits of decorum. And wit, with all its indirections, is an essential ingredient of these menus.

The food contents of literary menus are also tellingly linked with genre. To leave food out or keep it to a minimum is a more direct sign of good taste. Horace, for example, who

[22] *De Or.* 2. 58. 236: 'The territory of the ridiculous . . . involves a kind of grossness and deformity: the principal, or only, forms of humour are those that point out defects in a manner which is not itself gross' (*locus autem et regio quasi ridiculi . . . turpitudine et deformitate quadam continetur: haec enim ridentur vel sola, vel maxime, quae notant et designant turpitudinem aliquam non turpiter*); Quint. 6. 3. 89.

[23] 6. 3. 90.

[24] Quint 8. 6. 75 ff.

[25] e.g Quint 6. 3. 96: *si qua ambiguitate conditur* ('if it is seasoned with ambiguity'); Cic. *de Or.* 2. 255: *quod si admixtum est etian ambiguum, fit salsius* ('but if a dash of ambiguity is thrown in, it will be saltier').

was prepared to give an extended recipe for satire in the guise of Catius' culinary instructions (*Sat.* 2. 4), virtually eliminates food from his lyric invitations (wine and slender vegetables are acceptably ethereal). In one of his epistles, he sums up the menu in a nutshell (*modica . . . holus omne patella*: 'the whole meal is vegetables on a small plate'), before moving on to the more abstract concerns of the dinner.[26] In fact, the obsessive hygiene of this dinner, where the furniture is neat, the couches clean, the napkins spotless, the dishes self-revealing, and all stinking parasites banished, bears out Horace's unspoken decision to purge his letter of messy or perishable food.[27] Juvenal's menu, in this *Satire* 11, was, as we saw, similarly purified from external sources of corruption. There, however, the peerless apples, the prize kid, champion eggs, and mountainous asparagus needed to be hyperbolically described so that they would stand out luminously against the background of the adulterated, all-consuming city that Juvenal's satire tries to contain. For Pliny in his *Epistles*, on the other hand, to linger over a food-description will seem improper: a token summary has to suffice.

When a poet invites his reader to a *convivium*, that is also an invitation to the world of verbal frankness, humour, and indiscretion associated with the dinner party. The degree to which the reader chooses to enter into the spirit of impropriety is a matter of judgement, though we must be guided by the limits set out elsewhere in the author's writing. Is he spicy like Martial or po-faced like Pliny? Do we have the licence to read double meanings beneath the surface of propriety? Any author's *convivium* must be interpreted not only as a representation of his style, but also as evidence of the degree of candour he has allowed his work, and the flavour of his intimacy with his readers, both private and public.

[26] *Ep.* 1. 5. 2. On this poem, which I do not have space to discuss here, see Williams (1968: 7–10); McGann (1969: 44–6); Kilpatrick (1986: 61–5).

[27] *Ep.* 1. 5. 7: *munda supellex*; 22–4: *ne turpe toral, ne sordida mappa / corruget naris, ne non et cantharus et lanx / ostendat tibi te*; 29: *nimis arta premunt olidae convivia caprae*.

Catullus 13

Catullus 13 may seem like a perverse choice to begin with, as it contains hardly any food at all. It has been called 'a parody of invitation poetry',[28] and, while all classical invitation poems in a sense parody the genre of letter-writing which they imitate, this poem takes the host's traditional apology for poverty to extremes.[29] In fact, these extremes help to outline some of the oppositions that all invitation poetry contains. Catullus' guest is Fabullus, whose name ('little bean', 'bead', or 'pellet of dung'[30]) gives a foretaste of the tiny, worthless dinner for which he is destined. No sooner is he invited than Fabullus finds himself obliged to supply all the ingredients of the meal, a hazy affair, with no fixed date (*paucis . . . diebus*, 2). The menu is relayed in an absurd catalogue form, proceeding from the material to the insubstantial. Fabullus is required to bring a good big dinner, a blonde girl, wine, wit, and laughter:

> si tecum attuleris bonam atque magnam
> cenam, non sine candida puella
> et vino et sale et omnibus cachinnis. (3–5)

Catullus' sole contributions, meanwhile, are to be *meros amores* (9) and some *unguentum* (11) given by the gods to his *puella*, so delicious to smell that Fabullus will long to become 'all nose' (*totum . . . nasum*, 14).

Two main questions have interested critics of the poem, and they both concern the degree to which the dinner is real or metaphorical. The first is the identity of the mysterious *unguentum*. The usual interpretation is that this perfume is a symbol of Lesbia's beauty, whether or not it is real.[31] At the other end of the scale, Littman (1977) and Hallett (1978) think that it is a different sort of present: the smell of Lesbia's vaginal secretions or, alternatively, an anal lubricant, on the grounds that Catullus is too poor to afford the

[28] Williams (1968: 127).

[29] Cf. Philodemus, *Anth. Pal.* 11. 44; cf. 'shopping-list' poems: *Anth. Pal.* 5. 181, 183, 185. See also Nisbet and Hubbard (1975: 245) on *Od.* 1. 20.

[30] Cf. *OLD* s.v. *faba*; *fabula*.

[31] Defended by Witke (1980).

best sort of perfume. The second question is whether the whole poem is not in fact a metaphorical programme for Catullus' style. Race (1978), Marcovich (1982), and Bernstein (1985) are the chief advocates of this. Both kinds of interpretation call into question the validity of freely interpreting something that is not made explicit—in other words, reading between the lines. Each interpretation is made at the expense of its opposite. However, the fact that there is a controversy at all only emphasizes that this poem is constructed so as to create ambiguity. We can extract physical or metaphorical meanings from its vocabulary without having to reject either. And the vagueness of the last part of the poem may be intrinsic to the nature of the scent itself, which is both real and evocative.

The idea that the poem might be a metaphor for poetic programming is part of a growing recognition that many of Catullus' poems contain programmatic Callimachean vocabulary in the guise of some physical object.[32] More specifically, the poem has been read as concealed *recusatio*: modest poetic resources are offered in the urbane form of an apologetic invitation. Apparent poverty is taken to an extreme, then compensated for by *meros amores* (9), 'the epitome of suavity and sophistication which his charming (*venuste*, 6) friend will appreciate . . . the phrase *seu quid suavius elegantiusve est* could refer just as easily to Catullus' poetic style.'[33]

Explored in more detail, all the ingredients of the dinner can be ticked off as stylistic metaphors.[34] *Candida puella* represents the plain style in rhetorical terms;[35] *sal* is 'wit' as well as 'salt'; *venuste* means 'charmingly sophisticated'; *meros amores* can mean 'pure love poems' as well as 'unmixed

[32] First suggested in passing by Buchheit (1975: 43), who lists the relevant poems as follows: 'In nicht wenigen Fallen "verdeckt" er dagegen diese Tendenz hinter einem anderen Thema. Ich nenne nur die Gedichte 13. 14. 16. 35. 36. 44. 49. 50. Dabei ist für die meistern (13. 14. 35. 36. 44. 49) bezeichnend, dass das vordergründige Theme im Grunde belanglos ist, ja durchaus als Fiktion angesehen werden kann.'

[33] Race (1978: 187).

[34] By Bernstein (1985).

[35] See Ernesti s.v. *candidus*.

love'.[36] *Meros amores* might mean specifically Ἔρωτες, love-stories in the Hellenistic manner.[37] *Suavis* and *elegans* suggest literary stylishness.[38] *Unguentum* can refer simultaneously to Lesbia's beauty and to the source of Catullus' poetic inspiration. *Veneres cupidinesque* (12) are 'not strictly sexual, but stand as archetypes for erotic poetry and a life given over to sensual pleasure and love'; *nasum* (14) is not 'a joke intended to undercut the rest of the poem', nor a 'light-hearted throwaway',[39] but a metaphor for critical discernment. This continues the discussion of literary pleasure in sensual terms (*sal*—taste; *olfacies*—smell):[40] Fabullus will want to become 'all nose' to appreciate Catullus' poetry. The language of the poem begins to be recognizable as part of the much wider vocabulary Catullus uses to denote civilized and tasteful behaviour, in which he includes his own poetic work (*lepidus, facetus, salsus, sapiens, pulcher, aptus, formosus, venustus*, and so on).[41] In the words of one triumphant summary: 'All the terms fit a cohesive program which could easily describe Catullus' poetry: plain, witty and amorous'.[42]

The problems are not solved, however, by seeing the poem as entirely metaphorical. For a start, this ignores the gradual movement within the list of ingredients from solid and substantial—the dinner itself, into thin air—an ethereal scent. There is also a very clear-cut division between the things Fabullus is asked to bring and those that Catullus can supply on his own. The list begins with a *bonam atque magnam / cenam* (3), which would seem to be the solid and total essence of a dinner party (this is the point of the joke in the sequence *cenabis bene . . . si attuleris . . . bonam atque magnam cenam*).[43] Then follows the *candida puella*, the erotic element, both real, and symbolic of her erotic function as

[36] Cf. *Ecl.* 8. 23, 10. 54; Ov. *AA* 3. 343.

[37] Marcovich (1982: 135), citing Ov. *Tr.* 2. 361.

[38] Bernstein (1985: 129); cf. Witke (1980: 331): '*suavis* and *elegans* are well-known touchstones of aesthetic value in Catullus' polymetrics'.

[39] Bernstein (1985: 129).

[40] Cf. Hor. *Sat.* 1. 4. 8 (*emunctae naris*); Mart. 1. 141. 18; Plin *NH* pref. 7; Sen. *Suas.* 1. 6.

[41] See Ross (1969: 79): *suavis* as programmatic; Wiltshire (1977: 319–26).

[42] Bernstein (1985: 129).

[43] See Helm (1980–1: 213–14) for jokes παρὰ προσδοκίαν in the poem.

well.[44] *Vinum* is another material element, but it is also the catalyst of festive spirits and licentious speech.[45] Since these physical items are also semi-metaphorical, there is no point, for example, in stressing the importance of *sal* as a turning-point in the list: 'it looks backwards and forwards through its duality of meaning.'[46] The movement from substantial to symbolic is much more imperceptible. *Sal*, which was always a universal cultural metaphor (the slight but vital condiment that outweighed any staple in importance),[47] is also 'wit', the spice of dinner-party conversation.[48] The last ingredients of Fabullus's list, *omnes cachinni* (5, 'all the hoots of laughter'), of course leave the material plane entirely (hence the joke in *attuleris* (6), 'bring too'): convivial laughs are comically reduced to a collection or sum, as though such an insubstantial element could be quantified. Together, *sal* and *cachinni* stamp Catullus' own list with the mark of self-consciously convivial humour.

In particular, the first ingredient, the good big dinner (*bonam atque magnam cenam*, 3–4), is completely ignored by the programmatists. This is not surprising: it is antithetical to all the other stylistic metaphors. Could anything be more alien to Catullan (and Callimachean) aesthetics than an equation between what is good and what is large?[49]

[44] As Witke emphasizes (1980: 326).

[45] Cf. 12. 2, *in ioco atque vino*; 50. 6, *per iocum atque vinum*.

[46] Vessey (1971: 46).

[47] Cf. Plin. *NH* 31. 88: 'Civilized life cannot proceed without salt: it is so necessary an ingredient that it has become a metaphor for intense mental pleasure.' Ath. 10. 420c tells of a dinner where the host's request for bread became the 'spice' (ἥδυσμα) of the dinner. The joke lies in the temporary identification of opposites, staple and condiment. Cf. Aristotle *Rhet.* 1406a. 18. 19 using ἥδυσμα to describe the use of epithets. See Bramble (1974: 52–4) for other examples.

[48] *OLD* s.v. *sal*; cf. Hor. *Ep.* 2. 2. 60: *sale nigro* used in a literary sense in the context of a convivial metaphor.

[49] One exception is Cat. 64, where a lavish dinner is a figure for the most ostentatiously 'constructed' of all the poems: *large multiplici constructae sunt dape mensae* (304, 'the tables were generously piled with a multifarious feast'). *Struo, construo, exstruo*, originally building verbs, come to be associated with arranging feasts and constructing literary works (*OLD* s.vv.). *Struo* is found in many epic texts (e.g. Virg. *Aen.* 1. 704; Val. Flacc. 3. 570; Sil. 11. 277) in connection with feasts, as though these represented the laborious construction and imposing largesse of the texts in which they appear. In Catullus' feast the emphasis is on multiplicity rather than quantity: the feast is a miniature of the poem's *varietas*. *Multiplici* also suggests the structure of the poem, to which the central epyllion

Elsewhere, Catullus alludes to Callimachus' incarnation of Antimachus's *Lyde* as a παχὺ γράμμα (fr. 398 Pf.) and makes it clear that his own work is the antitype of all this:[50]

> parva mei mihi sint cordi monimenta . . .,
> at populus tumido gaudeat Antimacho.

If I can keep the modest monuments of my [friend] close to my heart, let the public delight in flatulent Antimachus.

If the list of prerequisites is to be read as a mock-banal prescription for Catullan poetry, it would be logical to include the substantial dinner itself.

The *candida puella*, we are then told, represents the 'plain' as opposed to the grand style. Catullus uses the word *candida* in another poem, however, with a less than ideal significance. Like Callimachus' *Lyde*, the irreproachable but charmless bulk of Quintia is a model of empty beauty, compared with Lesbia and her unique sex-appeal:

> Quintia formosa est multis. mihi candida, longa,
> recta est: haec ego sic singula confiteor.
> totum illud formosa nego: nam nulla venustas,
> nulla in tam magno est corpore mica salis. (86. 1–4)

Like Catullus' dinner, Quintia's qualities are reduced to banality by being presented in list-form: *candida, longa, recta* (blonde, tall, straight). The sum of these individual qualities (*singula*) is not a perfect whole (*totum illud formosa nego*). And while Lesbia is the quintessence of attractive beauty, Quintia lacks *venustas* and has no essential 'grain of salt' in her huge body: *nulla in tam magno est corpore mica salis*.[51] It is wrong, then, to attach special significance to the *candida puella*. Witke, for example, does when he claims that any 'crude' interpretation of Lesbia's *unguentum* diverts

depicted on the bedspread gives an extra fold. Cf. Martial's two poems on Ovid's *Metamorphoses*, a multiform poem (14. 192–3: *haec tibi multiplici quae structa est massa tabella / carminis Nasonis quinque decemque gerit*: 'this bulky volume, built up of many-leaved pages, holds the fifteen books of Naso's poems') and on confectioners, creators of mountains of sweets (14. 222–3: *mille tibi dulces operum manus ista figuras / extruet*: 'this hand builds you a thousand sweet edifices of pastry'; cf. 11. 31. 9).

[50] 95b. See Bramble (1974: 158) on *tumidus*.

[51] Cf. Mart. 7. 25. 3, where *nullaque mica salis* is re-applied to describe insipid epigrams.

attention from the other girl's radiant beauty, and turns Lesbia herself into an anticlimax.[52] *Candida puella*, like *candida* in 86, is part of a list whose sum is not total perfection,[53] Quintia's body, *tam magno corpore*, like the *bonam atque magnam cenam*, loses in ineffable *je ne sais quoi* what it makes up for in substance.

Another drawback of reading the list as the ingredients of Catullus' style is the fact that Fabullus is the one asked to provide them. Catullus could of course be displaying ironic *urbanitas*, pretending to be deficient in his own consummate qualities. But the banal list-form contrasts so much with Catullus' description of what he will offer in return that it is more likely that a contrast is being made between the bare essentials of a meal or poem and its intangible essence. This vital contrast has been ignored, as well as the progression from the most substantial element to the final secret ingredient which captures all the senses (*totum . . . nasum*).

What Catullus offers in return seems to be the clue to the meaning of the poem:

sed contra accipies meros amores
seu quid suauius elegantiusue est:
nam unguentum dabo, quod meae puellae
donarunt Veneres Cupidinesque,
quod tu cum olfacies, deos rogabis,
totum ut te faciant, Fabulle, nasum. (9–14)

Instead you will get the essence of love, or something more fragrant, more refined: I'll give you some ointment, which the gods of love gave my lady, and when you smell it, you will pray to the gods that they make you, Fabullus, all nose.

But what sort of substance these gifts represent is unclear. *Meros amores* is usually equated with the unguent that follows.[54] *Seu quid* etc. might mean 'if anything could be sweeter'[55] or 'or something that is sweeter'.[56] Whichever

[52] Witke (1980: 326).

[53] Cf. Cat. 35. 8: *candida puella* is used of a token mistress.

[54] e.g. Ellis (1867: 37): 'the pure spirit or quintessence of love . . . the *unguentum*, of course, is meant'; this interpretation goes back at least to Moretus, Venice 1554. *Meros* ('undiluted') of course introduces another παρὰ προσδοκίαν: we expect wine; instead, *amores* are substituted: see Bernstein (1985: 129), Vessey (1971: 46–7).

[55] Cf. 82. 4: *oculis seu quid carius est oculis*.

[56] Cf. 22. 12–13: *scurra / aut si quid hac re scitius videbatur*.

sense is divined, *unguentum* is at least roughly equivalent to *meros amores*.[57]

But how substantial is *meros amores*? Is it 'unmixed love', 'unmixed friendship',[58] or Alexandrian 'love-poems'?[59] Or is it something rather more sticky?[60] Martial, it is worth mentioning, reuses the phrase to describe a concrete object, a breast-band, at 14. 206:

collo necte, puer, meros amores,
ceston de sinu Veneris calentem.

Wind the essence of love round your neck, boy, a breast-band warm from the bosom of Venus.

A concrete object, that is, which smells of something more intangible. Marcovich argues that such a concrete use is not likely in Catullus' case, but the obviously double significance here can teach us a lesson about his *meros amores* too.

The quintessence of love, vaginal secretions, friendship, love-poems, the loved one herself: all these interpretations have been extracted from *meros amores*. And, as if this is not enough, *merus* and *amor* can have almost opposite meanings too: *merus* can mean 'mere' as well as 'sheer'; *amores* can be 'longings' or 'consummation'.[61] How generous, then, *is* Catullus' offer? Is he offering Fabullus the body of his mistress, to enjoy unknown sexual pleasure, or is he

[57] With the former meaning, *nam* (11) makes the *unguentum* identical anyway; with the latter meaning, *unguentum* intensifies all the qualities of *meros amores*, although, as will be seen later, it may in fact be even less of an offering.

[58] See Schuster (1925). Cf. Philodemus 23. 5 (= *Anth. Pal.* 11. 44. 5): the poet promises honest companionship as a compensation for a simple meal (ἑτάρους παναληθέας).

[59] Following Philodemus' promise of μελιχρότερα. See Marcovich (1982). Cf. Anacreon 96D, mingling together the muses and the splendid gifts of Aphrodite: ἀγλαὰ δῶρ' Ἀφροδίτης. Mixing (συμμίσγων) is another appropriately convivial metaphor, suggested by the mixing bowl (κρατήρ) in line 1. *Accipies* could mean *audies* (cf. Cat. 68. 13, 64. 325, 35. 6), and it would be more natural to describe both poetry and a scent as *suavis* or *elegans* than friendship: Marcovich (1982: 135–6); cf. *suavis odores*, 64. 87; *suave olentis amarici*, 61. 7; cf. Ross (1969: 79) on *suavis*.

[60] Littman (1977), who wants to read *unguentum* as 'vaginal secretions', is forced to read *meos amores*, taking *amores* to mean 'loved one' (cf. Cat. 15. 1; 21. 4): 'you will take my girl, or, what is better, her vaginal secretions'. Even with *meros*, this kind of interpretation is possible: 'Unmixed love (i.e. vaginal secretions) will put you in such heat that you will want to be made all nose.'

[61] *OLD* s.vv. *merus, amor*. Cf. Cat. 68. 69 (*amores* as sexual intercourse): *communes exerceremus amores*.

merely exciting a longing in him, in order to frustrate it? Or are the *amores* poems, the most insubstantial ingredients of the dinner, but worth more than the sum of Fabullus' list?

The same ambiguity affects the *unguentum* too. Traditionally, *unguentum* was seen as an actual unguent, an essential ingredient of Roman dining, or a symbol of Lesbia's beauty or fragrance. More crudely, Littman finds parallels for thinking that a natural lubricant is involved, Lesbia's vaginal secretions: Tibullus' description of *hippomanes*, the juice of mares on heat; an Aristophanic lubricant; Propertius' distinction between artificial perfumes and those created by Love himself, and Cynthia's 'smell of adultery'.[62] Along similar lines, Hallett (1978) suggests that the unguent is an anal lubricant.[63]

These flies in the ointment have been briskly swatted by Witke (1980), who shows that there are enough connections between perfume and the aura of beauty in Graeco-Roman culture and mythology for obscene suggestions to be quite unnecessary.[64] Once again, he over-compensates in his reaction, and underplays the specifically *sexual* associations of perfume. He concludes that it is a symbol of 'beauty', ignoring the encouragement that this poem and the rest of Catullus' work give us to think along other lines. We need to ask what there is about the tone of the poem that it elicits such crude interpretations.

To begin with, perfume in ancient literature had close associations with sexual foreplay, whether simply as a scent or as a lubricant as well. For Plato, perfumes, wine, and garlands were stimulants of harmful desires (*Rep.* 573a). In

[62] Tib. 2. 4. 57–8; Ar. *Lys.* 492–4; Prop. 2. 29. 17–18: *afflabunt tibi non Arabum de gramine odores / sed quos ipse suis fecit Amor manibus* ('wafting towards you will be not the scent of Arabian spices, but perfumes that the god of Love himself made with his own hands'); 2. 29. 38–9: *ut in toto nullus mihi corpore surgat / spiritus admisso notus adulterio* ('that no give-away breath of adultery rises from anywhere in my body').

[63] Via Martial's possibly obscene reinterpretation of Cat. 13 in 3. 12.

[64] e.g. Sextus Turpilius *Leucadia* (Ribbeck: 113): Venus rewards an old man with *unguenti alabastron: cum se inde totum ungeret, feminas in sui amorem trahebat* ('when he anointed his whole body with this, he made women fall in love with him'). Apul. *Met.* 6. 16: Venus sends Psyche to collect Proserpina's beauty in a *pyxis*.

the passage from Aristophanes quoted by Witke, Myrrhina (myrrh) frustrates her husband by producing the wrong sort of perfume, ῥόδιον μύρον, myrrh of roses, instead of scent that is ὄζον γάμων, proper for married love-making.[65] There need be no reference to natural lubricants here, and the smell of the female *pudenda* is never mentioned without disgust.[66] The Greek Adonia, a celebration of extramarital, unproductive love, and the Thesmophoria, a festival for married women, were similarly polarized by smells: the first by the scent of burning spices, the second by the foul-smelling breath of its fasting participants.[67] The association of scent with seduction and adultery would already explain Cynthia's protestation that there was no smell on her body to betray her.[68] In Plautus' *Casina*, again, it is the smell of perfumes that betray an old man's extramarital longings to his wife, though Casina (cinnamon), the object of those longings, remains only an elusive fragrance.[69] Perfume is also a prelude to seduction in Apuleius' *Metamorphoses*, where the matron anoints herself and the ass with balsam: *ut unguento fragrantissimo prolubium libidinis suscitaram* ('the fragrant perfume had stimulated my desire').[70]

[65] *Lys.* 942–4.

[66] Witke (1980: 327); cf. Richlin (1981: 116). Cf. Plaut. *Most.* 273: *mulier recte olet ubi nihil olet* ('a woman smells good if she doesn't smell at all'). This sort of abuse is usually directed at old and unattractive women. Even when arguments are made in favour of natural perfume, these assume that a pleasant smell is needed in connection with sex: e.g. Xen. *Symp.* 2. 3–4: young brides do not need perfumes as their bodies exhale their own perfume. Sometimes (this would support Littman's argument) the genitals are seen as natural alternatives to perfumed objects: e.g. *Lys.* 947: the husband calls his genitals ἕτερον ἀλάβαστον. Cf. Ar. *Lys.* 1004, Ruf. *Onom.* 111: myrtle, Venus' plant, is used to name the female genitals. But these examples only reinforce the theory that perfumes were preferred to natural smells.

[67] See Detienne (1977: esp. 61–6).

[68] Witke (1980: 328) argues that, as the phrase *toto corpore* is used, there is no specific reference to genitals here: she is probably referring to seductive pomades and perfumes.

[69] See MacCary and Willcock (1978: 126) ad 219: 'One has a strong suspicion that somewhere here (or in 225 [*qui quom amo Casinam magis niteo*; Schoell, *qui, quom amo, casiam magis inicio*]; cf. also 814 [*iam oboluit Casinus procul*]), there is an allusion to *casia*, "cinnamon", from which the girl's name is derived.' At 217 ff. Lysidamus extols *amor* as the spice of life. Cf. 277 for smelling as a metaphor for finding out: 'My wife has a sniff of what I am scheming' (*subolet hoc iam uxori quod ego machinor*).

[70] Apul. *Met.* 10. 21.

The idea that the unguent is redolent of Lesbia's 'beauty' or that it is the 'pure quintessence of love' clearly needs pepping up a bit. In Catullus 86 the point is made that Lesbia is not only beautiful but also uniquely sexy (she has the vital *sal* in her body). Hallett (1978) may be going too far in thinking that the unguent is an anal lubricant,[71] but Witke is still wrong to dismiss her claim that 'the whole idea of eating has sexual connotations'.[72] Implied links between the two kinds of appetite, for eating and for sex, permeate not only Catullus' own works but much of classical literature.[73] Catullus himself exploits similarities between the lunch invitation and the invitation to sex in 32, where he asks Ipsitilla ('the "it"-girl')[74] for a midday invitation to . . . nine continuous *fututiones* (32. 8). This initial confusion is replayed at the end of the poem, where Catullus contrasts his satisfied appetite with his hunger for sex:

> nam pransus iaceo et satur supinus
> pertundo tunicamque palliumque. (32. 10–11)

After lunch I lie on my back with my stomach full, and poke through my tunic and cloak.

In 21, Aurelius, parodically called father of starvations (*pater esuritionum*, 1), is accused of trying to corrupt Catullus' boyfriend with sexual voracity: *esurire . . . et sitire* (10–11).[75]

[71] Her interpretation depends on reading sexual innuendo into Martial's description of coming away hungry from Fabullus' dinner, where only unguent was provided. The poem would have a point even without these sexual meanings; and, if Martial wanted to exploit a possible redolence of Catullus' poem, there would be no point in doing that unless it was only 'latent' or 'under-suggested' in the original.

[72] Hallett (1978: 747); Witke (1980: 326).

[73] J. Henderson (1975: 47–8, 52, 60–1, 129–30, 142–4, 174) discusses food/sex references in Greek comedy; see MacCary and Willcock on food and sex in *Casina*, esp. 32: 'Erotic language is always full of implied comparisons with food.'

[74] Cf. Lucian, *Rhet. Praec.* 12: αὐτοΘαΐς, the essence of Thaïs.

[75] At 28 the companions of Piso get stuffed: *verpa / farti estis* (12–13). The etymological association of *mentula* with eating finds a parallel in a cooking metaphor at 94. 2: 'the pot chooses its own greens' (*ipsa olera olla legit*). Oral sex, too, is described in terms of eating: e.g. 80. 6, *grandia te medii tenta vorare viri?* ('that you gnaw the huge stiffness at a man's centre'); cf. 88. 8 *si demisso se ipse voret capite* ('if he bent down his head and devoured himself'). Cf. 2. 3–4; 8. 18; 57. 8; 59. 3–4; 74. 3; 91. 6; 99. 2. Oral appetite and verbosity: 98. 3–4. On oral imagery in 21, see Konstan (1979: 214–16); Richlin (1988: 149); on 28 see Skinner (1979: 139–40); on 29 see de Angeli (1969); on 83 Minyard (1971: 179); Scott (1971: 19); Skinner (1979: 145–7).

In short, Catullus is relying on innuendo, something that can be received or ignored, extended or limited, according to taste. This does not mean that, in every case where a word of double meaning occurs, the reader can justifiably extract both innocent and lubricious meanings.[76] But the 'maculate' genres of comedy, satire, and epigram simulate the atmosphere of licence, candour, and obscenity found in their social equivalents, the dinner party and the festival. The explicit connection made between food and sex elsewhere in Catullus allows us to read the first words, *cenabis bene*, as an anticipation of sex; this is reinforced by the dinner-party setting itself. Catullus' dialogue with Fabullus is noticeably joking and lubricious, unctuous in its intimacy: 'my Fabullus' (*mi Fabulle*); 'sexy fellow' (*venuste noster*); 'your Catullus' (*tui Catulli*). We have only to look at the blandishments of some Plautine cooks, pimps, and prostitutes[77] to see how Catullus, with his repeated promise *cenabis bene* and his own offerings, blurs the role of the host with that of the *leno* advertising his wares.

However, as Witke says, 'one of the poem's aims is to frustrate Fabullus'.[78] First he has to provide the dinner himself; then Catullus' promise of something in return evaporates. What might be 'nothing but sex', or, more intangibly, 'pure love poems', or simply 'mere longings', is

[76] Take e.g. the word *coliculus*, to which the phallus is compared at Petr. *Sat.* 132. 8 (cf. Lucil. 281M). When Plin. *NH* 19. 138 speaks of *coliculi*, sprouts, in an agricultural context, this does not allow us to assume that the obscene meaning could be read here as well (although we *can* assume that the secondary meaning was weak enough for Pliny to be able to use the word without fear of being thought obscene). But when Martial (13. 17) writes of plunging *coliculi* into green soda to make them less embarrassing (*ne tibi pallentes moveant fastidia caules*) (cf. 14. 101 *prototomis (pudet heu!) servio coliculis*), in the context of the Saturnalia (cf. 13. 1. 4 ff.), there are grounds for guessing why *coliculi* are embarrassing to look at. See below on Mart. 5. 78.

[77] e.g. *Pseud.* 881–2 (a cook): *nam ego ita convivis cenam conditam dabo / hodie atque ita suavi suavitate condiam* ('I shall serve your guests a seasoned dinner today, and I shall season it with such scrumptious succulence . . .'). *Bacch.* 1181 (a prostitute): *i hac mecum intro, ubi sit tibi lepide victibu', vino atque unguentis* ('Follow me inside, where everything is lovely with food, wine, and perfume'). *Poen.* 695–8 (a pimp): *edepol ne tibi illum possum festivom dare / siquidem potes esse te pati in lepido loco, / in lecto lepide strato lepidam mulierem / complexum contrectare* ('That's exactly the feast I'll lay on for you, if you can bear to be in a charming place, lying on a well-made bed with a charming lady to hug').

[78] Witke (1980: 329).

quickly withdrawn for an alternative, *unguentum*. But will Fabullus have Lesbia any more within his grasp? Witke says that, if *unguentum* meant vaginal secretions, Catullus would wish that Fabullus were another organ than the nose. But surely the nose is another blow to Fabullus' expectations, one part of the body which can never participate fully. It is wrong to see *totum . . . nasum* as a climax: 'A final extravagance offsets any unseemly over-commitment in the previous lines.'[79] This 'extravagance' is also a kind of meanness, which denies Fabullus the final consummation. The poem shares a plot with the *Lysistrata* passage and *Casina*: a woman emits a seductive perfume, redolent of the possibility of sex (in *Lysistrata* and *Casina* the women are named for their smell); but ultimately she frustrates her lover, and the expectations aroused by the smell evaporate.[80]

To pin the *unguentum* down as a physical secretion or lubricant is to misread the movement of the poem from substance (*bonam atque magnam cenam*) to ether, and to reduce the elusive but boundless possibilities of the ending. Martial's breast-band (14. 206), *meros amores*, still warm from Venus, is both physical and a memory.[81] The very open-endedness of Catullus' poem demonstrates the difficulties in defining suggestion or evocation as either exclusively physical or ethereal. It is not surprising, then, if we look elsewhere, that the sense of smell has recently been used as an analogy for symbolism:[82] neither smells nor symbols can be fully defined in linguistic terms or detached from their source; the meaning of both is constructed out of the individual's memory bank of associations and connotations. It is with some justification, too, that the *Oxford Latin Dictionary* calls *odor* a 'pregnant' word: its sense is either good or bad depending on the context. Smell in Latin is commonly a metaphor for suspicion, applied to situations

[79] Quinn (1970: 135).

[80] Vessey (1971) sees in the passage an allusion to *Od.* 18. 192, where Athene anoints Penelope with Aphrodite's κάλλος: Cat. is complimenting Lesbia by comparing her to Aphrodite; Penelope too was being beautified to frustrate the desires of the suitors.

[81] Cf. a vanished napkin: *mnemosynum* (12. 13).

[82] By Sperber (1975).

where something significant is hinted at, but cannot yet be defined.[83] In the same way, the sense of Catullus' *meros amores* and *unguentum*, though highly redolent, is never made explicit.[84]

Catullus leaves us, then, with a sense of having been eluded, and this is perhaps the quintessential flavour of the poem. If we look at Latin in general, we can see that rhetorical writers have a similar solution to the problem of defining qualities that cannot be taught, only recognized. They may suggest them negatively in terms of their opposites. *Sal* is 'not unsalty'; *urbanitas* is that which contains 'nothing jarring, nothing crude, nothing insipid, nothing outlandish'.[85] Otherwise they turn to semi-physical sensations as metaphors. Quintilian speaks of *urbanitas* as 'the taste of the city' (*gustum urbis*); or as a colour or flavour.[86] When Cicero's Brutus asks 'What is the colour of *urbanitas*?', Cicero replies: 'I don't know, but I know that it exists' (*Nescio . . . tantum esse quendam scio*).[87] Nothing could be more reminiscent of Catullus' epigram on the inexplicability of love: *odi et amo. quare id faciam, fortasse requiris? / nescio, sed fieri sentio et excrucior* (85, 'I hate and love at once. You may ask me why. I don't know, but I feel it happen and feel the pain'). Finally, and most significantly for our poem, Cicero chooses the phrase *odor urbanitatis* as an example of a metaphor taken from the senses.[88] As one modern critic has remarked: 'Cicero is certainly correct in pointing to the obscurity and vagueness of the picture presented by this combination, for nothing is more difficult to describe than a fragrance, let alone a "fragrance of urbanity".'[89]

83 e.g. Cic. *Ver.* 4. 53; *Clu.* 73; *Att.* 4. 18. 3.

84 A suggestively sexual meaning could be read into the phrase *venerem meram* in a statement made by a prostitute at Plaut. *Cist.* 314: *venerem meram haec aedes olent, quia amator expolivit* ('the house smells of the essence of love, since a lover has given it a lick and a promise').

85 Quint. 6. 3. 107: *quod non erit insulsum*; 6. 3. 19: *nihil absonum, nihil agreste, nihil inconditum, nihil peregrinum.*

86 6. 3. 17: *in toto* colore *dicendi, qualis apud Graecos atticismos ille reddens Athenarum proprium* saporem.

87 Cic. *Brut.* 171.

88 *De Or.* 3. 161.

89 Ramage (1973: 54).

The poem is also a guide to the balance of real and insubstantial in Catullus' work as a whole. In the space of his tiny *libellus*, he ranges between the extremes of body and ether. Even his love-poetry is made flesh, part of the arousing and pleasure-filled atmosphere in which it is created.[90] Catullus' diet of kisses in 48 can be seen as either insubstantial (a substitute for real food) or voluptuously physical. And his love-poems too are described as a thousand kisses: *milia multa basiorum legistis*.[91] For all these reasons it is difficult to say that *meros amores* is referring exclusively to either love-poetry or sex.

Catullus' literary polemic is also frequently inseparable from obscene or personal invective. Like the rhetorical writers, Catullus can only define his own ineffable qualities through their opposites—*illepidae atque inelegantes* ('charmless and unrefined'); *insulsa* ('insipid'); *infaceto est infacetior rure* ('more unsophisticated than the unsophisticated countryside')—or images of physical ugliness.[92] Evil smells,[93] deformities, mistakes in proportion, excessive appetite, all suggest the absent qualities of neatness, sweetness, and proportion that characterize good taste in behaviour or writing. Poem 43, for example, describes a girl with bad proportions: 'without a tiny nose, a pretty foot, dark eyes, long fingers, a dry mouth, or even a refined tongue' (*nec minimo . . . naso / nec bello pede nec nigris ocellis / nec longis digitis nec ore sicco / nec sane nimis elegante lingua*). Several of the terms here—*nasum, pes, os, lingua*—could be metaphors for rhetorical taste, metre, and delivery. But the insults are just as much tied to physical slanging in this contest between a provincial beauty and a sophisticated *femme fatale*.[94] The body's images of over-absorption, misproportion, and waste inform both literary insults (*cacata carta*: 'shitty sheets') and

[90] 16. 9: *quod pruriat incitare possunt*; cf. 50. 7 ff. Cf. Persius *Sat.* 1. 20–1 on sexually arousing poems.

[91] 16. 12–13.

[92] 6. 2, 16. 33 (cf. 17. 12), 22. 14.

[93] e.g. 69. 16, 71. 5, 97. 2.

[94] Cf. 86: the contest between Quintia and Lesbia, aesthetic on both a physical and a literary level.

personal abuse (*mentula magna minax*: 'big bad dick').[95] Even if these insults point indirectly to good taste and, by extension, to literary good taste, their obscenities still stain the collection as a whole.

It could even be argued that Catullus is using words like *suauius elegantiusue* ironically. Elsewhere, many of the so-called touchstones of Catullan aesthetics are used ironically to describe disgusting things: 'polished' (*expolitis*) of teeth cleaned with urine, which puts a different complexion on the veneer of the *libellus* (*arida modo pumice expolitum*); 'subtle' (*subtilis*) and 'light' (*levis*), of a fart; 'a fine fit' (*pulcre convenit*) of copulating *cinaedi*; 'rarefied' (*tenuis*) of a man worn out by incestuous sex.[96] Even the purity of excrement is used as an insult. It is so hard that even if you rub it between your hands you won't get them dirty: *id durius est faba et lapillis / quod tu si manibus teras fricesque / non umquam digitum inquinare posses.*[97] *Urbanitas* is not simply refined and rarefied wit: it is also, in Catullus' hands, the ironic juxtaposition of obscenity (bad taste) and neat composition (good taste). As Cicero says, it consists in describing grossness in a manner that is not gross.

Even so, Catullus escapes from Poem 13 without having named anything obscene. Witke claims that this poem cannot be sexually suggestive because 'other sexual references in Catullus are explicit, not veiled'.[98] This merely begs the

[95] 36. 1 *Annales Volusi, cacata carta*, either paper that *is* excrement, or paper to wipe away excrement; 95. 8, paper only suitable for wrapping fish in. Thompson (1964: 30–6) argues that the paper is used to cook fish in the kitchen rather than to wrap it in the market. 115: *non homo sed vero mentula magna minax;* Mamurra in 29 is voracious (*vorax*, 2) and overflowing (*superfluens*, 6).

[96] 39. 20; 1. 2; 54. 3; 57. 1; 89. 1.

[97] 23. 21–3.

[98] Witke (1980: 326). The example Witke chooses (1980: 331) to show how we might read anachronistic sexual connotations into an ancient text is peculiarly unfortunate. A medieval dialogue between a prostitute, Thaïs, and her customer (Hrosvitha, *Pafnutius*, in *Hrosvithae Opera*, ed. H. Homeyer (Munich, 1970), 3. 4. 337) turns out to be a conversation between two desert saints. This does not mean, however, that the writer is not exploiting the language of seduction when the two characters negotiate their meeting in a secret *penetral*. Cf. the tradition of prostitutes as heroines in Christian/Judaic mythology: Rahab of Jericho, Mary Magdalene; cf. the medieval image of the church as *mater misericordiae*, welcoming receptacle for many men. See Parker (1987: 9).

question: how explicit do they have to be to exist? Even Marcovich, who wants to abstract 'love-poems' from *meros amores*, concedes that the tone of the poem is erotic: 'the poem perspires with love (4 *candida puella*; 6 *venuste noster*; 9 *meros amores*; 11–12 *quod meae puellae / donarunt Veneres cupidinesque*) and with mirth (5 *et . . . et . . . et omnibus cachinnis*; 14 *totum . . . nasum*).[99]

Poem 13 does read as an evocation of Catullus' style, but the connections are more complicated than has previously been suggested. The catalogue at the beginning—food, women, wine, wit, and laughter—is not a programme for writing so much as equipment for reading, the licentious and pleasure-seeking spirits that the reader must bring with him (*attuleris*) in order to enjoy Catullus' poetry and see its different possibilities. *Candida puella* suggests the potential for sex; *vinum* invites drunken licentiousness;[100] *sal* could be spicy wit;[101] *cachinni* might be ribald laughter. After this prelude and the allusion to *Veneres Cupidinesque* as patrons of the *puella, meros amores* and *unguentum* can point in one direction only—though whether this is towards 'love', 'sex', 'love-poetry', or bodily secretions or organs is a question of taste.[102] In another poem, Catullus tries to entice his readers into approaching his poems as though they are tangible objects: 'If any of you are prepared to be my readers and not shrink from touching me with your hands' (*si . . . / lectores eritis manusque vestras / non horrebitis admovere nobis*).[103] But ultimately there are aspects of his style that are not meant to be pinned down. The essence of *urbanitas* is underplaying. *Meros amores* and *unguentum* combine those aspects of Catullus' writing that are both most wickedly suggestive *and* most elusive and intangible: the indescribable something that cannot be written down in a recipe.

[99] Marcovich (1982: 137–8).

[100] Cf. 12. 2. *in ioco atque vino*; 50. 6: *per iocum atque vinum*.

[101] Cf. *sal* used of sex-appeal at 86. 4: *nulla . . . mica salis*.

[102] Cf. the title of Bernstein's (1985) article: 'A Sense of Taste'—fitting for his tasteful conclusions about the poem.

[103] 14b. 1–3. Catullus brings his readers into contact with Furius' excrement in similar language: *tu si manibus teras fricesque*, 22.

Martial

With Martial, we are on much more solid ground: material things are less ambiguously the fodder of his writing. Martial himself uses the phrase *materiarum ingenium* (12 pref.) to describe his subject-matter, and his bundle of epigrams has been aptly called 'la poesia degli oggetti'.[104] The poems simulate all kinds of scrappy ephemera: labels, letters, lists, lavatory graffiti,[105] tags for presents (Books 13, 14), and of course their final, ignominious destination, wrappers for take-away food (13. 1. 1–2; 6. 61. 8). By concentrating on sordid minutiae—a dinner made up of leftovers (11. 65), stingy Saturnalian presents (4. 88), or a farm in which not even a cucumber could lie straight (11. 18)[106]—Martial draws attention to the surface worthlessness of the epigram, the most miniature verse-form. He summarizes Homer, Virgil, Ovid, and Livy in a nutshell, in preposterous two-line poems,[107] and is happy for epigram to stay at the bottom of the literary hierarchy.[108] Yet he also redeems the form with generous and inventive representations of meaness.[109] An oxymoron with which he dismisses one of his catalogues of rubbish, *haec sarcinarum pompa* (12. 35. 25: 'this procession of paraphernalia'), also betrays the importance that he has chosen to give it. Martial's success in portraying himself as a 'small' poet is reflected in the dearth of critical literature devoted to him. For example, the accumulated detail of his invitation poems is desperate for attention.

Here again, we need to recognize that Martial's invitations are projections of his literary persona, though the simple equation between a poet's invitation and his poetic style

[104] Salemme (1976).

[105] 12. 61. 8–10: *nigri fornicis ebrium poetam, / qui carbone rudi putrique creta / scribit carmina quae legant cacantes* ('the drunken poet of the dark brothel, who uses coarse charcoal and crummy chalk to scrawl poems for people to read in latrines').

[106] This is a parody of the normal poem of thanks to a patron (e.g. Hor. *Sat.* 2. 6). But it also helps to identify Martial as a poet for whom miniature presents are suitable.

[107] 14. 184, 186, 190, 192. Plin. *NH* 7. 20. 85 records that a parchment copy of the *Iliad* was once literally enclosed in a nutshell.

[108] See 12. 94.

[109] This sort of self-conscious experiment had already taken place in Greek epigram; for complaints about a miserly menu, cf. Rhianus 3256 ff. GP.

made by Race and others again needs modifying or redefining. Certainly, when Martial contrasts his own tiny flute-player (*parvi Condyli*, 3) with a *crassum volumen* (25, 'thick book') in 5. 78, he is using dinner-party entertainment to align himself with small-scale Callimachean poetics. And since he positively embraces material and lumber as subject-matter, the comparison between material feasts and (ethereal) poetry should be less strained. The fact that we have three different invitations from Martial shows, however, that there are ways of varying the presentation according to the poet's relationship with the recipient or recipients. Each private communication, when publicly aired, discloses a particular relationship between host and guest, or author and reader. And critics have not yet considered the vital role that food itself plays in embodying Martial's style of entertaining.

There are other intimations within Martial's poetry, ignored by Race, that a comparison between dinners and poetry is intended. First, Martial frequently identifies his work with the festive spheres of Roman life: the holiday, especially the Floralia or Saturnalia,[110] the games,[111] and the *convivium*. Individual books extend their arms in hospitality: Book 12 is a *cena adventoria*, a dinner to celebrate a homecoming, Book 13 (*Xenia*) an inventory of mottoes to accompany presents between guests and hosts. Some books are themselves personified as drunken revellers,[112] whose frank language sets the mood for Martial's epigrams.[113] Book

[110] Book 11 is especially identified with the Saturnalia (11. 2, 11. 6, 11. 15. 11–12: *versus / . . . Saturnalicios*); cf. 4. 14, 5. 84; Book 14, also, is a catalogue of mottoes for Saturnalian *apophoreta* or going-home presents; see 5. 18: the book as Saturnalian present (cf. *Silv*. 4. 9). Book 1 is compared to the games of the Floralia: (1 pref. 1): *Epigrammata illis scribuntur qui solent spectare Florales. Non intret Cato theatrum meum, aut si intraverit, spectet* ('My epigrams are written for people who watch the Floralia. Keep out of my theatre, Cato; if you must come in, you'll have to watch').

[111] The book of *Spectacula* epitomizes the displays of *monstra* in the circus.

[112] 5. 16. 9: *Conviva commissatorque libellus* ('my book is only a dinner-guest and party-goer'); 11. 15. 5 ff. *qui vino madeat nec erubescat / pingui sordidus esse Cosmiano, / ludat cum pueris, amet puellas* ('Let it be sodden with drink, and not blush to drip with ointments, frolic with the boys, make love with the girls'); 3. 68. 6: *saucia Terpsichore* ('sozzled Terpsichore').

[113] 1 pref.: *Lascivam verborum veritatem, id est epigrammaton linguam* ('Playful truth, that is, the language of epigrams'); cf. 1. 4. 8; 10. 18(17). 3; 10. 87. 7; *ioci*

2, for example, is seen as convivial entertainment for a guest between two stages of drunkenness: the limits of Martial's frankness are nebulous, as though the way we read his poems depends on how drunk we are. He teases prurient Catos and matrons who pretend to be shocked, but turn the pages faster when they see this secondary stage approaching.[114] The convivial or festive setting also helps to make Martial's poems seem trivial or ephemeral: it absolves him from moral responsibility and literary merit.[115]

Secondly, many of the more programmatic epigrams exploit the long tradition of using food-metaphors to advertise the poet's style.[116] It is characteristic of Martial to be alternately boastful and self-deprecating. In 10. 59 he scolds his reader for picking out the dainties among his epigrams and ignoring the solid fodder:

> dives et ex omni posita est instructa macello
> cena tibi, sed te mattea sola iuvat.
> non opus est nobis nimium lectore guloso;
> hunc volo, non fiat qui sine pane satur. (3–6)

A lavish dinner, stacked up from every market, is set before you, but you only like the choice bits. I can do without a gourmet for a reader: I want one who likes to fill up on bread.

Dives et instructa cena invests Martial's impromptu and haphazard work with a temporary, ironic dignity.[117] It may

(jokes): 4. 14. 12; *simplicitas* (candour): 1 pref.; *sal(es)* (wit): 3. 99. 3, 5. 2. 4, 8. 3. 19, 13. 1. 4.

[114] Catos: 1 pref., 10. 20(19). 21; 11. 15. 1; matrons: 3. 68; 5. 2. 1.

[115] 11. 15. 11–13: *versus . . . / Saturnalicios, Apollinaris: / mores non habet hic meos libellus* ('these are Saturnalian verses, Apollinaris: they do not reflect my morals'); cf. 7. 72. 16: *Non scripsit meus ista Martialis* ('My friend Martial didn't write this'). In 9. 89–90, the poems are imagined as part of a dinner-party competition for writing indifferent verse: '*lege nimis dura convivam scribere versus / cogis, Stella?*' '*Licet scribere nempe malos.*' ('Aren't your rules for making your guest write poems a bit strict, Stella?' 'You can always write bad ones').

[116] See Bramble (1974: 45–59); and, specifically on Martial, Preston (1920); and Pasoli (1970–2). They do not consider the role of irony in many of these programmatic statements.

[117] *Instruo* cf. 11. 31. 9: *multiplices struit tabellas* ('he constructs edifices of pastry'); 14. 222: *mille tibi dulces operum manus ista figuras extruet* ('this artist will make you a thousand types of confectionery'). NB the pun on *consumo* (to eat/use up (of writing)/spending) at 10. 59. 1: *consumpta . . . uno lemmate pagina* ('a whole page eaten up with one subject'): compare Juvenal's 'full margins' (1.5).

be relevant that in 12. 48 he asks a friend for *subitae ofellae* (17, 'instant meatballs', 'pot luck'), the sort of dinner he is able to return.[118] Epigrams are dishes in 9. 81 too: *cenae fercula nostrae* ('the dishes of our dinner'). Martial becomes a caterer to diners (*convivis*), popular readers, rather than professional poets or critics (*cocis*).[119]

Elsewhere, the satirical pungency of epigrams becomes the relish to more solid poetry.[120] At 9. 26 the small-scale muse appears in the shape of the humble olive:

> et parvae nonnulla est gratia Musae;
> appetitur posito vilis oliva lupo. (5–6)

Even the small Muse deserves our thanks; we have room for cheap olives after discarding the fish.

At 7. 25 Martial's salty and vinegary epigrams (*salis*, 3; *aceti*, 5) are contrasted with other poets' honey-apples and sweet figs:

> infanti melimela dato fatuasque mariscas
> nam mihi quae novit pungere, Chia sapit. (7–8)

Let children keep their honey-apples and saccharine figs. My Chian figs know how to pack a punch.

Varietas is promoted as a virtue at all costs: small and piquant epigrams alternate with uninspired ones; the work is a mixed bag of assorted goods (*ex omni macello*, 10. 59. 3). The lucky dip of generous and meagre going-home presents in Book 14 is an appropriate advertisement for the epigrams as a whole.[121] In 10. 45, Martial disowns his own anodyne eulogies as though they were bland food, conspiring with his readers in favour of piquancy and candour.[122] When it comes

[118] 10. 48. 15 contrasts *ofellae* with dishes that need a *structor*.

[119] For the poet as host or caterer, see p. 41, above.

[120] Plin *Ep*. 3. 21. 1 names as Martial's virtues salt and bile (*sal* and *fel*).

[121] 14. 1. 5. Cf. the *apophoreta* of Elegabalus (*SHA* 22) and Augustus (*Suet*. 75), lucky dips of vastly differing goods.

[122] *Lene* (1), *dulce* (1), *blanda* (2), *pingue* (3); Martial's rich loin of boar (*ilia Laurentis apri*, 4) and expensive wine are disdained by his guest (note the pun on *stomachus* (6), gastronomic or literary taste), who prefers to drink vinegary wine (*Vaticana*, 15). Martial exploits the contradictory connotations of different kinds of food: *acetum*, vinegar, is an ideal quality for epigram; in moral terms it smacks of excessive asceticism (cf. Hor. *Sat*. 2. 3. 117); similarly, *lenis* and *dulcis* are ideal in gastronomy, but anathema to satirical poetry. Martial may have chosen boar's meat because it had associations with luxury (cf. 7. 78. 3, 9. 14. 3, 12. 17. 4). But I have

to the reader's appetite, it is not surprising that he expects no more than light browsing;[123] and mocks anyone who can stomach a whole book at a sitting.[124]

These explicit analogies between food and different types of poetry suggest that the three invitation poems can also be read as literary programmes. Yet this is not a simple analogy between small-scale poetics and a frugal but decorous spread, as Race and Bramble suggest, but a far more dubious offering. Martial's menus exude impropriety: they are uneven, surprising, lascivious, joking. They are the licensed spirit of the *convivium* in bodily form; they play on the same discrepancies between material and metaphorical consumption, rules and liberty, which are part of the dinner party they pretend to be. Martial's modest invitations and suits for patronage may challenge the reader not to apply economic values to small-scale poetry,[125] but Catullus' intangible perfume is still replaced by gross and perishable food.[126] Exemplary but stereotypical food (lifted piecemeal from Virgil's and Ovid's pastorals) is mixed with comic and ignominious dishes. It is in Martial's menus as much as in his entertainments that we find the uneven lucky dip of his literary personality.

argued already that the Laurentian boar, fat on sedge and weeds, at Hor. *Sat.* 2. 4. 42, is an image of 'fat', that is, turgid and indigestible poetry. Similarly, the phrase used of insipid epigrams at 7. 25. 3, *nullaque mica salis* ('not a grain of salt'), comes from Catullus' epigram contrasting Quintia's bland beauty with Lesbia's essential spark (86), which I argued was a justification for small-scale but pungent poetry. Martial's borrowing of these phrases in an explicitly literary context supports my argument that they have a double meaning in their original setting.

[123] 13. 3. 8: *praetereas, si quid non facit ad stomachum* (readers are invited to skim through a list of food mottoes, passing by poems/dishes that do not take their fancy); cf. 10.1.

[124] 1. 118 (the last poem in the book): *cui legisse satis non est epigrammata centum, / nil illi satis est, Caediciane, mali* ('the man who isn't satisfied with a hundred epigrams, Caecidianus, will never get his fill of rubbish').

[125] Cf. 6. 48, where he criticizes excessive concern with food: it is not the host, but his food, that speaks to the guests (*non tu, Pomponi, cena diserta tua est*); cf. 9. 14: *amici* are suspect when they attend rich dinners; they are there for the food, not the host.

[126] 'Realism' is actually a misleading label for Martial's self-consciously grotesque style of writing. See Salemme (1976: 99): 'Possiamo noi parlare di "realismo" quando l'oggetto è colto nei suoi aspetti più groteschi e più deformati, quando cioè diviene più strumento di riso, piegato a un acrobatico gioco umoristico?'

1 Martial 5. 78

The first invitation draws on Catullus 13 for its urbane gestures of apology. Ironic deference to the guest[127] (*προπίνειν* (have pre-dinner drinks), *soles* (be in the habit), *domicenio* (eating in) invest Toranius' lonely dinner at home with blasé chic)[128] ushers in Martial's own 'starvation' menu (urbane litotes shapes *si* . . . / . . . *potes esurire mecum*, 1–2[129]). Toranius' grim lonely supper, *tristi domicenio* (1), promises *ioci* to come.[130] Martial's menu, a jostling queue of good and bad elements, pays lip-service to the structures of poetic variety and description. *Viles Cappadocae* (4) are, we are told, lettuces,[131] but they look at first sight like 'cheap Cappadocian slave-women', the dross of the Roman emporium.[132] Martial's meal is instantly linked with the bargain-price market: this is no unbought or purely Roman meal.[133] With *gravesque porri* (4), the reek of leeks wafts into the poem,[134] adding a token heaviness to the light and worthless lettuces. This word also begins to compromise any perfect match between light food and light entertainment (at 29 delicate flute-playing is *non grave*). The next dish is more artfully arranged: *divisis cybium latebit ovis* ('tuna lurks between chopped eggs'). *Divisis* and *ovis* are themselves divided in an oval structure to surround the cheap fish

[127] e.g. the repeated, mock-solicitous *si* (1, 3, 11, 17). Martial uses Toranius again at 9 pref. to parody the formulae of dedication: he greets Toranius but only to tell him that he is dedicating the book to another Toranius, who was kind enough to want Martial's bust in his library. His final words to Toranius there (*para hospitium*: 'be ready to play host') suggest that he is demanding a material return for his invitation poem.

[128] For dining at home as a sign of social failure, cf. Hor. *Sat*. 2. 7. 29–32, Mart. 5. 47, 11. 24. 15.

[129] *Λιτότης* of course originally meant 'meagre diet' (see Cic. *Fam*. 7. 26. 2); cf. Cic. *Orat*. 123 for *ieiunus*, thin hungry, of style, as opposed to *satur*.

[130] Cf. 29, *nec grave sit nec infacetum* ('neither serious nor humourless'). See Fraenkel (1932–3) for the principle that such adjectives are programmatic.

[131] Cf. Col. 10. 191.

[132] *Cappadox* (*OLD* s.v. 2) can be applied contemptuously to any Asiatic, especially a slave (Cic. *Flacc*. 61, Petr. *Sat*. 69. 2, Cic. *Red. Sen*. 14). Cf. Mart. 6. 77. 4, of slave litter-bearers. 9. 100. 5: Martial's toga is *vilis*

[133] Cf. 10. 59. 3: *ex omni macello*. Contrast Juvenal's homegrown menu at *Sat*. 11. 64–5: *fercula nunc audi nullis ornata macellis*.

[134] *Gravis* can mean 'strong-smelling'. Cf. 13. 18. 1: chives are described as *fila* . . . *graviter redolentia*.

lurking bashfully in the middle.[135] Like *grave, latebit* (lurk) is in pointed contrast with the programmatic section later, where the poet advertises the openness of his party (*sed finges nihil audiesve fictum*, 23). Even so, it also suggests the shrinking modesty of Martial's books (*timidumque brevemque libellum*, 12. 11. 7).

The *cena* proper contains more throwaway *varietas*, parodying the art of describing with banal epithets: *ustis* (6, burnt) and *algentem* (8, chilly); *nigra* (7, black), *virens* (7, green), *niveam* (9, snow-white), *pallens* (10, pale) and *rubente* (10, red). Another line mimics the arrangement of its contents: *nigra coliculus virens patella* (7, 'green broccoli on a black plate'). If we look more closely, however, these contents defy decorum. The black plate (*nigra patella*) looks like a variant on Horace's *modica patella* (*Ep.* 1. 5. 2), *puris catillis* (*Sat.* 2. 4. 75) and *angusto catino* (*Sat.* 2. 4. 77), which stood as images of a small poetic range, and, in the last two cases, alluded to the origins of *satura* as well. Here, the notion of modest writing (suggested by two diminutives) is taken to extremes of self-denigration.[136] In the context of Martial's other poetry, however, *niger* denotes not only humble blackware pottery, but also 'blackened', from the air in smoky, or dirty places (the word is also used of restaurant kitchens, coins, ovens, cookshops, smoke, and reputations).[137] The atmosphere of the *salax taberna* is suddenly piped into Martial's dining-room.

This already suggests that all is not what it might be. The menu to come—sprouting broccoli, beans and bacon, sausage and porridge—look innocent enough, the staples of the Roman peasant's diet, and a nostalgic gesture from a dis-

[135] For the pathetic fallacy of shame cf. Alexis *Crateias* (Ath. 3. 107c = 2. 335K): a liver is wrapped in its own caul, ashamed of its colour (presumably the colour of blushing); cf. Crobulus (Ath. 3. 107e = 3. 381K): the shamed liver of a dung-eating boar.

[136] Other diminutives: *botellus* (9), *cenula* (22, 31), and Martial's references to his books as *libelli* (11. 15. 3, 4, 13; 11. 16. 7), in the tradition of self-deprecation (cf. Cat. 1. 1, Juv. 1. 86).

[137] *Culina*, 3. 2. 3, 10. 66. 3; *monetae*, 1. 99. 13; *fornex*, 12. 61. 8; *popina*, 7. 61. 8; *fama*, 10. 3. 9; cf. 2. 90. 7, *fumi*. Black salt (*sale nigro*) is used of coarse poetry at Hor. *Ep.* 2. 2. 60.

enchanted city-dweller.[138] So far they have only been read as examples of simple-meal description or, by extension, as metaphors for the plain style.[139] Book 5, we find, is dedicated to matrons and children, but this is quite clearly with tongue in cheek: the book is twice linked with the Saturnalia, for example (5. 18, 84). The usual link between Martial's poems and riotous dinners, together with the programmatic rejection of *tristi domicenio*[140] and the smoky suggestions of *nigra*, give us enough of a licence to read innuendo into the dishes that follow. We have seen that Martial portrays his books as revellers lapsing gradually into verbal frankness. Terpsichore, for example, gets drunk and begins to utter obscenities in the place of euphemisms or *double entendres* (*schemata dubia* 3. 68).[141] Book 5 may be for matrons, but the central dishes, apparently innocent, are still laden with innuendo, *schemata dubia* typical of a lascivious poet and designed to tickle ribald readers. According to taste, they can be decorous or obscene.

To start with, *coliculus*, broccoli, could be a phallic metaphor. Adams lists it as such along with *caulis*, but he objects that the words are not current enough to be classified as common slang.[142] This only highlights the difficulties in pinning down a sexual vocabulary when innuendo is by definition only suggestive.[143] Elsewhere in Martial, *coliculus*

[138] Cabbage: Hor. *Sat.* 2. 1. 74, 2. 2. 117, 2. 6. 64, *Ep.* 1. 5. 2; Virg. *Georg.* 130, Ov. *Met.* 8. 647; bacon: Hor. *Sat.* 2. 6. 64, 85; Ov. *Met.* 648; beans: Hor. *Sat.* 2. 6. 63; porridge: Varro, *LL* 5. 105, Val. Max. 2. 5. 5, Juv. 11. 58.

[139] See Bramble (1974: 47–8) on these types of food in Horace, Ovid, and Juvenal.

[140] Cf. 29, *nec grave sit nec infacetum*; 10. 18(17). 3: *sollemnesque iocos nec tristia carmina*.

[141] Cf. 11. 15. 5–10.

[142] Adams (1982: 26 ff.): Adams sees Lucil. 281M = 305W, *praecidit caulem testisque una amputat ambo* ('he chopped off his brussels sprout and both his balls with one blow'), as the only unambiguous occurrence of the word; Celsus' medical use of *colis* is only a calque on Greek *καυλός* (though this itself is an organic metaphor meaning 'stalk'); Petr. *Sat.* 132. 8 *languidior coliculi*, in Encolpius' ('the crotch') address to his wilting organ, is only a simile. But cf. also Plaut. *Cas.* 911: *num radix? . . . num cucumis?* ('was it a radish, was it a cucumber?')

[143] For food suggesting sex elsewhere in Martial, cf. 3. 82. 21, *turturum natis* (pigeons' rumps); 9. 2. 3, *siliginei cunni* (fondant fannies); 14. 70, the *inguina* (privates) of a pastry Priapus (cf. Petr. *Sat.* 60. 4). See Adams (1982: 138 ff.) on Martial's eating-vocabulary for sex, e.g. 9. 63. 2: Phoebus is asked to dinner so that the company of *cinaedi* can dine on his *mentula*.

is curiously connected with feelings of embarrassment. At 13. 17 he advises people who are ashamed of looking at pale *coliculi* to boil them in nitrate to make them green.[144] At 14. 101, he sighs (*pudet heu!*) when telling us that one can use a rather special mushroom-boiler for cooking sprouts. The *coliculus* in 5. 78 is green and apparently less suggestive, but *virens* can also mean 'flourishing, at the height of its (sexual) powers'.[145] And the phrase *digitis tenendus ustis* ('to be held in scorched fingers') tempts the salacious reader further.[146] In the light of this, the next dish, sausage and porridge, is no less suggestive: *et pultem niveam premens botellus* ('and a sausage pressing hard on snowy porridge'). Martial uses *premere* elsewhere of (male) sexual embraces.[147] Here a (male) sausage squeezes snowy (female) porridge.[148] Finally, it comes as no surprise to find that Catos and matrons are themselves brought in as reluctant voyeurs at this bawdy display, in the shape of the third dish: *et pallens faba cum rubente lardo* (10, 'and pale beans with blushing bacon'). *Pallens* and *rubens* may look like gratuitously 'poetic' or 'ecphrastic' epithets, but they also personify the foodstuffs: the beans and bacon are a prudish couple, one pale, the other blushing to see the vigorous cabbage-phallus and the sausage embracing the porridge.[149]

The *secunda mensa* is a hotchpotch of mean and pompous food. *Marcentes uvae* (12, 'tired or withered grapes') is an unflattering periphrasis for raisins. The pears are saddled with a bathetic grandeur: *nomen . . . quae ferunt Syrorum* (13). Syrians, like Cappadocians, are usually slaves and litter-bearers in Martial:[150] like their namesakes, the pears are bearers—of an ignoble name. Chestnuts are infused with

[144] *Ne tibi pallentes moveant fastidia caules / nitrata viridis brassica fiat aqua.*

[145] *OLD* s.v. *vireo*, 2b.

[146] Cf. 11. 46. 3: *truditur et digitis pannucea mentula lassis* ('the shrivelled cock is tickled by jaded fingers').

[147] e.g. 3. 58. 17, 4. 4. 4, 12. 96. 2.

[148] For *niveus* of purity: cf. Cat. 64. 36. 4, *niveos . . . virginis artus*; Mart. 8.73. 2, *nivea simplicitate*.

[149] Martial's readers at 6. 60. 3: *ecce rubet quidam, pallet . . .* ('Look, they're all blushing, turning pale'). The revelling book at 11. 15. 5 has no such shame: *nec erubescat*.

[150] 10. 6. 2, 9. 2. 11.

Virgilian vapours: *quas docta Neapolis creavit, / lento castaneae vapore tostae* (14–15, 'chestnuts produced in learned Naples and roasted on a slow fire');[151] the wine is plonk: *vinum tu facies bonum bibendo* (16, 'You will improve the wine as you drink it').[152]

After this list of dubious goods,[153] Martial anticipates his reader's satiety, but presses on, overloading the poem with throw-away material:

> post haec omnia forte si movebit
> Bacchus quam solet esuritionem (17–18)

After all this, if Bacchus does his usual trick of stirring your hunger again . . .

Forte si movebit seems to invite vomiting or defecation,[154] but the list marches on relentlessly. Noblemen clash with the humble litter-bearers: with *nobiles olivae . . . et fervens cicer et tepens lupinus* ('noble olives, heated chick-peas, and warm lupins'), the auxiliary troops are marshalled, and the host sweeps a bow to the needs of his imperious guest.[155] *Fervens* and *tepens* provide more *varietas* within a small range, an almost haphazard allotting of adjectives (at 1. 103 it is *cicer* that is *tepidum*); *tepens* suggests a rather damp fading-out after the heat of the rescue.

Martial's final question is provocative: *Parva est cenula—quis potest negare?* (22, 'The supper is small—who can deny it?'). In one sense, given the meagreness of its individual ingredients, the dinner *is* small; at the same time it has distended the thin epigram with its length. In the list of entertainments, the clichés of Callimachean poetics become clear: the small flute-player contrasted with the large book,

[151] Cf. *Ecl.* 1. 81, *castaneae molles*; 1. 4, *lentus in umbra*.

[152] Cf. 1. 105: Martial's Nomentan wine adapts itself to any label (a parallel for his own versatility?). Apologies for wine: cf. Hor. *Od.* 1. 20. 1, *vile potabis*; Petr. *Sat.* 39. 2, *suave faciatis*; 48. 1, *vos illud oportet bonum faciatis*.

[153] Like Martial's *apophoreta* (11. 1. 5) the dishes are a mixed bag or lucky dip.

[154] *Movere* (of bowels): Mart. 11. 52. 5 on the uses of lettuce (*ventri lactuca movendo / utilis*); cf. Col. 6. 30. 8.

[155] Matron's *Deipnon* (Ath. 4. 134d–137c) is the *locus classicus* for the confusion of military and gastronomic vocabulary.

the claims of truthfulness.[156] Flamboyant flamenco dancers are banished from the dinner.[157] But are they really? Martial himself is Spanish and lascivious.[158] The lewd description is a parody of innocent disclaimers; Martial has his cake and eats it:

> nec de Gadibus inprobis puellae
> vibrabunt sine fine prurientes
> lascivos docili tremore lumbos. (26–8)

There won't be girls from naughty Cadiz, tickling endlessly, wiggling their hips with practised manœuvres.

The final summary, *haec est cenula* ('here you have the supper'), has a throw-away 'take it or leave it' flavour. Martial keeps the link with Catullus by referring to female guests. But instead of being the climax, the true spice of the dinner, the women appear as something of an afterthought, a series of almost arbitrary couplings: *Claudiam sequeris. Quam nobis cupis esse tu priorem?* (31–2, Toranius is invited to choose Martial's escort), which takes its lead from the arrangement of the menu. We are thrown back on the supper itself as the real substance of the poem.[159]

2 *Martial 10. 48*

In another poem, the lone guest is exchanged for a crowd of poets and patrons:[160] L. Arruntius Stella and five other recipients of Martial's poems, Nepos, Canius, Cerealis, Flaccus, and Lupus.[161] Martial returns his obligations to his

[156] Cf. 8. 3. 21: Martial contrasts his own constricted pipe (*angusta avena*) with epic trumpets (*tubae*); cf. Virg. *Ecl.* 1. 2: *tenui avena*.

[157] See Pliny. *Ep.* 1. 15 and Juv. 11. 162–4 for these as tasteless forms of entertainment.

[158] On dancers and sexuality, see Adams (1982: 194) and cf. *Copa* 1 f.

[159] The meaning of these last lines is obscure: Greenough (1890: 191–2), argues that *quam* is interrogative; Jackson (1883: 25–6), had suggested that ring-composition was likely, and proposed *haec est cenula. Claudiam sequeris, / quam noris. cupis esurire mecum?* ('Here you have the supper. Claudia, whom you know, will be your escort. Are you still willing to starve *chez moi*?').

[160] Cf. Hor. *Ep.* 1. 5. 29 on the undesirability of crowds at dinners: *nimis arta premunt olidae convivia caprae* ('smelly goats at a dinner cramp one's style'). Varro (ap. Gell. 13. 11) sets the ideal number at between three and nine.

[161] Nepos: 6. 27; Canius: 1. 69; Flaccus: 8. 45, 11. 27; Lupus: 5. 56, 10. 40, 11. 18, 11. 88; Cerialis: 11. 52.

supporters in one mass party, squeezing six guests on to one couch. The poem is a series of jokes on the theme of excess and distortions of scale: numbers and sizes which are immoderate in one context become inadequate in another. The public world of Egypt and Rome, temples, soldiers, and lawcourts, is banished to the furthest margins of the poem (1–2, 24); this larger world is cramped compared with the liberty of the dinner couch.[162] The time of the dinner party, for example, the eighth hour, marks the ideal temperature of the neighbouring baths of Nero (*temperat haec thermas*, 3), by contrast with the excessive heat of the sixth and seventh: *nimios vapores* (3), *inmodico Nerone* (4): Nero's baths loom over the dinner like an immoderate tyrant. The number six is hardly sufficient, though, when it comes to dinner-guests: *septem sigma capit, sex sumus, adde Lupum* (6, 'The S-shaped couch takes seven, we are six: add on Lupus'). The sixth verse becomes a figure for its subject: the S-shaped *sigma* is suggested by sibilants within the line; there are six words in line 5, seven in line 6, where the seventh guest is added, but the pentameter is a shortened version of the hexameter before, so numbers are being confused, rather than ordered in any scheme. These alphabetic and arithmetical games introduce the spirit of convivial behaviour into the poem.[163]

The poem has been described as 'sober and unlikely to mislead'.[164] This is on the basis of the information it contains about the productivity of Martial's Nomentan farm (the bailiff's wife in 7; the Nomentan wine in 19). Elsewhere (for example in 7. 31, and of course in 5. 78), Martial claims to buy provisions in the Roman market; 11. 18 is a series of variations of the sterility and smallness of the suburban farm given to him by Lupus.[165] With Stella, his most important benefactor, present, it is in Martial's interests to look like a prosperous landowner. The small garden, mocked in 11. 18, now becomes an image of variety:[166] *varias quas habet hortus*

[162] Cf. *turba* (1), *pilata cohors* (2).

[163] Cf. *SHA Antoninus Geta* 5. 8 for dinners according to the letters of the alphabet.

[164] Saller (1983).

[165] The butt of jokes in 10. 48 as well (6, 14).

[166] See above p. 47 on *Moretum* 62: *exiguus spatio, variis sed fertilis herbis*.

opes ('all the varied produce of the garden'). In other ways, however, the poem is anything but sober. Convivial licence bubbles up at the end in Martial's remarks about the conversation:[167]

> accedent sine felle ioci nec mane timenda
> libertas et nil quod tacuisse velis (21–2)

add jokes without bile and freedom to speak without regrets tomorrow and no secrets anywhere . . .

The menu, too, contains this spirit, though it has in the past been read only literally. The capacity of food to invoke a mood or to give clues has once again been underestimated.

The first dish, for example, laxative mallows (*exoneraturas ventrem . . . malvas* 7[168]), has made one commentator rush to conclude that the Romans were obsessed with bowel movement.[169] He compares Trimalchio's mutterings about his constipation, and the picture from Ostia of philosophers at dinner uttering scatological aphorisms.[170] But the fact that the subject appears so frequently in Latin literature does not mean that it was any more acceptable than it is today.[171] Both texts cited are assuming a licence in defiance of social convention. Both are parodies, of polite manners or moral philosophy. Martial's use of the subject, too, is a sign of the licence he has assumed. All three passages have a *convivium* as their setting, which suggests that the dinner party was a possible, if not entirely appropriate place for the discussion of bodily functions. The word *exonerare*, to unload, can also be applied to the person defecating,[172] and the ambiguity of *exoneraturas* (whose stomachs are the mallows going to unload?) suggests that the vegetables are themselves acting like rude guests who defecate before or during the dinner.[173] Bowel freedom is part of the *moral* reasoning behind the

[167] See Race (1978: 187).

[168] Cf. Hor. *Epod.* 2. 57–8: *gravi / malvae salubres corpori.*

[169] Kay (1985) ad 11. 52. 5.

[170] Petr. *Sat.* 47. 2–4; illustrated in Meiggs (1973: 429).

[171] Cic. *Fam.* 9. 22. 5 comments on the Stoics' acceptance of farting and belching as prerequisites of personal freedom; this is clearly in conflict with polite behaviour.

[172] e.g. Sen. *Ep.* 82. 12: *ad exonerandum ventrem secessit.*

[173] Cic. *Fam.* 7. 26. 2 personifies the mallows and beet that gave him diarrhoea: *a beta et a malva deceptus sum.*

meal; it is convivial *libertas* taken to an extreme.[174] It is also a comic subject. Opening the menu, the defecating mallows are a token of licensed behaviour and of licentious conversation. No sooner have the guests been squeezed on to the couch (*capit*, 6) than they are going to be unloaded.

All the other ingredients can be interpreted as guests too. *Lactuca sedens* (9) is usually translated 'squat'. But once again the adjective also suggests the laxative qualities of lettuce, about which Martial is more explicit elsewhere (*sedere* is one of his euphemisms for *cacare*).[175] The active use of the verb suggests that the lettuce is itself squatting before dinner.[176]

Tonsile porrum (9), 'the shaven leek', is a novel way of describing *porrum sectile*, cut leeks or chives. But is there also a point to 'shaven'? It is apparently not known when the Romans shaved, though they may well have waited until late in the day like modern Mediterraneans.[177] Two classical references do suggest that shaving was part of the general process of washing and anointing before dinner.[178] So *tonsile* looks like another joking reference to dinner-guests performing their toilet. *Ructatrix mentha* (10), 'mint to make one belch', or, more literally, 'belching mint', again personifies the spirit of bodily freedom and freedom to talk about the body that characterizes the frank *convivium*.

The next item, *herba salax* (10), eruca or rocket, is a well-known aphrodisiac;[179] the principal meaning of *salax* is lascivious, highly-sexed (from *salire*, to leap; a contrast with *sedens*). The preoccupations of this guest are obvious. At 11. 25 Martial describes the *mentula* as potentially *salax*:

[174] Cf. 21–2: *nec mane timenda / libertas*.

[175] Cf. 11. 52. 5–6: *ventri lactuca movendo / utilis*; 3. 89. 1, where it is coupled with mallows: *utere lactucis et mollibus utere malvis*. *Sedere*: Mart. 11. 77. 2.

[176] *Sedere* can have another obscene meaning, 'to take up one's position as a prostitute' (see Herescu (1960) and Booth (1980)): the lettuce might be another type of person associated with the more permissive *convivium*.

[177] See Balsdon (1969: 20).

[178] Varro ap. Non. 179M: *sunt circumtonsi et terti atque unctuli* ('they are clipped, scrubbed and anointed'); Quint. 1. 6. 44: *lavamur et tondemur et convivimus ex consuetudine* ('we wash, shave, and dine as usual').

[179] Cf. Ov. *Ars Am*. 2. 422. Martial links it at 3. 75 with other aphrodisiac plants, bulbs and savory: cf. 13. 62. 2: pigeon as an anaphrodisiac.

illa salax nimium nec paucis nota puellis
stare Lino desit mentula. lingua, cave.

That saucy old cock that all the girls know—Linus can't get it up any more: watch out, tongue.

Mentula, as we have seen, has the status of a taboo, the ultimate forbidden word which emerges in drunken convivial conversation,[180] served up on a plate in the disguise of a *virens coliculus* at 5. 78. 7. As well as picturing belching and lecherous diners in miniature form, this line sketches out a teasingly euphemistic reference to the salacious *mentula* with the words *mentha . . . salax*. Cicero, in his letter on euphemism, tells us that *menta* (mint) is an acceptable word, while its diminutive is not.[181] Once again, Martial narrowly protects himself against the matrons' disgust: the menu can be read innocently on the surface.

By line 11, the party is in full swing; a dish of mackerel and chopped eggs depicts a banqueter in mid-carouse: *secta coronabunt rutatos ova lacertos* ('chopped eggs garnish mackerel with rue'). *Coronabunt*, used here of a garnish,[182] suggests the garlands associated with the *convivium*.[183] *Lacertus*, mackerel, adds to the confusion between foodstuffs and diners with its pun on *lacertus*, arm.[184] Rue (*rutatos*) is a byword for bitterness,[185] which seems to offer a pointed contrast with the rejection of bitter conversation at 21 (*sine ioci felle*).[186] Once again, *secta* and *ova* (chopped eggs) are

180 11. 15. 8–10; 3. 68. 7–10.

181 Cic. *Fam.* 9. 22. 3.

182 Cf. 13. 35. 2: *grata corona* (for sausages).

183 e.g. 3. 68. 5, *rosas*; 3. 65. 8, *madidas nardo passa corona comas* ('a garland that has felt hair drenched in ointment'); 9. 90. 6, *frontem sutilibus ruber coronis* ('a forehead rosy with entwined garlands').

184 For *lacertus* as a meagre fish, cf. 11. 52. 7, *tenui lacerto*; 11. 27. 3, *tenuem lacertum*; 7. 78. 1, *coda lacerti* (as take-away food at the baths, cf. 12. 19. 1). Cf. Crassus' joke (Cic. *de Or* 2. 59. 240): *comedisse eum lacertum Largi* ('he has eaten Largus' arm/mackerel'): Cicero calls this joke *salsa*, another pun in the context.

185 e.g. at Cic. *Fam.* 16. 23. 2.

186 The expression *in rutae folium conicere* is usually translated 'to knock into a cocked hat'. Smith ad Petr. *Sat.* 37. 10, 58. 5, where it occurs, argues that there is no evidence for Friedländer's claim (1886: ad loc.) that *in rutae folium* was proverbial of a small space, instead, he translates: 'to beat up with a leaf of rue'. However, in Martial at least, rue does seem to be proverbially small. At 11. 18. 4 it is a plant in a tiny garden, parodically described as a grove of Diana; at 11. 31. 17 it is a comic

split from each other in the line: more of Martial's graphics.[187]

The final dish is the apogee of these banqueting images: *et madidum thynni de sale sumen erit* (12, 'a sow's udder will drip with tuna brine'). Sow's udder (*sumen*) is perhaps the most unexpected dish served here. Elsewhere, Martial lists it with other luxurious dishes.[188] At 7. 78, for example, it is specifically contrasted with *lacertus*, a mean fish; here, the *sumen* is dripping with brine (*muria*, classed as a mean present at 13. 107).[189] The puzzling combination suggests a dinner for all tastes and means.[190] The *sumen* can also be taken as an image of generosity or *ubertas*;[191] it may well also be a gesture towards Stella's grandness.[192] But the most significant aspect of the line is the way in which the foodstuff embodies conviviality. *Madidus* is linked with the *convivium* through three of its meanings. It can mean 'wet or dripping', of food;[193] 'drunk, inebriated';[194] or 'dripping with ointment'.[195] In some cases the different meanings can be blurred; for example, at 1. 41. 6 it is unclear whether a *madidus conviva* is wet with ointment or drunk on wine.[196] There is also a fourth, figurative meaning: 'steeped' (in

expression for the exiguous wrappers used to hold perfume (*[condat] in rutae folium Capelliana*). *Ova*, incidentally, are mentioned after *herba salax* at Ov. *Ars Am.* 423.

[187] Cf. 5. 78. 5 (*divisis . . . ovis*).

[188] 2. 37. 2; 7. 78. 3, *sumen aprum leporem boletos ostrea mullos*; cf. 9. 14. 3, 12. 17. 4, 11. 52. 13, 12. 48. 9. Cf. Pers. 1. 53, *calidum sumen*; Philodemus, *Anth. Pal.* 11. 44, οὔθατα.

[189] Martial describes it by periphrasis as *sale thynni*. Cf. Quint. 8. 23: an orator apparently used *duratos muria pisces* (fish preserved in brine) as a euphemism for some other less elegant word (possibly *salsamenta*).

[190] Cf. 14. 1. 5 on the *Apophoreta*: *divitis alternas et pauperis accipe sortes* ('Draw these lots, rich and poor by turns').

[191] Cf. 13. 44–5: *esse putes nondum sumen; sic ubere largo / et fluit et vivo lacte papilla tumet* ('You would think this was not yet a dish of sow's udder, so generously do the teats swell and drip with living milk').

[192] 11. 52. 13: *sumina* are among those foodstuffs that even Stella would only serve occasionally (*Stella solet rara nisi ponere cena*). The single *sumen* could be seen as the most precious offering from Martial's Nomentan farm.

[193] e.g. 3. 2. 4, *cordylas madida tegas papyro*; 5. 39. 3, *Hyblaeis madidas thymis placentas*; 1. 41. 6, *madidum cicer*.

[194] 1. 70. 9, 6. 89. 2, 9. 22. 11, 9. 73. 5, 9. 6. 11, 11. 15. 5, 14. 64. 1.

[195] 3. 65. 8, *madidus nardo passa corona comas*.

[196] Cf. 14. 1. 9 on the Saturnalia: *madidis diebus*.

learning or art). Horace uses this in another convivial context (*Socraticis madet / sermonibus*) and Martial puns on the double meanings of *madidus* and *ius* when he describes someone steeped in the law (*iure madens*).[197] The udder is drenched with salt fish (*sale thynni*), and the point of this humble ingredient now becomes clear: *sal* is of course the standard Latin metaphor for wit,[198] and, again and again, the chief flavour of Martial's own witty poetry.[199] *Madidus* and *sal* occur together in figurative senses at 6. 44. 2, where Martial accuses Calliodorus of thinking he is steeped in wit: *solum [te] multo permaduisse sale*. So the fourth sense of *madere* is also in play in the description of the *sumen* at Martial's party. The dish is drunk, anointed, drenched in salt, and also steeped in wit, summing up the material and abstract aspects of the convivial spirit, and the flavour of Martial's writing.[200]

The description of the *gustus* (*gustus in his*, 13) can be read, then, not only as a catalogue of simple food, but also as a metaphorical *gustus* or foretaste of the dinner.[201] Not only will the food excite a festive mood in the diners: that mood is actually rooted in the food itself, jokingly characterized as a party of dinner-guests, squatting, shaven, belching, lecherous, garlanded, drunken, and witty. These grotesque descriptions invert the ancient belief that humans absorb the characteristics of their food or its environment,[202] and they also anticipate the promise of *ioci* at 21. For the second time, Martial skates over the embarrassment of describing his own food by making it humorous. The food catalogues in 10. 48

[197] Hor. *Od*. 3. 21. 9; Mart. 7. 51. 5; cf. 1. 39. 3, *Cecropiae madidus Latiaque Minervae artibus*; cf. 6. 44. 2 (see below).

[198] e.g. Cic. *de Or*. 1. 34. 159; and Horace's *sale nigro* (*Ep*. 2. 2. 60, again in the context of a convivial metaphor for writing); Quint. 6. 3. 18–19.

[199] e.g. at 13. 1. 4, *novos sales*, 3. 99. 3, *innocuos sales*; 5. 2. 4, *sales nudi*. Cf. 7. 25. 3 on insipid epigrams: *nullaque mica salis*; 3. 20. 9: Martial's salty epigrams contrasted with Attic wit; 8. 3. 19, *at tu Romano lepidos sale tinge libellos*.

[200] Cf. e.g. 4. 14. 2, *lascivis madidos iocis libellos* ('books drenched in naughty jokes'), 8. 3. 19, *Romano lepidos sale tinge libellos* ('dip your charming books in Roman salt').

[201] *Gustus* as foretaste, sample: e.g. Plin. *Ep*. 4. 27. 5; Val. Max. 3 pref; Col. 11. 1. 2.

[202] See Gourévitch (1974).

and 5. 78 vindicate themselves on a close reading; there is more to them than at first meets the eye.[203]

There have been six lots of food so far, so the meal is not yet complete, if we follow the principles of line 6: *sex sumus, adde Lupum*. Lupus is added here too, in the shape of a mocking reference to pastoral meals: *haedus inhumani raptus ab ore lupi* (14, 'a kid snatched from the jaws of an inhuman wolf').[204] Here animal and man are confused: Lupus, teased as a glutton, is indeed 'inhuman'.[205] Meatballs are to follow: *et quae non egeant ferro structoris ofellae* (15, 'meatballs which need no carver's knife'). The contrast is between elaborate and impromptu meals (*subitae ofellae* is Martial's term for potluck at 12. 48. 17), but we can also see a comparison between types of writing. *Ferro* suggests the military epic; *structor* also has associations with grand designs.[206] Martial's throw-away epigrams have no need of pompous epic organization[207] (there is also a contrast between the small informal dinner and the ominous military presence at the beginning of the poem). Coarse food is added in the next line; but there is self-conscious wordplay in *faba fabrorum* (16). Martial allies himself with humble craftsmen, not heroic carvers: *faba*

[203] A similarly trivial-seeming, but significant catalogue in the 'Ithaca' section of *Ulysses* is discussed by Tucker (1984: 134). It describes the contents of a kitchen dresser opened by Bloom on the night after Molly has slept with Boylan. Not only are the half-finished pots and bottles the remnants of Molly and Boylan's feast (and as such a token of their presence there), but verbal clues within the description betray Molly's act of adultery. Tucker comments: 'It purports to be a straightforward, objective, denotative listing of items, but in fact the words used are highly connotative to the reader. She cites 'a half-empty bottle of . . . port, half disrobed of its swathe of coral-pink tissue paper' and 'a quarter of soured adulterated milk'. Another example is 'an oval basket bedded with fibre'.

[204] Cf. Hor. *Ep*. 2. 60, *haedus ereptus lupo*; cf. Prop. 4. 4. 54, *inhumanae lupae*.

[205] *Lupus* is also the name of a large fish (the bass); Martial seems to use his patron as the antithesis of his own small self (it was Lupus who gave him the small garden at 11. 18).

[206] *Ferrum* as metonym of war: Lucan, 9. 245; Stat. *Theb*. 4. 145. *Struere* of writing a grand work: Hor. *Ep*. 1. 3. 6, Ov. *Pont*. 2. 5. 1 (also a military image); of arranging lavish dinners: Mart. 3. 45. 3, 11. 31. 9, 14. 222. 1.

[207] *Instructa* is used parodically at 10. 59. 3 of Martial's feast of epigrams. Cf. the ironically chic *mattea* (a small dainty), used of Martial's *bons mots* in the next line, instead of the pejorative food images used to describe scraps of learning taken away from lectures or Senecan *sententiae*: *esca, frustum, prunuli, globuli* (see Bramble (1974: 143 ff.)).

fabrorum prototomique rudes ('blacksmiths' beans and backward sprouts').[208] A more mundane kind of survivor, a ham that has lasted three dinners,[209] parodies the romantic picture of the kid saved from the wolf's jaws (17).

Martial's dinner is unabashedly makeshift; he piles up dishes without discrimination, jumbling the relics of pastoral with the remains of yesterday's dinner into a muddle of surprises and inconsistencies.[210] The feasters are offered food even when they are full: *saturis mitia poma dabo* (18, 'when you are full I shall ply you with sweet apples'). The salt and rue of the *gustus* are varied with sweet flavours; there is a similar parallel between the wine, *sine faece* (19, 'without sediment'), and the jokes, *sine felle* (21, 'without bile'). *Saturis mitia poma* conflates two lines from the *Eclogues*: 1. 80 (*mitia poma*) and the last line, 10. 77: *ite saturae . . . capellae*. Martial evokes the notions of due limits contained in Virgil's book, only to prolong the list with further additions, treating abstract convivial qualities as though they were part of the catalogue: *accedent sine ioci felle nec mane timenda / libertas et nil quod tacuisse velis* (21–2). Ring composition recalls the threatened liberty of the outside world. The guests are free to discuss the circus, a subject removed from serious political slander, and a good match for the festive atmosphere of the epigrams.[211]

de prasino conviva meus venetoque loquatur,
nec faciunt quemquam pocula nostra reum. (23–4)

My guest will chat about the Greens and the Blues: my wine-cups will not send anyone to court.

[208] Cf. 13. 13. 1, *fabrorum prandia* (artisans' lunches); Col. 10. 113, *quaeque fabis habilis fabrilia miscet* ('all the crafty things a clever cook mixes with beans'); Pers. 5. 18, *plebeia prandia* (plebeian lunches). Cf. Mart. 9. 11. 11, *versu . . . non rudi* ('not unsophisticated verse'): Martial can also admit to urbanity in another context.

[209] Cf. Petr. *Sat.* 66. 7.

[210] On surprise or παρὰ προσδοκίαν as a feature of the *convivium*, cf. Petr. *Sat. passim*; the first recipe in Apicius is for *conditum paradoxum* (1. 1).

[211] Cf. the *spectacula*, epitomized in Book 1. Disclaiming malice, see e.g. 1 pref.: *salva infimarum quoque personarum reverentia ludant . . . absit a iocorum nostrorum simplicitate malignus interpres* ('the playfulness of my books does not compromise even the lowest classes . . . any spiteful interpreter of my candid humour should keep his distance'); cf. 5. 15, 7. 72, 10. 3, 10. 5, 10. 33. See in general Bramble (1974: 190–204).

This last section of the poem is more obviously a programme for Martial's frank and playful style of writing.[212] But the menu itself is just as much a foretaste of the festivity, lewdness, and jokes that characterize his dinners and his writing.

3 Martial 11. 52

The invitation in 11. 52 is different again: it is from a poet to a poet, Julius Cerealis. Appropriately enough, Martial uses the invitation to conjure up a series of jokes on different themes which link food and paper. Among these is a well-known poetic cliché, derived from Catullus and used by Martial of his own food-containing mottoes, the *Xenia*: that the fate of failed poems is to end up wrapping take-away food, or food baked in the oven. Real and literary dishes are stirred up together. The perishable and meagre food of urban bath-complexes and *popinae* finds a place with the perennial bucolic feasts of Virgil and Ovid.

Cenabis belle (1) is of course an ironic variation on Catullus' *cenabis bene*.[213] With *belle*, Martial flaunts tastefulness, and disclaims it as a laughable virtue. But he soon gives the lie to his own pretensions: *conditio est melior si tibi nulla, veni* (2; Cerealis will only come if it is his last resort). Martial sets the dinner in the city, next to a bath-house (*scis quam sint Stephani balnea iuncta mihi*, 4);[214] he ends with a reference to pastoral poems (*Rura*, 18). The *gustus* in this menu looks interchangeable with take-away food. At 12. 19. 1 Martial lists *lactucas, ova, lacertum* as typical of the kind of food sold at the baths, and lettuces, eggs, and an old sardine (*cordyla*, indeterminately redeemed by being described as 'bigger than a small *lacertus*') are the first items on this menu. At the same time, *tenuis lacertus* faintly suggests, for example,

[212] Cf. 10. 47–9: *vires ingenuae, salubre corpus, / prudens simplicitas, pares amici, / convictus facilis, sine arte mensa* ('hearty stock, a healthy body, candour with tact, equal friends, relaxed and spontaneous meals').

[213] Cf. 2. 7, where *belle* is repeated *ad nauseam* to describe the superficial accomplishments of a dilettante (contrasted with *bene*); contrast Petr. *Sat*. 46. 2, *inveniemus quod manducemus . . . belle erit* for a non-ironic use of the word to describe a meal.

[214] Cf. Sen. *Ep*. 56 on the commotion of living above a bath-house.

Propertius 4. 3. 23, *teneros lacertos*, used of a woman's arms. Martial is parodying 'literary' erotic language. The *cordyla*, discreetly covered (*tegant*), smacks of the food-wrapper cliché with which Martial introduces and damns his own *Xenia* in Book 13.[215] The irony is that the redeemed paper is immediately put to use as a container for paper descriptions of food. An even closer parallel is 3. 2, where Martial condemns his poems to the same fate:

> ne nigram cito raptus in culinam
> cordylas madida tegas papyro
> vel turis piperisve sis cucullus (3–5)

lest you are snatched away to some blackened kitchen to wrap sardines in sodden papyrus, or knock incense and pepper into a cocked hat.

Incense and pepper are lifted here from Horace's *Epistle* 2. 1. 269–70, and they are also the first items in Martial's *Xenia*. In his menu for Julius Cerealis, Martial wraps the tuna in eggs and rue-leaves: *sed quam cum rutae frondibus ova tegant* (8). Rue-leaves are also an alternative wrapping for perfume at 11. 31. 17.[216] There are other eggs too (*varietas* within individual courses), *altera [ova] non derunt tenui versata favilla* (9, 'more eggs to come, gently roasted on slow embers'), recycled from Ovid's rustic feast for Baucis and Philemon.[217] The cheese (*massa coacta foco*) is a pastoral coagulate (from Ov. *Met*. 8. 666: *lactis massa coacti* and Virg. *Ecl*. 1. 81: *pressi copia lactis*). The olives that follow exploit the exaggerated clichés of describing food fresh from the tree.[218] Elsewhere, olives, too, are among Martial's poem-

[215] Cat. 95. 8: *laxas scombris saepe dabunt tunicas* (loose wraps for mackerel); Hor. *Ep*. 2. 1. 269–70; Pers. 1. 43; Stat. *Silv*. 4. 9. 12; Sidon. 9. 321; Martial 3. 2. 3–5; 4. 86. 8: *nec scombris tunicas dabis molestas* (lit. cumbersome tunics for mackerel: a reference to inflammable tunics worn by criminals burnt to death in the arena; cf. 3. 50. 9); 6. 61(60). 9: *redimunt soli carmina docta coci* (cooks recycling poems). See Pasoli (1970–2); Thomson (1964: 130–6) argues that the paper was used to cook fish in the kitchen, rather than wrap it in the market.

[216] See n. 186 above for the evidence that rue leaves were associated with exiguousness.

[217] Ov. *Met*. 8. 667: *ovaque non acri leviter versata favilla*.

[218] Cf. Mart. 5. 78. 8: *algentem modo qui reliquit hortum* ('which has just left the chilly garden'); 19–20: *olivae, / Piceni modo quas tulere rami* ('olives, ripe from Picenian trees'). See Mennell (1985: 347) for modern parallels.

wrapped take-away dishes: *ne . . . paenula desit olivis* (13. 1. 1, 'to provide a mackintosh for olives').

Martial next imposes Callimachean limits on the hors-d'œuvres: *haec satis in gustu* ('enough of the appetizers').[219] He frustrates Cerialis' curiosity to taste more of the meal. For the next course, Martial lays on a mock-exposé of poetic invention. This luxurious food is not 'real', but only exists on paper:

Cetera nosse cupis?
mentiar, ut venias: pisces, conchylia, sumen,
et chortis saturas atque paludis aves,
quae nec Stella solet rara nisi ponere cena.

Do you want to know the rest? I'll lie to make you come: fish, seafood, sow's udder, fattened birds from farmyard and marsh, which even Stella saves for special occasions.

The progression to the more insubstantial part of the meal is by way of food that is itself insubstantial.

The final part of the poem deals with entertainment, which in this case is of course poetic recitation, a last link between literature and the *convivium*.[220] It comes as something of a surprise to find Martial inviting Cerialis to read from his *Gigantes*. Normally, he never wastes an opportunity to inveigh against lengthy poems and lengthy recitations, the antithesis of his own fragmented and rapid turnover. In 3. 45, even a long menu—turbots, a 2-pound (or two-volume?) mullet (*mullumve bilibrem*), mushrooms, and oysters—is no consolation for dreary recitation, while in 3. 50 a huge book (*ingens liber*) is served up as an hors-d'œuvre betwen the lettuces and the fish-sauce. In 7. 51, a pun on *satis* spans listening and eating at another literary *convivium* without limits: *et cum 'Iam satis est' dixeris, ille leget* (14, 'even when you say "I'm full", he goes on reading').[221] As Kay remarks: 'one wonders how Cerialis would interpret Martial's offer in

[219] An echo of the end of Horace *Sat.* 1. 1 and Virg. *Ecl.* 1. Here, there is a pun on *satis* in the context of eating.

[220] Kay (1985) ad 16: 'considering the literary circles in which M. and Cerialis moved . . . and the preceding reference to Stella, a recitation was definitely on the horizon.'

[221] Cf. 5. 78. 25 (*crassum volumen*); cf. Philodemus, *Anth. Pal.* 11. 44.

view of his tirades against stale mythological epic elsewhere.' He concludes that the offer is sincere: 'Again it is selfless of Martial to offer to sit through something which might well open limitless vistas of boredom. That is friendship.'[222] It seems much more likely, of course, that Martial is being ironic.

The alternative is for Cerialis to read from his *Rura*, Bucolics or Georgics, which Martial places second only to Virgil's: *Rura vel aeterno proxima Vergilio* (18, 'or rustic poems, second in place to immortal Virgil's'). *Rura* can also mean 'country estate', and the language used jokingly suggests a geographical site for Cerialis' *Rura*, next to Virgil's own farm (*aeterno proxima Vergilio*). There is also a contrast between the physical location of Martial's dinner, next to the baths of Stephanus (*Stephani balnea iuncta mihi*, 4), and Cerialis' more abstract literary proximity to the writings of Virgil. The final antithesis of the poem is between Virgil's eternal rustic poems and the ephemera of the small urban poet.

PLINY, *Epistle* 1. 15

The lubricious and playful tone that can be teased out of Martial's epigrams is completely suppressed in Pliny. Admittedly, he does once come out of his shell to reveal his weakness for writing obscene hendecasyllables, but this is only a temporary and calculated lapse.[223] If Pliny were a meal, he would be a humane dinner party of sympathetic friends, where the food was served smoothly and unobtrusively, and was secondary in importance to the relaxed but high-minded conversation.

This is in fact exactly the analogy Pliny himself makes for his letters, which he designs as a careful balance of varied activities, scholarly, political, and recreational, both in the arrangement of the collection as a whole and in the internal contrasts of each letter. The balance of the ideal meal, with its literal and metaphorical mixture of flavours, and the

[222] Kay (1985) ad loc.
[223] Plin. *Ep*. 4. 14. 2–3.

obligations of an attentive host provide Pliny with many similes for his own judicious ordering of his life and letters. At 2. 5. 8, for example, he compares one of his own varied speeches to a meal that has something to please everyone, even if individual dishes are not universally liked:

Nam in ratione convivorum, quamvis a plerisque cibis singuli temperemus, totam tamen cenam laudare omnes solemus, nec ea quae stomachus noster recusat, adimunt gratiam illis quibus capitur.

If you take people at a dinner party, you will find that we all praise the whole dinner, even if individually we are put off eating several of the dishes, and that even those dishes that we do not like do not detract from the pleasure of the ones we find inviting.

Pliny, the ingratiating author of the *Panegyricus*, presents himself both as the all-accommodating host and, within the simile, as the suave guest, who is polite enough to conceal his dislikes and compliment the host on the whole dinner. He demonstrates his affability again in 7. 3, where a life divided between town and country is recommended with a convivial simile. This simile reveals 'incidentally' that Pliny's own style of entertaining is based on the same principles of *variatio* that govern his life and writing:

si cenam tibi facerem, dulcibus cibis acres acutosque miscerem, ut obtusus illis et oblitus stomachus his excitaretur, ita nunc hortor ut iucundissimum genus vitae non nullis interdum quasi acoribus condias.

If I gave you dinner, I would make sure I mixed sharp and savoury dishes with the sweet ones, so that once your appetite was dulled by the sweet dishes it would be stimulated again by the sharp ones: in the same way, I advise you to season your pleasant life every so often with a dash of piquancy.

A well-judged mixture of sweet and savoury flavours prevents his guests from ever feeling jaded. *Stomachus*, 'taste' (literal and otherwise), and *condire*, to flavour (literally and metaphorically), reinforce the comparison.[224]

Pliny's ideal dinner is glimpsed in 3. 1, when he spends a day in the country with his role-model, the veteran *novus*

[224] Cf. *palatus* and *sal* as metaphors at Hor. *Ep.* 2. 2. 60, 62; Quint. 6. 3. 19.

homo, Spurinna. Pliny makes a careful note of all the details. Spurinna delays dinner in order to listen to some light reading; his guests are free join in, or do as they please; his dinner is elegant but frugal (*cena non minus nitida quam frugi*); his silver solid and antique (*argento puro et antiquo*); his bronze is for use, not display, and he is not over-enthusiastic in his passion for it (*delectatur nec adficitur*); the dinner is punctuated by performances of comedy, and prolonged sociably into the night (8–10). Spurinna's restraint, affability, and easy high-mindedness suggest the confident *sprezzatura* of the born aristocrat rather than the nervous zeal of the new man (even his silver has the casual appearance of an heirloom). We are also told that the food is not so important that it cannot be delayed (Pliny's deliberately undetailed description of it subscribes to this ethic), and that the comic performances that divide up the meal are the real spice of the dinner, which inverts the expected sentiment, that study is the staple to which pleasures are the spice;[225] *comoedis cena distinguitur, ut voluptates quoque studiis condiantur*. A perfect balance is achieved between intellectual pursuits and the necessary light-heartedness appropriate to the dinner party (*remissius aliquid et dulcius*; *comoedis*; *comitate*).

This conjunction of pleasure and study at the *convivium* is mirrored in Pliny's image of the ideal of balance and variety in both life and literary activity:

> Ut in vita sic in studiis pulcherrimum et humanissimum existimo severitatem comitatemque miscere . . . qua ratione ductus graviora opera lusibus iocisque distinguo . . . Ita solemus . . . satietatis periculum fugere. (8. 21. 1–4)

> In literature as in life, I think it is fine and civilized to mix the serious and the sociable aspects . . . on these grounds I intersperse my more solemn pieces with light-hearted and humorous ones . . . this is how we avoid satiety.

The opposite extreme, however, is just as instructive. At 9. 17 Pliny 'reveals' his own good taste and the gentlemanly tolerance that overrides all questions of taste when he

225 Cf. e.g. Quint. 5. 14. 35.

chastises a friend for sneering at hosts who use clowns for entertainment, instead of the trio of readers, lyre-players, and comedians that gentlemen prefer.[226] A bad-mannered host is one who distinguishes between his guests and serves them different food according to rank: *sibi et paucis opima quaedam, ceteris vilia et minuta ponebat* (2. 6. 2, 'he gives himself and a few friends the richest food, and serves the rest cheap scraps'). Such a dinner is the antithesis of Spurinna's: *sordidum simul et sumptuosum* (2. 6. 1, 'stingy as well as extravagant'); *luxuria specie frugalitatis* (6, 'luxury in the guise of thrift'). Pliny contrasts it with his own egalitarian hospitality: *eadem omnibus pono, ad cenam, non ad notam invito cunctisque rebus exaequo, quos mensa et toro aequavi* ('I serve everyone the same food: I am entertaining them, not means-testing them, and, when I invite people as equals to the same table, I give them equal treatment'). Of course, in this area, Pliny's philosophy does not square with the style of the letters: rich food for a few, cheap scraps for the rest sums up his own treatment of his different correspondents. Later in the same letter, he describes slumming it with his freedmen: *liberti mei non idem quod ego bibunt, sed idem ego quod liberti* (4, 'my freedmen don't drink what I drink: I drink what my freedmen drink'). Pliny's *liberalitas* or *humanitas*, it is worth noting, extends only to bringing himself down to the level of his inferiors, not to elevating them to his own heights. At 9. 23. 4 he makes a point of describing his gracious politeness to a *municeps* he met at a dinner party: *noblesse oblige*.[227]

Another convivial image which imitates the structure of Pliny's philosophy of life and art is a rare description of an alfresco meal on the edge of a fountain at his Tusculan villa (5. 6. 37). Around the basin, he tells us, are placed the *gustatorium* and the heavier dishes, while the lighter dishes float on the water in the shape of little boats or birds: *gustatorium graviorque cena margini imponitur, levior naucularum et avium figuris innatans circumit*. This contrived

[226] Clowns: 9. 17. 1: *scurrae cinaedi moriones mensis inerrabant*; Pliny prefers *lector aut lyristes aut comoedus*: cf. 9. 36. 4; 1. 15. 2.

[227] D'Arms (1984: 347–8) sees in this passage evidence that the political notion of *hospitium* was put into practice at meals.

contrast between the serious and the playful, weighty and figurative, is the perfect visual image of Pliny's own style.

Apart from these convivial metaphors and sideways glances, any detailed picture of Pliny's own taste in entertaining is elusive. This is itself significant: to reproduce a recipe would go against the assumption that a good meal can only be recognized, not prescribed. We learn about Pliny's tastes only from his descriptions of other people's meals.

In *Epistle* 1. 15, however, we do find a menu, or at least part of one. This is a prose variation on the invitation theme, a retrospective suit for damages from a friend, Septicius Clarus, who failed to turn up to dinner and spurned Pliny's own simple but genial meal for a more opulent feast elsewhere. The intimate, mock-angry tone of the letter places it among the more playful sections of his *œuvre*, what he calls *ludus iocique* (8. 21). The fact that both Pliny's 'convivial' letters, 1. 15 and 3. 12 (in which he *accepts* a dinner-invitation), share a bantering tone suggests that a facetious style was naturally chosen for representing the meal: its jokes recreate or evoke the convivial mood.[228] In many ways, 1. 15 is an exception to Pliny's usual harmoniously balanced style. The potential for embarrassment in describing one's own food makes smooth and balanced exposition impossible: Pliny resorts to the stratagems of irony, urbane wit, parodic unevenness and bad taste, litotes and exaggeration, in order to suggest, playfully and indirectly, his own sense of proportion.

Septicius Clarus is also the informal dedicatee of Pliny's letters: in 1. 1, he is offered this makeshift, impromptu collection, bundled together supposedly at random: *epistulas, quas paulo curatius scripsissem . . . ut quaeque in manus venerat* (1. 1. 1, 'any letters that were composed with a little care . . . arranged just as they came to hand').[229] The grandest

[228] See Sherwin-White (1985: 43).

[229] Statius' *Silvae* (lit. 'unfinished material') has often been suggested as an influence (e.g. by Peter (1873: 689 ff.)). This aspect of the letters can be generally compared with other forms of writing that simulate the appearance of formlessness: epigram and satire. On the underlying artfulness of the letters see Sherwin-White (1985: 5).

and most ambitious of these letters is reserved for the historian Tacitus, with the humble envoi *aliud est enim epistulam aliud historiam, aliud amico aliud omnibus scribere* (6. 16. 22, 'it is one thing to write letters, another to write history, one thing to write to a friend, another to write for the whole world'); but Pliny does not presume to offer the collection itself to anyone so distinguished. Nevertheless, this choice of a young equestrian protégé on the brink of his political career is an advertisement for Pliny's shrewdness as well as for his modest pretensions. Clarus was destined to become praetorian prefect under Hadrian (119), and was the dedicatee of another perennial history-book: Suetonius' *Lives of the Caesars*.[230] The fact that he is the overall recipient of the *Letters* suggests that *Ep.* 1. 15, whether or not it records a real event, is included for symbolic reasons.[231] It is not simply a 'filler', a short, mundane note that gives body to the whole.[232] And it is not, despite appearances, any more of a functional document than Martial's invitations. In fact, the central opposition of the piece, between Pliny's simple vegetables and serious entertainment, and his rival's luxurious dishes and flamenco dancers, may well be modelled on Martial anyway, or at least be exploiting the cultural oppositions that Martial and others had fixed in this particular form.[233]

Alternatively, of course, we can read the piece as another convivial manifesto for the plain style, with the opposition between fuss and simplicity as a double reference to styles of dining and styles of writing, in the manner of Horace and Martial.[234] Two important qualifications need to be added.

[230] See Sherwin-White (1985: 85).

[231] See Beard (1985) on the *Acta Fratrum Arvalium* as 'symbolic' writing.

[232] See Sherwin-White (1985: 47); Booth (1961); Watt (1976: 9–37, 104–51); Bryson (1981: 1–28) on the inclusion of insignificant detail in fiction or art as a device for creating a sense of the real.

[233] Mart. 5. 78, 10. 48; cf. also Philodemus, *Anth. Pal.* 11. 44; Hor. *Ep.* 1. 5; Juv. 11. See also Sherwin-White (1985: 17): 'the two "invitations to dinner" . . . exploit the stock comparison of the plain and the extravagant meal.' Edmunds (1980), as we have seen, persists in regarding it as a real control for Martial's poems.

[234] Race (1978: 186) compares it to the contrast between *Persicos apparatus* and *simplici myrto* in Hor. *Od.* 1. 38, concluding that 'Pliny is talking more about the style of his letters than about some real dinner.'

First, Pliny is not simply alluding to convivial models for their own sake. He is projecting the literary commonplace on to a simulation, at least, of real life. Pliny selects information that fits both his literary stance and his chosen autobiography. The old opposition between simple and luxurious food is included because it has cultural connotations much wider than its literary resonances alone: it defines Pliny's place in the political world (removed from the scramble of opportunist dinner parties); it identifies him, more nebulously but no less pertinently, as a man of taste; it shows us that he does have a convivial life, and joking relationships with his contemporaries.[235] Secondly, it is not enough simply to extract the material and abstract images of simplicity and luxury and dismiss them as a code, literary or otherwise. An extra dimension is added when we appreciate that to recite one's own menu is conventionally a mark of bad taste. Pliny dissolves his embarrassment by exaggerating bad taste into a joke.

Once again, then, the connections that the letter suggests between a dinner party and the writer's *œuvre* are far more involved. We should expect to find that this very specific social exchange is something of a complement to the dedicatory epistle. The elder statesman's invitation may be imbued in experience of the world of politics and social climbing, but what it offers is a casual escape from it.[236] In the same way, the collection announces itself as a casual alternative to grander writing (*neque enim historiam componebam*, 1. 1. 1: 'after all, I am not writing history'): informal, ephemeral, and infused with *comitas*, the spirit of social affability. More and more, we can see that this invitation was a necessary ingredient of the epistles.

Pliny's complaint is lodged as a parody of a lawsuit claiming damages; Clarus is required to pay costs in full. Pompous legal vocabulary reinforces the impression that such a complaint is petty-minded and bad-mannered in the context of a dinner party. *Dicitur ius*, 'the court is in

[235] The exposé of the dedicatee's bad manners marks the informality of the book.

[236] Cf. 4, *nusquam incautius*. Sherwin-White (1985) suggests a possible reference to spies (cf. 4. 9. 6, 9. 13. 10; or 8. 4. 8 for a more general use).

session',[237] is another 'convivial' pun on *ius* (law) and *ius* (sauce), reviving the inappropriate analogy between law-court and *convivium*.[238] The absurd comparison has a point to make: Pliny's dinner could not have been further removed from the political world of risk and responsibility (cf. *nusquam incautius*, 4); to enumerate the items on the menu parodies the system of values based on crude calculation that drove Clarus elsewhere.[239]

Paratae erant lactucae singulae, cochleae ternae, ova bina, halica cum mulso et nive (nam hanc quoque computabis, immo hanc in primis, quae perit in ferculo), olivae betacei cucurbitae bulbi, alia mille non minus lauta. audisses comoedum vel lectorem vel lyristen vel, quae mea liberalitas, omnes. at tu apud nescio quem ostrea vulvas echinos Gaditanas maluisti. (1. 15. 2)

On the table were: one lettuce each, three snails each, two eggs, emmer-gruel with honeyed wine and snow (you had better take this into your calculations too—in fact it ought to be one of the most expensive items, given that it melts away on the plate), olives, beetroot, gourds, onions, and a thousand other delicacies. You could have listened to a comic actor or a reader or a lyre-player, or, such is my generosity, all three. But you chose to go to somewhere where you could have oysters, sows' wombs, sea-urchins, and Spanish dancing-girls instead.

Only the *gustatio*, the first course, is described in full.[240] Pliny is giving Clarus just a *gustus* (foretaste) of his dinner, just as the letter-form can only stretch to a *gustus* or digest of

[237] Merrill (1935) thinks this is an action for damages under the Lex Aquilia, quoting *Dig.* 9. 2. 27. 5; Sherwin-White (1985) thinks the metaphor is not so specific.

[238] See p. 77 n. 98 above.

[239] Elagabalus (*SHA* 29. 9) is reported to have loved hearing the price of his food exaggerated at the table: he called it the 'appetizer' for the dinner; cf. Amm. Marc. 28. 4. 13 on hosts who have scales to weigh food ostentatiously in front of their guests. Cf. Stat. *Silv*. 4. 9. 6: *licet ecce computemus* ('let's add up our accounts'); *CIL* 9. 2689 (a restaurant bill): *copo: computemus*.

[240] Merrill (1935) seems to imply that the final dishes are part of the *cena* proper: 'Pliny's enumeration begins in normal order with hors d'oeuvres, washed down with *mulsum*, and proceeds to the regular *fercula*.' There are no grounds for this. All the dishes are conventionally part of the *gustus*—lettuce: Martial 13. 14 on the fashion for beginning meals with lettuce (cf. Plin. *NH* 19. 38. 127); snails: Mart. 13. 53; eggs: Hor. *Sat*. 1. 3. 6, *ab ovo usque ad mala* (cf. Cic. *Fam*. 9. 20. 1); olives: Mart. 13. 36. 2 on olives as first and last course; gourds: Apic. 3. 4. 1 for an appetizer made of gourds (*gustum de cucurbitis*).

experience; it is only a sample or specimen of biography. At 4. 27. 5 Pliny recites some verses as a foretaste of the whole *œuvre* of another poet: *ad hunc gustum totum librum repromitto* ('I promise that this is an accurate foretaste of his whole work'); at 9. 18. 1 he sends Sabinus a selection of his speeches, so as not to overload him: *per partes tamen et quasi digesta* ('only in fragments, and predigested, as it were').[241] In the same way, the list is also only a sample of the complete menu. According to one critic, Pliny is very successful in suggesting the difference between his own simple but elegant meal and the expensive but banal food Clarus preferred.[242] In fact, Pliny makes his own meal seem banal too. There is none of Horace's fastidious detail or Martial's throw-away embellishment: for Pliny, it would not be urbane to continue his satirical καθ' ἕκαστα inventory beyond a certain point. A mock-tedious recitation is narrowly prevented from actually becoming tedious; Pliny expresses his own disdain for such literal itemization by jumping abruptly to the end: *alia mille non minus lauta*.

Within the menu, Pliny manœuvres with tiptoe precision. The first section uses numerals to suggest mean-spirited exactness; Pliny is taking the costs 'to the letter' (cf. *ad assem*, 1): *lactucae singulae, cochleae ternae, ova bina*.[243] The fact that the numbers involved, one, three, two, are not in normal order makes the list seem more realistically, or rather randomly, ordered. This is something that they have in common with Pliny's letters, which are arranged *ut quaeque in manus venerat* (1. 1. 1, 'as they came to hand'). The list seems straightforward enough: vegetables and eggs are shorthand for simplicity. But the fourth item, *halica cum mulso et nive*, has puzzled commentators.[244] *Halica*, emmer-gruel, and

[241] Cf. 2. 5. 12: *quia existimatur pars aliqua etiam sine ceteris esse perfecta*.

[242] Guillemin (1934: 15): 'Pline indique avec autant de finesse que d'ironie la différence entre la réception simple, mais distinguée et intelligente qu'il préparait à Sept. Clarus . . . et le festin somptueux et banal qui lui a été préferé.'

[243] Merrill (1935: ad loc.) 'the numerals are doubtless meant to give the effect of precision in the reckoning of damages, and only incidentally of frugality in the supplies'. But we need to be aware of all the statements, direct and indirect, that Pliny is making.

[244] Vegetables: e.g. Hor. *Sat*. 2. 1. 74, 2. 2. 117, 2. 6. 64; *Ep*. 1. 5. 2; Ov. *Met*. 8. 647; eggs: Petr. *Sat*. 46. 2; Ov. *Met*. 8. 667.

mulsum, honeyed wine, seem to be chalk and cheese, as one of Martial's epigrams suggests: *nos alicam, poterit mulsum tibi mittere dives* ('we can send you emmer-gruel, a rich man can send you mead').[245] And snow, as an addition to drinks, was a stock example of expensive superfluity.[246] So Sherwin-White concludes: 'Pliny here betrays his ignorance of the simple meal.'

It is presumptuous, however, to call any ancient author, least of all Pliny, ignorant of social codes. The mixture needs to be seen as an example of flexible compromise. The smart dinner could be neither completely rustic nor extravagant.[247] Pliny's balance of simple and expensive ingredients makes his dinner the most nearly irreproachable according to all the various codes, placed impregnably in the middle region of simple elegance.[248] Besides, it is the less substantial items that are more expensive: this reinforces the impression of discreet stylishness. The point of having snow is not only to suggest over-literal mean spirits. Pliny charges most of all for it because it is perishable.[249] Like Catullus' *unguentum*, the snow is only semi-substantial; its value lies also in its power to evoke the more abstract qualities of the *cena*, in this case its fleeting, irrevocable quality.[250] The letters are presented as ephemera, snatches of momentary life; Pliny contrasts Tacitus' histories at 6. 16. 2: *scriptorum tuorum aeternitas* ('your own eternal writings'). And Clarus has missed his opportunity.

The inventory continues with four more items, then suddenly tails away: *alia mille non minus lauta*. In both content and style, the dinner has begun to be *misrepresented*: too much of the *gustatio*, distended as a banal and random list; nothing of the central *fercula*. *Alia mille non minus lauta* makes the remainder seem infinite, but also puts a tasteful

245 Mart. 13. 6.

246 See e.g. Plin. *NH* 19. 19. 54; Sen. *Ep*. 95. 25.

247 See also Shero (1929) on Lucilius' satirical description of a rustic dinner, with its long list of herbs (Charisius *GLK* 1. 100. 29: *enumeratis multis herbis. Enumeratis* recalls the banal itemization in Pliny's letter).

248 Cf. Spurinna's meal, 3. 1. 9: plain but elegant.

249 2: *nam hanc quoque computabis, immo hanc in primis quae perit in ferculo*.

250 And possibly, indirectly, summer heat. Sherwin-White (1985) thinks the meal took place in summer because the vegetables mentioned are summer crops.

limit on the joke.[251] It shares this technique with some other famous inventories. Ovid, *Amores* 1. 5 contains another comic catalogue of something that cannot really be itemized, female beauty, which is curtailed with the words *singula quid referam?* (23, 'I can't list each thing'). In the context, this drawing of a veil is also a euphemism: Ovid had just reached the girl's middle. In *Twelfth Night*, Olivia says: 'I will give out diverse schedules of my beauty—It shall be inventoried and every particle and utensil labell'd to my will: as—item, two lips indifferent red; item, two grey eyes with lids to them—item, one neck, one chin, and so forth . . .'. This list too gives the flavour of an inventory, but does not go on for too long, fading out with 'one neck, one chin'.[252]

Pliny's irony can be detected in the rhetorical structure of the list as well. He arranges his clauses as (fulsome) tetracola rather than (more decorous) tricola: *lactucae singulae, cochleae ternae, ova bina*; *olivae betacei cucurbitae bulbi*. Moral or representational excess is suggested through the structures of rhetorical excess.[253] The vegetables swell in size as far as the *cucurbitae*,[254] then collapse bathetically with *bulbi*,[255] another contrived breach of logical classification. A further tetracolon bundles together the alternative list of pleasures: *ostrea vulvas echinos Gaditanas*. *Gaditanae*, Spanish dancers, may well refer to Martial's dancers at 5. 78. Here they are jokingly included in a list of food, as though they were among the sensory pleasures of the meal.[256] Four is often an excessive number in gastronomy, as well as rhetoric. Pliny's uncle calls four wines superfluous;[257] the so-called *tetra-* or *pentapharmacon* is cited as the monstrous favourite dish of

[251] Cf. 3. 12 where Pliny asks to be entertained moderately: *sit expedita sit parca . . . nostrae tamen cenae ut adparatus et impendii, sic temporis modus constet*.

[252] Ov. *Ars Am.* 1. 5. 23; *Twelfth Night*, I. v. 228 ff.

[253] See Sen. *Con.* 9. 2. 27 on a tetracolon of Murredius: *serviebat cubiculo, praeter meretrici, carcer convivio, dies nocti*. Seneca criticizes the final clause for being included clearly only to complete the metre and fill up space: *novissima pars sine sensu dicta est, ut impleretur numerus*. See Norden (1898: i. 289–90).

[254] Prop. 4. 2. 43 on gourds' swelling stomachs: *tumidoque cucurbita ventre*.

[255] For comic meanings of bulbs, see Mart. 3. 75. 3; Pers. 4. 36.

[256] Cf. Ar. *Ach.* 1091, where prostitutes are inserted into a list of sweetmeats; Ar. *Plut.* 190 ff.

[257] *NH* 14. 16. 97.

two late emperors.[258] An ancient board-game found in Rome imitates an inn-sign with four alliterative items, suggesting infinite excess: *abemus in cena pullum piscem pernam paonem* ('for dinner we've got chicken, fish, ham, and peacock').[259]

After all this, Pliny's menu is disappointing for the antiquarian or social historian.[260] But that in itself is informative. Unlike Martial's lingering parodic descriptions, which revel in their own bad taste, this list is bare and comparatively flavourless, diverting attention away from itself to the less material aspects of the meal.[261] The tricola that follow, *quantum nos lusissemus risissemus studuissemus* ('what fun, laughter, and civilized entertainment we would have had'), *nusquam hilarius simplicius incautius* ('never with such simple and relaxed good spirits'), signal a return to sincerity and 'straight' representation.[262]

Pliny's letter demonstrates urbane manners through ironic imitations of bad taste, which are always redeemed by the limits that he puts on them. By satirizing his own discrimination and generosity as 'exhibitionism' (*nec id modicum*, 1: 'and rather a substantial one at that'; *non minus lauta*, 2: 'no less dainty'; *quae mea liberalitas*, 2: 'such is my generosity'),[263] he has it both ways, communicating the casual elegance of his hospitality and disclaiming it as something trivial. The 'crude' enumeration of the hors-d'œuvres lets Pliny suggest the blatant simplicity of the meal, with all its social, moral, and literary connotations.

[258] Aelius Verus (*SHA* 5. 4) and Hadrian (*SHA* 21. 4): as the name suggests, this was a joke version of a medical panacea with four or five ingredients.

[259] See Ihm (1890); and *Bullettino della Commissione Archeologia Municipale* (1876), 188, table. XXI for a sketch.

[260] Sherwin-White (1985: 120): 'it adds nothing to the stock of information on the subject'.

[261] Cf. Hor. *Ep.* 1. 5. 2, where the meal is summarized on one plate: *modica holus omne patella*.

[262] Similarly, the sentence *audisses comoedum vel lectorem vel lyristen vel . . . (quae mea liberalitas) omnes* defeats expectation. Instead of further (and perhaps more vulgar) entertainments (like the jesters of 9. 17), Pliny restricts his generosity to a total of three entertainers. This may be a private joke: at 5. 19. 3 Pliny praises a slave who is equally versatile as *comoedus, lector*, or *citharista*. The triplet is varied at 9. 17: *lector aut lyristes aut comoedus*; at 1. 16. 1 Pliny approves of a poet whose style is adaptable: *quam varium quam flexible quam multiplex*.

[263] Cf. Theophr. 20. 9: the ill-bred man boasts about his own water, vegetables, and cook.

Indirectly, he drives it home that the qualities of an ideal meal are too elusive to be listed or reduced to a formula. Plutarch writes: 'The Romans are fond of quoting a clever and gregarious man who said, after eating alone, "I have eaten, but not dined today," implying that a dinner always needs the company of friends for its seasoning.'[264] *Comitas*, the sociable combination of *ludi* and *studia*[265] that gives this letter its flavour, is the essential ingredient of Pliny's meal—which Clarus has failed to provide.[266]

[264] *Mor.* 69c.

[265] Cf. 8. 21. 1, *graviora opera lusis iocisque* ('serious works mixed with light-hearted and humorous ones'); 1. 15. 3, *quantum lusissemus risissemus studuissemus* ('what fun, laughter, and civilized entertainment we would have had').

[266] Cf. Plut. *Mor.* 697c: 'Homer calls salt "divine" . . . but the most truly divine seasoning at the dinner-table is the presence of a friend or close acquaintance—not because he eats and drinks with us, but because he takes part in the conversation.' Winniczuck (1966) argues that Pliny replaced Cicero's term for the ideal civilized virtue, *urbanitas*, with *comitas*. For another summary of the difference between tasteful and tasteless meals, see Pliny's praise of Trajan's highbrow dinners (*Pan.* 49. 8): 'We did not come to admire gold and silver, or elaborately devised feasts, but your own charming and pleasant manner, of which, since it is always sincere and genuine and adorned with dignity, we can never get our fill. The royal table is not surrounded by the ministers of foreign cults or indecent jesters, but with warm hospitality, tasteful humour, and respect for learning.'

5
Garlic Breath
HORACE, *EPODE* 3

Garlic is one foodstuff that has inspired a love–hate relationship throughout history.[1] In Anglo-Saxon England, its haters have always been the more strident party. John Evelyn banishes it in ignominy from his *Acetaria*:

> Whilst we absolutely forbid it entrance into our Salleting, by reason of its intolerable Rankness, and which made it so detested of old; that the eating of it was (as we read) part of the Punishment for such as had committed the horrid'st crime; 'tis not for ladies' Palates, nor those who court them, further than to permit a light touch on the dish, with a clove thereof, much better supply'd by the gentler Roccombo.

It is one of Smollett's chief grouses about France: 'For my own part I hate the French cookery, and abominate garlick, with which all their ragouts, in this part of the country, are highly seasoned.' And Mrs Beeton allows it grudgingly into one Indian recipe: 'The smell of this plant is generally considered offensive, and it is the most acrimonious of the whole alliaceous tribe.'[2]

Horace's *Epode* 3 is the original, compact model for virulence against the herb; it is the source that Evelyn takes literally for the connection between garlic and 'the horrid'st crime'. In this poem, garlic not only fuels the poet's dyspeptic rage: it is also the object on which he vents it. Such a

[1] Kenney (1984: 43): 'Praise and blame of, and jokes about, garlic have always abounded.' See Hicks (1986) and Rios (1986) on the mythology of garlic in later periods.

[2] Evelyn, *Acetaria* (London, 1961), 28; Smollett, *Travels through France and Italy* (1766), Letter 8; Isabella Beeton, *Household Management* (London, 1861), 392.

vicious circle has great metaphorical potential, potential which has, as usual, been underrated. This single foodstuff from one tiny poem can serve to illustrate a more general hypothesis: that we can only begin to understand the meaning of food in a literary text if we assume that it was chosen specially, whether for its unique flavour or for its distinctive properties. In this case, the manifold and contradictory qualities of garlic make it a uniquely appropriate 'food' for iambic anger, as well as a complex metaphor for the peculiarly symmetrical conflict between the iambist and his victims.

The garlic poem is usually regarded as the slightest of the *Epodes* in a book which also deals with the effects of civil war and the destruction of Rome.[3] Horace launches into a hyperbolical tirade against the after-effects of garlic, equating the plant preposterously with witches' poison and snakes' blood, and finishing with a vengeful curse on Maecenas involving bad breath and sexual rejection. According to Fraenkel, the poem is a parody of the stance of the iambic writer: 'a good deal of mocking pathos and quasi-Archilochean indignation delivered with feigned grandiloquence.'[4] This observation is important, and we can take it much further. Horace seems to be using the incident to serve up a 'physiology' of iambic anger; he makes it a poisonous or evil-smelling afflatus which emanates from the poet's raging entrails. Small-scale and light-hearted, the poem is also an epitome of iambic writing.

'Epode' comes from the Greek word ἐπῳδή, meaning a spell or incantation.[5] Although scholars are now sceptical about the ancient theory that iambics and comedy grew straight out of a social ritual (the phallic songs improvised in

[3] Cf. e.g. Büchner (1970: 50): 'Im Raum der Epoden begegnen eine Fülle von Erscheinungen der Wirklichkeit. Grösstes—das Schicksal Roms—steht neben Geringstem—der Unbekömmlichkeit von eines Kräuterklosses.' It is often classed as a *jeu d'esprit* in the manner of Catullus: e.g. by Fraenkel (1957: 68); Fedeli (1979: 112); Shackleton Bailey (1982: 6).

[4] Fraenkel (1957: 68).

[5] Cf. the triple meaning of *carmen*—'song', 'poem', or 'charm': *OLD* s.v.

Greek cities to ward off malevolent influences[6]), it is clear that these songs may at least have provided a model for literary invective, especially when the phallic amulet was so prominent in Roman culture.[7] Many of Horace's *Epodes* make allusions to their apotropaic origins. Horace's conflict with the witch Canidia, for example, harks back more to the supposed function of iambics as an antidote to witchcraft and enchantment than to any real-life feud. There is a certain symmetry, even sympathy, between these enemies already: Horace the writer of epodes, and Canidia, author of magic spells.

The iambic form is also associated with invective against enemies within normal society. In Greece, a cluster of legends sprang up accusing Archilochus and Hipponax of trying to murder their real-life enemies through their verses. However, as one critic concludes: 'The three most prominent characteristics of Archilochean blame turn out to be a distance from the object, a consciousness of function, and a manipulation of convention . . . Anger is generalized and the audience is reminded that attack is a form of artistic activity.'[8] Even Horace's predecessors, then, produced iambics of a paradoxical kind: virulent abuse which seemed to be born in the heat of the moment; formal artefacts which alluded to the conventions of the model. Horace's *Epodes* share this formal quality. The polite world of abstracts is apparently abandoned for grossly physical images: disgusting descriptions of repulsive old women; animal abuse; and references to the bodily functions of the author (eating, nausea, and sex).[9] At the same time, an overwhelmingly physical treatment of abuse is also an allusion to the conventions of the form; the angry *Epodes* are hemmed in by formal patterns of attack and retaliation.[10]

[6] e.g. Aristotle (*Poet.* 1448b–1449a); see Richlin (1981); Elliot (1960: 4–6), following Cornford (1914: 49–50) regards the descent as comparatively uncontroversial.

[7] See Richlin (1981: 63).

[8] Burnett (1983: 60).

[9] *Non descendat in ventrem meum* (2. 53); *fluentem nauseam* (9. 35); *minus . . . languet fascinum* (8. 18). See J. G. W. Henderson (1987: esp. 110, 115–16).

[10] See e.g. 2. 1–66, 67–79; 3. 1–18, 19–22; 5. 1–82, 83–102; 15. 1–23, 24; 17. 1–52, 53–81; and Carrubba (1969).

In the metaphorical language that surrounds the production of iambic poetry in Greece and Rome, the iambist is traditionally described in violently physical terms. He is the generator of bitter or pungent verse, even, crossing the boundaries of edibility, of poison itself. The exact physiological origins of this bitterness or poison are never clear or consistent. Either the iambist is thought to have absorbed the 'poison' of the aggressor into his veins, or else his vitriolic nature gives him the power to summon up an equivalent bodily response (the liver was thought to be the organ of anger, and excess bile created a surplus of bitterness in the body).[11] So the poetry is ambiguously either the poison, the reaction, or the antidote. This emerges from the varied chemical explanations of Archilochus' poetry, for example. In every case, iambics are poisonous (the Greeks derived the word from *ἰός*, poison). Callimachus analyses this poison into bitter dog's gall and a sharp wasp's sting.[12] According to another version, Archilochus' iambic verse is born from a complex physiological reaction: the bitter rage of iambics generated by a tongue bitter with gall.[13] Another etymological poem, deriving 'iambics', from *ἰός*, poison, and *βάζειν*, to dip, shows Archilochus dipping his bitter muse in viper's bile.[14] Snake-bites, poisonous pens, and venomous breath are commingled in Latin descriptions of iambics as well. Tacitus speaks of bitter iambics; Ovid of biting poetry; Martial of viper's poison. And Quintilian lights on *acerbitas* as their chief flavour.[15]

[11] See Onians (1954: 84–9).

[12] Fr. 380 Pf.

[13] *Anth. Pal.* 7. 69.

[14] Ibid. 7. 71. Cf. *AP* 9. 185: *ἰός*.

[15] Tacitus, *Dial.* 10. 4: *iamborum amaritudinem*; Ov. *Trist.* 2. 563: *mordaci carmine* (cf. 2. 565–6: *candidus a salibus suffusis felle refugi / nulla venenato lettera mixta ioco est*). Mart. 7. 12. 6–7: *Lycambeo sanguine . . . vipereumque vomat nostro sub nomine virus*; 7. 72. 13: *atro carmina quae madent veneno*. Quintil. 10. 1. 96; cf. 10. 1. 94. Lucilius' satires contain *acerbitas et abunde salis*; Babrius (prol. 1. 19): *πικρῶν ἰάμβων*. Metaphors of venom and gall also characterize malice in speech or literature outside specifically iambic writing. Cat. 44. 12: a speech *plenum veneni et pestilentiae*; Hor. *Sat.* 1. 7. 1: *Rupili pus atque venenum*. *Venenum* and *virus* are also used of abstract qualities: Cic. *Amic.* 89: *odium quod est venenum amicitiae*; Fronto, *Aur.* 1. 258 (168N): *virtutibus ceteris iracundia venenum ac pernicies fuit*; Cat. 77. 5–6: *nostrae crudele venenum / vitae*. Cf. *virus*: Cic. *Amic.* 87: *aliquem, apud quem evomat virus acerbitatis suae*; Ov. *Pont.* 4. 6. 34: *verba velut tinctu singula virus habent*; Sil.

Iambics are also conventionally pugnacious and vengeful,[16] born out of heat, anger, and youthful impetuousness. Ovid speaks of pugnacious iambics; Catullus turns iambics into vicious swords; Statius makes these swords the instruments of revenge.[17] Horace unites war and anger in his description of Archilochus at *Ars Poetica* 79: *Archilochum proprio rabies armavit iambo*. At *Odes* 1. 16. 24 he apologizes for his youthful epodes, angry young man poems, dashed off in the heat of the moment:

me quoque pectori
temptavit in dulci iuventa
fervor et in celeres iambos
misit furentem . . .

The heat of passion assailed me in my tender youth, and drove me to hasty iambics.

Plutarch, in his life of Cato, invents a similarly histrionic scenario for Cato's iambics against Scipio: 'Angrily and impetuously he turned his energies to the writing of iambics, in which he made a violent attack upon Scipio, using the bitter style (τῷ πικρῷ) of Archilochus and allowing himself to exaggerate and make childish jokes.'[18]

Another persistent theme is the link between iambics and virility, harking back to the original phallic songs.[19] Barefaced frankness was part of iambic *libertas* (Quintilian notes that words considered improper elsewhere are praised when they occur in iambics or old comedy[20]). Quintilian's own description of Archilochus' iambic style (10. 1. 60) concentrates all these metaphors, phallic, warlike, and violent, into one display of bristling masculinity: *Summa in hoc vis*

11. 557: *virus futile linguae*; Apul. *Apol.* 67: *ibi omne virus totis viribus adnixi effundere ibi maxime angebantur*. For other uses of poison and bitterness in connection with iambic or satirical writing see Burnett (1983: 55 ff.); Bramble (1974: 190–204): 'The Disclaimer of Malice'. On literary 'flavours' in general, see above p. 41.

[16] Burnett (1983: 55 ff.); Bramble (1974: 190–204).

[17] Ov. *Ib.* 519: *pugnacis iambi*; Cat 36. 5: *desissemque truces vibrare iambos*; Stat. *Silv.* 2. 2. 115: *minax ultorem stringit iambon*.

[18] Plut. *Cat. Min.* 7, quoted by Burnett (1983: 55).

[19] See Elliott (1960: 4–6); Burnett (1983: 77), referring to J. Henderson (1975: 14–18).

[20] Quint. 10. 1. 9.

elocutionis, cum validae tum breves vibrantesque sententiae, plurimum sanguinis atque nervorum ('His language had extraordinary force, his ideas were tough, punchy, and scintillating, full of blood and muscle').

One consistent feature of all the metaphors for iambics is that they involve physical or physiological reaction to an external stimulus. Iambic aggression always seems to be prompted by aggression from another source. In other words, the iambist duplicates the malevolence and vindictiveness of an aggressor with a kind of symmetry. In Horace's poetry, as in the social practices which are its models, the identities of victim and aggressor are actually confused and made interchangeable by the 'sympathetic' nature of retaliation.

We can understand the nature of iambic poetry better if we consider another peculiarly symmetrical ancient belief which supplied some of its images: belief in the evil eye. The Latin word for the evil eye, *fascinum*, is also the name of the phallic amulet worn against it. If we are to believe the ancient sources, phalluses were everywhere in the ancient world, standing outside Pompeian houses, scaring away the birds in vegetable gardens, hanging round the necks of babies and famous generals: in fact, a great deal more exposed than they are today, aggressively masculine symbols protecting society from the evil eye.[21] However anachronistic this popular belief was for educated Romans, it survived to give an aggressively masculine tone to Horace's *Epodes*. In the *Satires*, where Horace to some extent dismantles his iambic aggression, the poet speaks in the voice of a statue of Priapus, who protects the gardens of Maecenas from the witches Canidia and Sagana with his obscene groin (*obscenum inguen*, 1. 8.). Instead of using his weapon to rape them, he insults them with his powers of deflation, letting out a monstrous fart.

Again, the physical shape of the amulet or Priapic statue suggests that the type of aggression it warded off was

[21] Plin. *NH* 19. 50 on statues of Priapus, which warded off evil spirits. Generals and babies: Plin. *NH* 28. 39. See Richlin (1981: 63) on the power of the *fascinum* in the ancient world. See also Ling (1990: 51–5) for pictures of the phalluses

specifically a *physical* assault. *Invidia*, the Latin term for 'spiteful envy', implies the possession of an offensive or aggressive gaze (*offectus oculique venena maligni*[22]). Ovid's portrait of Invidia seethes with Archilochean poison: an evil squint, tartarous teeth, a heart streaming with bile, and a tongue dripping with venom.[23] To the modern sceptic, of course, the evil eye is a myth. The aggression really begins with the 'normal' members of society, who are using this label to exclude some more abnormal-seeming member, perhaps someone deformed or diseased: the so-called possessor of the evil eye is in fact the victim of social aggression.[24] Similarly, powers of fascination or poisoning are often ascribed to women, because they cannot participate in the normal aggressive activities of men. The phallic amulet or statue implied a retaliation in kind to the witch's *effascinationes*, but it also reinforced male supremacy over women.[25] To accuse someone of stinking, an example relevant to the *Epodes*, is again to imply that they have initiated an offensive attack, but it is also to exclude them from normal society. In the case of *literary* abuse, the writer himself has the power to initiate the idea that the aggressor or victim is to be identified with malice and cruelty. Take this example from a rhetorical textbook of a malevolent man stalking the forum like a snake with poisoned fangs in search of victims:[26]

iste qui cotidie per forum medium tamquam iubatus draco serpit, dentibus aduncis, aspectu venenato, spiritu rabido, circum inspectans huc et illuc si quem reperiat cui aliquid mali faucibus adflare, ore adtingere, dentibus insecare, lingua aspergere possit.

The man who slinks every day through the forum like a crested snake, with fanged teeth, a poisonous stare, and foaming breath, glancing about him this way and that, looking for someone to blast with his venomous fumes, taint with his mouth, slice with his teeth, and spray with his tongue.

ubiquitous at Pompeii, particularly in connection with food (outside restaurants, food-shops, ovens, etc.).

22 Grattius *Cynergetica* 406.

23 *Met.* 2. 776–7: *nusquam recta acies, livent rubigine dentes, / pectora felle virent, lingua est suffusa veneno.*

24 See Siebers (1983: 39–40).

25 See Richlin (1981: 58 ff.) on Priapus as a threatening male figure.

26 *Rhet. ad Herennium* 4. 62. Cf. Dem. *Adv. Aristogeit.* 1. 52.

The example is there to illustrate how similes can be used to excite hatred (*ut in odium adducat*). In other words, it shows how to blight someone's character by accusing them of blighting others.

The evil eye, the symmetrical conflict between protector and aggressor, sparring displays of male potency: all these give the *Epodes* their thrust and bite. Virility and strength, concentrated in words like *vis, vires, potior, posse, valere, efficax, virtus*, are the issues at stake in each conflict.[27] Horace's own iambic potency, however, is at risk throughout. The rage of frustrated love and impotence is as much a part of his persona as straightforward aggression. The language of phallic machismo is used to set in play some of the discrepancies between traditional Roman masculinity and the effeminacy of weakling poets. Despite all Horace's bravado, we are left in some doubt about his own masculine virtues, especially as he makes no secret in his poems of his *cognomen* Flaccus,[28] and tells Maecenas in *Epode* 1 that he is too weak to go to war, unlike more virile men (*non mollis viros*, 1. 10). The limp *fascinum* of 8 and 12 is, paradoxically, the greatest gesture of abuse towards repellent women, especially when Horace is 'known' to be potent with another mistress.[29] Horace is left with words as his instruments of revenge for emasculating or making offensive the powerful men and women who attack him.[30] The final poem of the book (17), however, is a stalemate between the *vires* of Horace and Canidia: he admits that her knowledge is efficacious (*efficax*, 1), and makes his own poems defer to her books of spells (*libros carminum valentium*, 4); she deplores

[27] *Vis*, 6. 6, 5. 94; *vires*, 8. 2; *potior*, 15. 13; *valere*, 5. 62, 5. 87, 11. 11, 17. 4; *posse*, 6. 3, 12. 15, 17. 45, 78, 79; *efficax*, 3. 17, 17. 1; *virtus*, 15. 11, 9. 26, 16. 39. A comprehensive survey of this theme has recently been provided by Fitzgerald (1988).

[28] 15. 12: *nam si quid in Flacco viri est* ('if limp Horace is man enough').

[29] 12. 14–16: *Inachia langues minus ac me; / Inachiam ter nocte potes, mihi semper ad unum / mollis opus* ('You are stiffer with Inachia than with me. You manage her three times a night; you always go limp at the first attempt with me'). Cf. 8. 17.

[30] e.g. in 7, Roman soldiers rushing into war are characterized as the victims of divine possession or infatuation: blind fury (*furor caecus*, 14), bitter violence (*vis acrior*, 14) and cruel fate (*acerba fata*, 17) drive them on, stupefying their minds: *mentes . . . perculsae stupent* (16).

the fact that her own magic has no strength against him (*artis in te nil agentis*, 81).

Horace's stress on the symmetry or sympathy between himself and his enemies often amounts almost to an identification of the two. In *Epode* 2, for example, he chooses to satirize a hypocritical moneylender who idealizes country life without being able to leave the town, and we know from *Epode* 3 that Horace himself cannot stomach country life despite playing at it on his Sabine farm. Until the last lines of the poem, when the speaker is identified (*haec ubi locutus faenerator Alfius*, 67), we assume that it is Horace himself, especially as the speaker pictures himself with a bronzed Sabine or Apulian wife (*Sabina qualis aut perusta solibus / pernicis uxor Apuli*, 2. 41–2).[31] Similarly, in *Epode* 4, Horace victimizes a military tribune with servile origins (*contra . . . servilem manum*, 19); yet he himself was once a military tribune and the son of a freedman.[32] Again, in *Epodes* 8 and 12, he anatomizes the decaying bodies of two lecherous old women, comparing them to cows with diarrhoea and black elephants.[33] A novel explanation of this is that these women are bodily figures for the *wrong* kind of literary style: ugly, putrid, and misproportioned.[34] That is, however, to apply to the *Epodes* the same criteria Horace demands for his more decorous poetry. It fits better with the consciously 'horrid' character of this kind of writing to say that the grotesque bodies that exhibit black teeth (*dens ater*, 8. 3) and bad odours (*malus odor*, 12. 7–8) are fleshly versions of the physical ugliness, mordant tone, and unseductive aura of the iambics themselves.[35]

Horace's conflict with the witch Canidia is his most persistent one, and here the closest symmetry can be found: Horace responds to her *venena* (5. 22, 87; 17. 35) with his own verbal venom; meets her spells and incantations with his

[31] Horace as an Apulian: see *Sat*. 2. 1. 34.

[32] See J. G. W. Henderson (1987: 111); Fitzgerald (1988: 182).

[33] 8. 6: *crudae bovis*; 12. 1: *nigris barris*.

[34] See Clayman (1980: 79–80) on *Epode* 8, and (1975) on *Epodes* 8 and 12 as embodiments of bad literary taste.

[35] J. G. W. Henderson (1987: 116) on *Epode* 8: 'The poem does "incarnate" iambic writing as a pharmacological brew of *carmen*.'

own ἐπῳδαί and *carmina*; bites back when she gnaws with her savage teeth (*saeva dente livido*, 5. 47).[36] If the ugly old women's bodies are the flesh of the iambic style, Canidia's malignant afflatus is its inspiration.[37] The poem that most graphically illustrates this circular quality in iambic malice is *Epode* 6, with its self-conscious allusions to Archilochus and Hipponax:[38]

> cave, cave: namque in malos asperrimus
> parata tollo cornua,
> qualis Lycambae spretus infido gener,
> aut acer hostis Bupalo. (11–14)

Beware, beware: my horns are out against my foes, like Archilochus, spurned by his treacherous father-in-law, or Hipponax, the fierce enemy of Bupalus.

Here Horace hounds a savage dog that pursues him with poisoned teeth (*atro dente me petiverit*, 15), and promises to return bite for bite (*me remorsurum*, 4) in his own vicious way.[39]

All this dog-eat-dog savagery gives the book a rather unprepossessing character. The *Epodes* could well be described as a book of anti-poems, where Horace wishes *bon voyage* to someone and hopes he drowns, raves about the countryside and turns out to be a cynical moneylender, dissects an old woman's putrid body and tells her she disgusts him, and curses his kind patron with bad breath. We begin to wonder who is the real aggressor, these hideous poetic characters or Horace himself.

This sympathy between two bitterly opposed enemies becomes significant when we come to consider the role of garlic in *Epode* 3. The traditional view is simply that Maecenas fed Horace on garlic in order to explode his humbug about the joys of country life.[40] *O dura ilia messorum*

[36] At 5. 87 ff. the *puer* victimized by Canidia responds with his own curse on her.

[37] Cf. *Sat.* 2. 8. 95–6: *velut illis / Canidia adflasset* ('as though Canidia had breathed on them').

[38] See E. A. Schmidt (1977).

[39] Compare the teeth of the repulsive old woman (*dens ater*, 8. 3) and the fierce teeth of Canidia (*saeva dente livido*, 5. 47).

[40] See Setaoli (1981: 1701); Hermann (1953): Horace ate the garlic at his Sabine farm; Turolla (1957): Horace ate a *moretum* during an excursion to the country with Maecenas, who offered jokingly to eat garlic himself.

(4, 'Harvesters with your tough guts') appears to be an allusion to an idyllic scene in *Eclogue* 2 where Thestylis pounds up garlic and thyme (aromatic herbs: *herbas olentis*, 2. 11) into a *pesto* for her weary harvesters. *Dura ilia* sets up an unspoken contrast with urban effeminacy (*mollitia*), the kind Alfius was guilty of in *Epode* 2.

This opposition may be in play, but the possibilities of garlic are far wider than that. In fact, the complexity of the Roman's attitude to garlic pinpoints a confusion in their cultural identity as a whole. Garlic's most inescapable quality was its smell. As one Roman etymologist puts it, garlic is called *allium* in Latin *quod oleat*, 'because it smells'.[41] All the ancient proverbs that use garlic suggest that contact with the herb involves some unpleasant undertaking or duty: the responsibility lingers on the person who has touched it. Pomponius, for example, gives us *aleo* [*si*] *ludam, sane meae male olant manus* ('If I play with garlic, my hands are bound to stink').[42] And in Terence's *Phormio* a rustic caution is apparently derived from garlic cookery: *tute hoc intristi: tibi omnest exedendum* ('You crumbled it up: you've got to eat it').[43]

On any cultural scale in antiquity, garlic was at the opposite extreme from sweet perfumes.[44] At the Greek festival Skirophoria, married women fasted and ate garlic in a symbolic break from their husbands which reaffirmed their marital status; the Adonia was a festival for prostitutes where seductive perfumes and quickly withering pot-herbs symbolized the charms and sterility of transient love.[45] Garlic, for the Romans, smacked of the honest but unsubtle habits

[41] Isid. *Etym.* 17. 10. 4.

[42] *Atell.* 6.

[43] *Phorm.* 318. Donatus ad loc. says this refers to a rustic garlic *pesto*: *Hoc autem inter rusticos de alliato moretario dici solet*. Henisch (1976: 107) quotes a medieval French recipe along the same lines: 'The mortar always smells of garlic' (perhaps a reference to humble origins); Rios (1986) quotes a Spanish proverb which seems to be using garlic as a social leveller: 'Garlic and pure wine, and then you will know who each man is.'

[44] Greek *σκόρδον* can be used of human excrement: see *Papyrus Graecus Holmiensis* ed. O. Lagercrantz, Uppsala 1906. Presumably this is on the grounds of its smell. Cf. Latin *oletum*: human excrement (Paul. Fest. 203M).

[45] See Detienne (1977: 80, 101).

of their ancestors, who, as the satirist Varro once said, were still men of good spirit, however much onion and garlic they had on their breath (an untranslatable pun on *anima*, 'spirit' or 'breath').[46] Perfume, at the other extreme, had the sinister and seductive aura of Greece and the Orient. For Plautus, as we have seen, garlic is always a sign of the provincial aspects of Roman taste (usually contrasted with exotic perfumes or spices): the reek of a rustic slave, the whiff of the Roman galleys, or the trademark of an incompetent cooking-school.[47] Garlic-eating was also a characteristic that the Romans transferred to their own barbarian inferiors: a late Roman bishop, Sidonius, shudders at the thought of living among Burgundians belching out garlic fumes at nine in the morning; and the Romans depised the Egyptians for worshipping this humble plant as a god (in the anthropological tradition begun by Herodotus in which foreigners were seen to invert one's own practices).[48]

Civilized Romans emphasized their ideal separation from both their crude ancestors and foreign effeminates by despising both extremes. The emperor Vespasian once said to a soldier who smelled of perfume: *maluissem alium oboluisses* ('I'd rather you smelled of garlic').[49] Because garlic compromised social activity, it was shunned by people of taste. Pliny and Columella give hints on when to plant it and how to cook it so that it will not taint the breath, or what to eat with it to disguise the smell.[50] Garlic is conspicuously absent from Apicius' cookery book, included in only a handful of recipes (just as Mrs Beeton only includes it in her recipe for Indian chutney, a permissible foreign dish).[51] That is hardly surprising. If one function of such technical books was to disseminate inherited knowledge from the élite to the self-taught bourgeoisie, the latter was the class which, in its striving for social acceptance, was most likely to eschew

[46] Varro *Men.* 63: *avi et atavi nostri, cum alium ac cepe eorum verba olerent, tamen optime animati erant.*

[47] Plaut. *Most.* 39; *Poen.* 1314; *Pseud.* 814.

[48] Sidon. *Carm.* 12. 14–15; Plin. *NH* 19. 32. 101.

[49] Suet. *Vesp.* 8. 3.

[50] Plin. *NH* 19. 34. 111, 113; Col. 11. 3. 22.

[51] Apic. 4. 1. 3; 9. 13. 3.

garlic.[52] Another clue about the civilizing of appetite comes from Pliny. While talking about the flatulent powers of radishes, he says that they are a *cibus inliberalis*, not food for a gentleman.[53] So garlic marked a sharp divide in the social stratification of cuisine, as so often in later Europe. Roman élite cooking generally lacked garlic; peasant cooking reeked of it. In the *Moretum*, for example, a soft cheese is flavoured with four heads of garlic: the earthy peasant Simulus digs up the bulbs from his garden, cursing the fumes as he pounds them in the mortar: *inmeritoque furens dicit convicia fumo*.[54] But it has no place in the genteel *moretaria* recipes of Apicius and Columella.[55]

The issue of flavouring has an extra dimension in literary works, where the mere mention of the word 'garlic' gives the work a flavour which cannot be expunged. Garlic gives roughness and bite to the slender *Eclogues*.[56] It is hard to imagine it appearing in the works of Jane Austen or Henry James, or not appearing in Shakespeare or Smollett. Aristophanes, in *Peace*, has a figure who represents the spirit of war mashing up garlic, leeks, cheese, and honey in a mortar: a μυττωτός, the Greek equivalent of a *moretum*, symbolizes the peace between the Greek states.[57] Each ingredient represents a particular state: leeks are a pun on the name of Prasiae, Sicily was famous for its cheese, Athens for its honey, while Megara was Greece's largest exporter of garlic. Megara was always badly treated by Athens; in *Acharnians*, a poor inhabitant of Megara pleads to be given just a clove of garlic, a pathetic reminder of his former prosperity.[58] Garlic is partly making a political point, then, but the fact that it appears at all among the ingredients of this comic μυττωτός is also an indication of how far the author will go to give his

[52] See Goody (1977: 129–45): 'The recipe, the prescription and the experiment', esp. 136, 143.

[53] *NH* 19. 79.

[54] *Moretum* 108.

[55] Apic. 1. 21; Col. 12. 59.

[56] Virg. *Ecl*. 2. 11 (garlic is the equivalent as a taste to *raucus* ('harsh, grating') at 2. 12 as a sound).

[57] Ar. *Peace* 228–9.

[58] Ar. *Ach*. 813.

work a flavour of pungent realism and comic licence. When garlic appears in Aristophanes' plays, whether as part of the cheesy mess which symbolized the peace, infusing a warlike spirit into a politically minded sausage-seller, or making a rebellious wife repellent to her husband, it is one of the wickeder ingredients of what Plutarch called the Greek comic poets' 'wicked pickles'.[59] The pseudo-Roman meal served by the French in the middle of Smollett's *Peregrine Pickle* (where there is a more obvious connection between the feast and the title of the book) has one dish so garlicky that it causes great offence to one of the guests. As we saw, Smollett had a very English prejudice against Gallic garlic-eating, and here he seems to be misrepresenting Apicius' smart Roman cooking as yet another filthy foreign cuisine.

When Horace complains about the effect of the virulent herb on his stomach, he is partly saying, then, that he is not a man with guts, an old Roman or a peasant, but an effete and over-civilized hypocrite. This is one dimension of the poem, but the fact that garlic is mentioned at all also gives the book an uncompromisingly earthy aroma. Garlic had other properties besides, not usually noted by the poem's critics, that make it more than just a bad smell in the *Epodes*. For the Romans, as for us, it was a plant of contradictions, and this is what gave it its symbolic power. Above all, garlic had deep-rooted links with iambic poetry.

The first thing that Pliny says about garlic is that it has a *magna vis*, great potency.[60] The concept is typical of Greek and Roman natural history, which analysed and classified the plant world as an analogue of human power struggles: each plant had its peculiar *vis* or *virtus* against other plants (an idea which persists in the archaic English word 'virtue', describing some efficacious property). One anonymous prayer addresses 'every powerful herb' (*omnes potentes herbas*).[61] In Horace's *Epodes*, Canidia is *efficax* (17. 1) because she knows about poisonous plants (5. 21, 67–8), while the *puer* claims that her poisons are powerless (*non valent*, 5.

[59] *Mor.* 68c.
[60] Plin. *NH* 20. 50.
[61] Minor Latin Poets (Loeb pp. 344 ff.).

87).[62] This *magna vis* is also highly reminiscent of Quintilian's description of Archilochus' iambic fury: *summa in hoc vis elocutionis*. And in all our sources, the pungent or bitter taste of garlic conjures up the same adjectives that are used elsewhere of iambics: *asper* (pungent), *acer* (acrid), and *mordax* (biting).[63] Horace, in particular, uses these same adjectives to describe the flavour of his own iambic style: *acer hostis* (6. 14, a fierce enemy); *remorsurum* (6. 4, bite back), *dens* (8. 3, 5. 47, 6. 15, tooth), and, at *Sat*. 1. 4. 93, *lividus et mordax* (bruised and biting), of the vicious poet.

Garlic also inspired or reinforced a passionate, hot-headed temperament. According to Galen, it heated the body and ought to be avoided by people of a choleric disposition.[64] That is, garlic had a natural affinity with a bilious disposition, and caused a physiological imbalance when given to people who were already hot by nature. As we have seen, the iambic *persona* was traditionally fervid and bilious. Horace describes himself as sun-loving and irascible (*solibus aptum, / irasci celerem*).[65] At *Sat*. 2. 1. 34, we learn that he comes from the borders of Apulia (linked with suntanned people at *Epod*. 2. 41–2: *perusta solibus / pernicis uxor Apuli*). Again, another of the many *vitia* Pliny ascribes to garlic is that it excites thirst: *sitim gignit*.[66] In *Epode* 3, Horace's overheated body is compared to thirsty Apulia enduring the heat of the dog days: *siderum insedit vapor / siticulosae Apuliae* (15–16).[67] *Vapor* is used not only of meteorological heat, but also of

[62] Galen writes on the different *powers* of foodstuffs: *Περὶ Τροφέων Δυναμέως*.

[63] Cels. 3. 22. 11, *cibus . . . acer ut alium*; 4. 4, *quae sunt acerrima, id est in sinapi alium, cepam*; Plin. *NH* 19. 111; Col. 9. 14. 3, *fetentibus acrimoniis alii vel ceparum*; Prud. *Perist*. 10. 260, *venerare acerbum cepe, mordax alium*.

[64] *De Alim. fac*. 2. 71.

[65] *Ep*. 1. 20. 24–5. He also describes himself as *praecanum*—prematurely grey—another link with Canidia.

[66] *NH* 20. 57.

[67] Dog days = *canicula* (cf. *sidus fervidum, Epod*. 1. 27) reminds us of other vicious dogs in the *Epodes*: 6. 1, *canis*; 5. 23, *ieiunae canis*; 17. 12, Hector's body left to *canibus*—and a possible pun on Canidia (despite the difference in vowel-length), who appears symmetrically opposite to these lines on the dog days (15–16) at 3. 7–8. Cf. Pers. 3. 5, where the heat of the dog-day sun (*siccas canicula messes / iam dudum coquit*) is the background to an explosion of bile within the poet's body (*turgescit vitrea bilis*, 8).

heat within the body, caused by fever or violent emotion.[68] Horace the Apulian is a microcosm of the temperament or temperature of his own native land.

Garlic was also known in antiquity for its protective powers. How exactly it was meant to be efficacious, whether it imparted a warlike spirit or gave apotropaic protection against evil, is not clear. There was presumably some connection with the unmistakable pungency of the plant: if it visibly warded off human beings, what might it not do against invisible powers? Perhaps it is most realistic to accept that all these powers were subsumed under the nebulous phrase *magna vis*: this might be both an aggressive and a protective power. Garlic was eaten, at least by *Greek* soldiers,[69] before they went into battle, and smeared on to cocks before a cockfight. The sausage-seller in Aristophanes' *Knights* is fed garlic to make him a better political fighter.[70] Cicero quotes a Greek proverb whose full version is *ἵνα μὴ ποτε φάγῃ σκόροδα, μηδὲ κυάμους* ('May I never eat garlic or beans').[71] This is the Greek equivalent of English 'anything for a quiet life'. As the scholiast explains, garlic was eaten by soldiers going to war, while beans were eaten by jurymen to keep them awake. It may not be a coincidence, of course, that the smell of garlic and the flatulent properties of beans are ideal examples of the sort of alimentary choices that have to be lived with. After *Epode* 1, *Ibis liburnis*, Horace's apology for not going to war (or not writing epic) and *Epode* 2, where he appeals for a quiet life *procul negotiis*, away from war and the city,[72] we are perhaps meant to assume that Maecenas gave Horace the garlic to induce him to go to war or write epic, and that his innate weakness or irritability made it too much for him to stomach. In other words, the poem is

[68] e.g. Sen. *Phaed.* 640–1: *pectus insanum vapor / amorque torret* ('the heat of love scorches my heart into insanity'); Apul. *Met.* 10. 2: *fluctuare . . . vaporibus febrium* ('toss with the heat of a fever'). *Insideo* can also be used of physiological oppression by fever (cf. Plin. *Ep.* 7. 19. 3, *insident febres*) or bile (Sen. *Ep.* 55. 2, *sive bilis insederat faucibus*).

[69] And by Plautus' Roman oarsmen (*Poen.* 1313–14).

[70] *Eq.* 494: *ἵν' ἄμεινον ὦ τᾶν ἐσκοροδισμένος μάχῃ.*

[71] *Ad Att.* 13. 42. 3; see Ar. *Lys.* 690. Leutsch-Schneidewin i. 421. Cf. Xen. *Symp.* 4. 9.

[72] 2. 5–7.

another concealed *recusatio*. Perhaps it also suggests that he does not even have the stomach for full-blooded iambics or that iambic rage is necessarily crippling for the poet.

Garlic also had a reputation as an antidote to poisons or evil spirits. Pliny lists as its natural enemies snakes, scorpions, or perhaps all beasts, shrew-mouse bites, poisons like aconite and henbane, and dog bites.[73] He attributes garlic's protective powers to its smell, and uses *valere* (to be strong) and *debellare* (to combat) to describe its antidotal effect. In Western folklore, the plant is, of course, infamous as an amulet against witches and vampires. Persius mentions the oriental custom of eating three cloves of garlic every morning as a protection against evil spirits; again, whether because of its smell or its vaguer 'powers' is not clear.[74] The Romans also left dishes of garlic at the crossroad shrines of the witch-goddess Hecate; and Odysseus ate a plant which was probably a kind of garlic to avoid being turned into a pig by the witch Circe. It may be that garlic was ascribed this power against witchcraft not because its chemical composition was the *opposite* of a witch's potions, but more because it was somehow similar or sympathetic.

Vis and *virtus*, of course, conjure up ideas of masculine vigour, and the distinction between sexual and warlike vigour was often blurred. While Roman babies were wearing tiny phalluses round their necks, their Greek and Egyptian counterparts were wearing cloves of garlic. This leads to another contradiction: the *vis* of garlic makes it both apotropaic and aphrodisiac. According to Pliny, garlic mixed with green coriander made a man lecherous (*venerem stimulare*); and it has kept its reputation for being the food of love.[75] Chaucer's Summoner relished garlic, onions, and leeks, and was as hot-blooded and lecherous as a sparrow.[76] Garlic, in Greek at least, was one of many metaphors for the phallus,

[73] *NH* 20. 50. [74] 5. 188. [75] *NH* 20. 57.

[76] 'A Somonour was ther with us in that place, / That hadde a fyr-reed cherubynnes face, / For saucefleem he was, with eyen narwe. / As hoot he was and lecherous as a sparwe / . . . / Wel loved he garleek, oynons, and eek lekes, / And for to drynken strong wyn, reed as blood.' *Canterbury Tales*, 623–36.

and is the focus of innuendo in several scenes in Aristophanes.[77] At the same time, its smell was its most unavoidable and off-putting characteristic.[78] Discussions of this smell, and the smell of other pungent vegetables, often involve sexual innuendo of a negative kind.[79] In Aristophanes' *Thesmophoriazusae*, rebellious women eat garlic in order to disappoint their husbands.[80] Another comic philanderer abstains from leeks in order to be able to kiss his lover.[81] Martial, too, recommends only limited sexual contact after a dose of leeks (13. 18–19, *fila Tarentini graviter redolentia porri / edisti quotiens, oscula clusa dato.*) In Xenophon's *Symposium*, Niceratus is accused of eating onions to pretend to his wife that no one has kissed him.[82] Garlic, as we have seen, was also eaten at festivals of abstinence as a physical repellant.[83] So the herb is ambiguous in the context of love. It fills the eater with lust, but is anaphrodisiac for the beloved; it produces masculine vigour, but is off-putting because of its fetid smell. Like the protective phalluses of the ancient world, it was prophylactic as well as erogenous.

Garlic and iambic poetry seem, then, to be tied together by the same contradictory or self-defeating properties. Iambics are traditionally dashed off in a mood of heat and passion, typical of the effects of garlic; like garlic, they bristle

[77] See Seager (1983) for innuendo surrounding garlic in Aristophanes: e.g. *Ach.* 164–6; *Eq.* 494, 946; *Vesp.* 1172. He suggests that when the women admit at Ar. *Thesm.* 493–6 that they have been eating garlic to disappoint their husbands, this conceals another confession: they are trying to hide the taste of another kind of garlic, the phallus.

[78] Garlic coupled with onion: e.g. Plin. *NH* 2. 16, *fetidas cepas, alia et similia*; Cels. 3. 20. 1, *odore foedo movent, qualis est . . . acetum alium cepa*; Col. 9. 14. 3, *fetentibus acrimoniis alii vel ceparum*; Sidon. *Carm.* 12. 14–15, *cui non alia sordidaeque cepae / ructant mane novo decem apparatus*; Var. *Men.* 63; Plaut. *Most.* 39. Contrast *Moretum* 99–100, *fragrantia . . . alia.* Kenney (1984: ad loc.) takes this to imply a pungent or repellent smell: 'this seems to be the earliest example connoting a repellent smell (not registered by *OLD* s.v. *fragro* b)'. But it is worth comparing this fragrant smell with that of the harvest *pesto* mixed by Thestylis in *Eclogue* 2 (*olentis herbas*): fresh garlic might have had a pleasantly pungent smell, temporarily, at least.

[79] See Watson (1983: 83).

[80] Ar. *Thes.* 493–6.

[81] Ath. 13. 572c = K2. 385.

[82] Xen. *Symp.* 4. 8.

[83] See Seager (1983: 140); *Et. Mag.* 769. 1.

with priapic vigour and ward off witch's poisons, snakebites, and evil vapours. At the same time, they are the opposite of sweet and sympathetic antidotes: the flavour of iambics is 'bitter', 'acerbic', and 'biting'. So garlic seems to be a uniquely suitable fuel for Horace's iambic rage: iambics are the unprepossessing, vile-smelling exhalations that protect society from its enemies.[84]

Horace's poem is less a description of an amusing incident, then, than a physiology of iambic anger. The most pungent metaphors for iambic rage and abuse, venom and bile, are restored to their original bodily sites, reincarnated. The venomous style is explained in appropriately circular terms: garlic, ingested, becomes the flavour of iambic poetry as well as the object of iambic abuse. Horace exploits the double nature of the relationship between food and its eater: food is both an external object, to be consumed, and an incorporated part of the eater's identity. He gives a specifically iambic flavour to the neutral metaphor of the poet's 'diet'.[85] His hyperbolic claim that the garlic was *venenum* (5) is itself an example of 'venomous' Archilochean invective; his rage at being poisoned is the result of physical bile *and* an outburst of verbal 'bile'.[86] To make this literally true, Horace actually breathes out the word for bile in the course of the spell, concealing it in the middle of *fefellit* ('tricked', 7).

Two more physiological traditions are in play: first, the belief that an eater absorbed the qualities (the innate 'mood' or 'temperament') of the foodstuffs he ate or of their environment; [87] and secondly the belief that certain organs, the *praecordia* and the *iecur*, were the seats both of the digestion and of the emotions, especially anger and love.[88] Horace's

[84] The fragments of Hipponax contain references to a *μυττωτός* (= *moretum*, 35. 2) and a *κυκεών* (mixed barley drink, 26. 2), both ancient garlic dishes, so the specific connection with garlic may have antedated Horace.

[85] See Bramble (1974: 51) on this metaphor, and the use of *alere* to describe the rearing of an orator: e.g. Cic. *Brut.* 126; *de Off.* 1. 105; Hor. *AP* 30.

[86] See Bramble (1974: 29 ff.) on the 'rejuvenation' of stylistic metaphors.

[87] See Gourévitch (1974).

[88] See Onians (1954: 40–3), arguing that the *praecordia*, regarded as one of the seats of consciousness, were the lungs; ibid. 88–9 on the liver as seat of love and anger.

verbal poison is conceived as the result of food poisoning, which simultaneously involves an imbalance in temper. We cannot know the exact chemical basis of his anger, as Graeco-Roman physiology ascribed multiple functions to lungs and liver, and different explanations were given for the interactions between body and mind: it is either an excess of bile produced as a reaction to bad digestion or, more directly, poison converted from the food into a bodily or emotional fluid or *afflatus*.

The transformation of emotion into the facts of eating and having indigestion carries its own comic and undignified connotations.[89] Although Horace equates garlic with poison in the poem, this is of course a joke. Pliny does say that, if taken in large quantities, garlic harms the stomach, but this does not make it into a deadly *venenum*.[90] It is Horace himself, then, who makes the transformation. He is implying that, however much potency it may give a potential warrior, garlic is still overpowering for a weaker man; it can still knock one back in large doses. In the central part of the poem, Horace equates garlic with a savage poison, vipers' blood (*viperinus cruor*, 6), and with food infected by the witch Canidia.[91] His garlic-filled body becomes an inverted microcosm of the ideal land described in *Epode* 16, free of vipers (52), Colchian witches (58), contagious diseases (61), and astral blight (61–2); a land which is neither too wet nor too dry:

> nulla nocent pectori contagia, nullius astri
> gregem aestuosa torret impotentia. (61–2)

No contagion blights the flocks, no sweltering astral distemper scorches the herds.

In other words, he transforms garlic into the very sources of poison to which it is traditionally the antidote: the snakes,

[89] Fraenkel (1957: 68) notes a similarity with Cat. 44. There the poet complains of a cold he has caught from reading a particularly 'frigid' book and decides to get revenge by buying up *omnia venena* from the booksellers and poisoning his friend.

[90] *NH* 20. 57.

[91] *An malas tractavit Canidia dapes* (7–8, 'or did Canidia taint the feast?'); a witch's brew has been suggested already by *incoctus herbis* (7 'cooked with herbs').

scorpions and dogs, aconite, and henbane listed by Pliny, and the evil spirits mentioned by Persius.[92]

Watson insists that all the comparisons in the poem are made with heat-producing qualities as the point of connection.[93] It could be argued that, pharmacologically, garlic was an antidote because it was heat-producing and its enemy poisons were cold (hemlock, mentioned in line 2, was the archetypal cold poison). Heat is certainly one of the connections being made in the comparisons to come between garlic and the thirsty dogdays (*siderum . . . vapor*, 15) and Nessus' shirt (*inarsit aestuosius*, 18). However, garlic may equally have been regarded as an antidote on homeopathic grounds, because the bite or poison of serpents, beasts, or plants was similar to the taste of garlic: both were *mordax, acerbus,* or *acer*. The fact that Dioscorides names two varieties 'adder-garlic' and 'viper-garlic' [94] shows how snake-like qualities became part of the identity of the antidote. In other words, a sympathy between poison and antidote was there to be exploited.

This sympathy emerges in the Greek and Latin words for poison, *φάρμακον*, which can also mean a remedy, and *venenum*, which can also mean a love charm.[95] *Venenum* (5) is an appropriately ambiguous expression for the sensation in Horace's entrails. The *praecordia* are the seat of the emotions as well as of the digestion, of love as well as anger. Horace's description of heat or poison in his body could almost be read as a description of love-sickness, which elsewhere in the *Epodes* is cast in terms of burning heat or lethargy. At 11. 15–16 Horace's anger and jealousy towards a sexual rival causes an eruption of bile: *quodsi meis inaestuet praecordiis* /

[92] Plin. *NH* 20. 23. 50; Pers. 5. 187–8.

[93] Watson (1983: 84).

[94] Vind. C. 249 v. 117: ἀλιουμ κολυβρινουμ, ἀλιουμ βιπερινουμ. Cf. ὀφιοσκοροδον, 'snake garlic'.

[95] Mauss (1954: 127 n. 101) compares Eng. gift/German Gift; Latin and Greek *dosis*—gift/poison; Latin *venenum*; Greek φίλτρον, φάρμακον. Derrida (1981) discusses the ambiguities of *pharmakon* with reference to Plato's philosophy; Goldhill (1986: esp. 34) discusses these ambiguities in the erotic context of Theoc. 11. The κυκεών mentioned by Hipponax is to be a φάρμακον πονηρίης (fr. 39. 4), and φάρμακον is a repeated word in fr. 5.

libera bilis ('once free-flowing bile is seething in my intestines').[96] In *Epode* 14, love is described as though it were a *venenum*, a love-potion or a poison, seeping through the veins. Love is the poison that weakens and breaks the iambic spell: *deus, deus nam me vetat / inceptos, olim promissum carmen, iambos / ad umbilicum adducere* (14. 6–8, 'It is the god who thwarts my iambics in their progress, my once-promised spell, and keeps them from their end'). The antagonism between Horace and Canidia, too, is based on the manipulative magic of love. *Epode* 5 describes the attempts of the *venefica* to brew a love-potion (*amoris poculum*, 38; cf. *potionibus,* 73, *poculum*, 78) from herbal poisons (*venena*, 21–2, 62) and the liver of a young boy (*iecur*, the seat of love, 37).[97] Under the influence of Canidia's potions in *Epode* 17, Horace prays for release from burning pain:

ardeo
quantum neque atro delibutus Hercules
Nessi cruore nec Sicana fervida
virens in Aetna flamma. (17. 30–3)

I am on fire, more so than Hercules doused in Nessus' dark blood, or the leaping flame on Sicilian Etna.

These similes bring us back full circle to Horace's description of his burning entrails in *Epode* 3. 15–18: the shirt of Nessus and the geographical reference to southern Italy. The metaphors of love-sickness and bewitchment feed Horace's description of his burning *praecordia*, a confusion made possible by the double function of the organs of love and anger. Bewitchment is used in the *Epodes* as a physical explanation of those metaphors: it identifies the malevolent agent and vindicates malevolent revenge.

The mythological *exemplum* at the heart of the poem—the story of Medea anointing Jason to protect him from the fiery

[96] Love as an affliction: 11. 2, *amore percussum gravi* (*percutere* can also be used of a snake's bite); as burning: *ardor*, 11. 27; cf. 14. 9–10, *arsisse*; 14. 13–14, *ureris ipse miser*; as a narcotic: 14. 1–4, *mollis inertia cur tantam diffuderit imis / oblivionem sensibus / pocula Lethaeos ut si ducentia somnos / arente fauce traxerim* ('why faint inertia has poured forgetfulness down to the depths of my senses, as though I had drained a potion down my burning throat to bring on the sleep of the dead').

[97] At 5. 81–2 Canidia prays that her victim may burn with love like burning pitch: *quam non amore sic meo flagres uti / bitumen atris ignibus*.

bulls and sending poisoned gifts to incinerate his mistress—tells us more about the mysterious identity of *venenum*. Lines 11–12 have presented the chief dilemma: *ignota tauris illigaturum iuga / perunxit hoc Iasonem* ('This is what she smeared on Jason to help him yoke the unsuspecting bulls'). Watson argues that the context demands another illustration of the heat-producing and destructive effects of garlic, yet the salve which Medea gave to Jason was obviously heat-resistant.[98] If Horace were deliberately making the story comic by introducing garlic as a joke repellant on the grounds of its smell, this would spoil the surprise ending of the poem; in any case, it would be pointless for Jason to be able to repel the bulls when he wanted to yoke them.

For a solution, Watson appeals to the dual powers of many ancient herbs which were both protective and deadly.[99] Pliny provides a good illustration of this when he tells us that some powerful poisons can cancel out the deadly effect of other ones: 'Scorpions become numb when they touch aconite, and go into a pale stupor, accepting their defeat. But white hellebore comes to their aid, dispelling their torpor with its touch; the aconite yields to two evil agents, one peculiar to itself and one common to all creatures'.[100] As Watson says, 'There is no difficulty in ascribing ambivalence of effect to a single *pharmakon*'. Medea's remedy is as much an example of the herb's natural *vis* as a bathetic joke about its pungent smell.[101]

This tiny episode is also typical of Horace's treatment of enmity in the *Epodes*, which makes us see the nemesis or antithesis of every conflict, the symmetry between antagonists. Until now, the poet has seemed to be the victim of a

[98] Watson (1983: 84 ff.). Cf. Pind *Pyth*. 4. 221–2, *σὺν δ' ἐλαίῳ φαρμακώσαισ' ἀντίτομα / στερεᾶν ὀδυνᾶν / δῶκε χρίεσθαι*; Ap. Rhod. 3. 1305, *κούρης δέ ἑ φάρμακ' ἔρυτο*.

[99] p. 85. His parallels are: the herb *prometheion* used by Medea both for her protective salve (Apollonius 3. 851–3) and for her poisonous gifts to Glauce (Sen. *Med*. 706–9); the gorgon's blood, a deadly poison (Eur. *Ion*. 999 ff.) or a revivifying agent (Apollod. 3. 10. 33); and, significantly for the *Epodes*, the *genitalia*, associated both with love-magic (Apul. *Apol*. 34: fish used in love-magic with names sounding like words for the genitals), and, in the form of the *fascinum*, with the apotropaic magic of amulets (Plin. *NH* 28. 39).

[100] *NH* 27. 6.

[101] I have argued above that the 'powers' of garlic were indirectly associated with its smell anyway.

witch's spell, bound by bitter pains.[102] Jason's conflict with the fiery bulls illustrates a victory over an enemy. He himself becomes the one who binds and gains control: *ignota tauris illigaturum iuga* (3. 11). But the lines could also be read in another way. The witch Medea, who is love-struck by Jason (*candidum / Medea mirata est ducem*, 9–10), devises this salve not only for Jason's protection but also as her own *venenum* or love-spell, so that she herself can bind Jason, and fix on him unawares the yoke of marriage (different *ignota iuga*). Even this *pharmakon* contains its own inversion.

The second half of the story, which describes Medea's present of gifts smeared with garlic to send to Jason's mistress (*hoc delibutis . . . donis*, 10. 13), uses the mythological motif of the poisoned gift (for example Pandora's Box, or the Wooden Horse).[103] In fact, the whole of this later saga has been foreshadowed in the earlier episode of smearing and yoking. As if to emphasize the sympathy between garlic and its enemy poisons, Horace puts in what looks like a decorative detail about Medea's winged dragon. But not only is *serpente . . . alite* (14, 'a winged snake') a contradiction in terms (literally, 'a winged creeping thing'): the choice of *alite* also reminds us of *alium*, garlic (3). The deadly enemies, garlic and snakes, antidote and poison, are tied together as familiars.

The theme of the poisoned gift recurs in the final comparison, with the shirt of Nessus which burned Hercules to death (17–18). In the legend, Nessus was killed by Hercules, but gave Deianira his shirt, telling her that it was a love-charm. After sending it to Hercules to win him away from another woman, she discovered too late that it contained a deadly poison. Again one kind of *venenum*—a love-charm—was revealed to be the other—poison. The same confusion is made in Greek. In her last speech in *Trachiniae* before discovering the true nature of the shirt, Deianira refers to it as a *pharmakon* (*Tr.* 485): the ambiguity of the word

[102] For the image of binding, cf. 5. 71, *solutis ambulat*, of someone freed from a witch's power.

[103] Which of course provides a linguistic pun for Mauss (Gift = 'poison' in German, 'present' in English).

anticipates the final *peripeteia*. *Efficax*, the epithet attached to Hercules (17), is an important word in the *Epodes*, signifying power and resourcefulness (compare *efficaci scientiae*, 17. 1). In this context, it is versatile and provocative. It may seem ridiculous that Horatius Flaccus should compare himself to Hercules, the hero of the labours, but we are reminded that this was the one occasion when Hercules had no resources against the *vis* of a violent poison. With the imprecation which follows, Horace seems to be on top again. Yet the extent of his efficacy is in doubt in the poem, both in and out of the illusion that garlic is a poison.

The final section of the poem, the sting in its tail, implies that Maecenas is responsible for the garlic Horace has eaten. Until then, the garlic itself seemed to be the sole agent and victim of Horace's virulence.[104] Horace turns the tables on his patron by inflicting a cruel curse, sexual rejection caused by bad breath:[105]

Si quid umquam tale concupiveris,
 iocose Maecenas, precor
manum puella savio opponat tuo,
 extrema et in sponda cubet. (19–22)

If you ever have this sort of desire, Maecenas, you practical joker, I hope your girlfriend fights back your kisses, and gives you a wide berth in bed.

Past interpreters of the poem have tried to reconstruct the background to the situation described, and are particularly concerned that Maecenas does not himself appear to have suffered from food poisoning, when he presumably shared the same *dapes*. Perhaps the fact that Maecenas is unaffected is meant to imply that he is one of the real men (*non mollis viros*) who can stand pungent food. It also concerns them that Maecenas' punishment is not the same as Horace's.

[104] For the personification of foodstuffs which harm the body and are treated as enemies cf. Plaut. *Pseud.* 814 ff. where pungent herbs including *alium* are likened to screech-owls disembowelling the guests (820–1, *non condimentis condiunt, sed strigibus / vivis convivis intestina quae exedint*); mallows and beets personified after they have given Cicero diarrhoea: Cic. *Fam.* 7. 26. 2.

[105] Watson (1983: 80–3).

Watson, for example, asks: 'Would it not have been more appropriate for the poet to wish the same pangs rather than a sexual rebuff?'[106]

Nevertheless a sexual punishment *is* appropriate in a poem that is full of sexual undercurrents, suggested both by the aroma of garlic and by the iambic frame. Horace's pangs are phrased in the language of love-sickness; he contrasts his own effeminacy (*mollitia*) with the tough guts or loins (*dura ilia*) of peasants (*ilia* can mean the groin as well as the entrails[107]); in the central myths, scenes of sexual jealousy are enacted. It seems appropriate that, in the disputed clause *si quid umquam tale concupiveris* (19), the object of desire should be obscure. Watson believes that the object in question is garlic (compare *hoc*, 12, and *hoc*, 13).[108] But *tale* could also mean 'a joke of this kind', and *concupiveris* could be taken ambiguously, linking Maecenas' desire for garlic or for making trouble with the mythical tales of poison and sexual jealousy, and with Maecenas' own punishment, which thwarts his own *cupido*.

The Medea and Hercules myths each have a trio of protagonists; Horace, Maecenas, and the *puella* correspond to these, but it is not at all clear who should take which role. Medea was, of course, famous for her sinister cooking exploits, so it looks initially as though Horace is casting Maecenas as the witch. He holds him personally responsible for cooking a garlic dish which tasted of viper's blood and poison, and, by implication, for sending poisoned gifts. But Maecenas, the naval hero of Actium (*candide Maecenas*, 14. 5)[109] could also be Jason (called *candidum ducem*, 9–10). Horace, on the other hand, the enemy of Canidia, writer of (ἐπῳδαί) and *carmina*, is a perfect parallel for Medea: he taints the meal verbally, turning garlic into evil poison and snake's blood and tampering with myths by 'cooking' garlic

[106] p. 81.

[107] e.g. Cat. 11. 20, *omnium ilia rumpens*; Virg. *Ecl.*: 7. 26, *invidia rumpantis ut ilia Codro*.

[108] p. 81; he cites Cels. 3. 6. 1 for *concupisco* used of desire for food.

[109] Cf. 1. 11–14 where Horace swears to take Maecenas as his leader: *te . . . sequemur*.

into them.[110] In *Sat.* 2. 8, Horace's farewell to satire, the guests leave Nasidienus' party avenged (*fugimus ulti*, 93) and taste nothing of the exquisite last course, as though Canidia had blasted the food with snake-poison: *velut illis / Canidia adflasset peior serpentibus Afris.* In the central myth of *Epode* 3, Medea sends poisoned gifts, as revenge, to Jason's mistress, then flees on a winged serpent: *ulta . . . paelicem / serpente fugit alite*, 13–14. Horace uses a similar technique in both poems: it is his own comparisons that blight the food and make it poisonous. His foul afflatus is flavoured by the snake-breath of his satirical Canidia.[111] Horace, like Medea, gets revenge on a *paelex*;[112] and Medea, we remember, is the one who binds Jason in the end, just as Horace takes revenge on Maecenas at the end of the poem.[113] So there is a delicious ambiguity in the relevance of this central myth to the main story. It may not be so grotesque after all for the feeble Horace to compare himself with *efficacis Herculis.* But it looks as though we will have to live with the uncertainty.

In the final episode, Horace curses Maecenas with putrid breath and untouchability. That is to say, we are not told about bad breath itself, only about its consequences. But the evil smell of garlic is present as an unspoken oxymoron in the word *savio*, kiss (21, from *suavis*, sweet). To accuse someone of smelling is like accusing them of the evil eye, of being contagious, or of being a witch: the other party is made to seem guilty of being offensive or aggressive, and becomes society's scapegoat. Horace's epithet *olentem Maevium* (10. 2, 'stinking Maevius') distils the whole apotropaic message of that poem; by reviling the evil smell issuing from the old woman in 12, he can put the blame on her for offending his senses. The final curse, if we continue to see Horace's

[110] Cf. *incoctus* (3). Cf. the metaphorical use of *recoquere*, which is often an allusion to Medea's experiments with rejuvenating soups: see Cat. 54. 5, *Sufficio seni recocto*; Hor. *Sat.* 2. 5. 55–6, *recoctus / scriba ex quinqueviro*; Quint. 12. 6. 7, *se . . . Apollonio . . . rursus formandum ac velut recoquendum dedit.*

[111] *Afflatus* can be used of snakes, witches, or planets. Snakes: Larg. 165 on remedies for snakes' bites or poisonous breath; witches: Horace *Sat.* 2. 8. 95; planets: Petr. 2. 7, *veluti pestilenti quodam sidere afflavit*; Plin. *NH* 2. 108, *afflantur alii sidere.*

[112] Acron identifies her as Terentia, Maecenas' concubine.

[113] Cf. 15. 24, *ast ego vicissim risero* ('but I shall have the last laugh').

emotion as a physiological process, is itself a consequence of his garlic-eating: it is his own offensive exhalation.[114] So Maecenas' punishment is directly symmetrical to that of the person who wishes it on him. Horace's afflatus is the effect of the poison he has received and also its antidote. On the bed where Maecenas has made his final tryst (*extrema sponda*, 22, the edge of the bed, looks like a pun on *spondere*, to pledge) perches a baleful incubus (*in . . . cubet*, 22).[115] He will be shunned like the parricide at the beginning (1–3), for whom garlic was the imagined penalty. The symmetry of the two wishes, *edit* (3) and *opponat* (21), and the two gestures involved—the hand raised against the throat or mouth (*guttur* in 2 is chosen for its versatile meaning, both throat and gullet)—emphasizes the similarities and differences between an impious crime and bad breath.[116]

The reference to parricides also makes this poem a miniature of the larger intestinal conflicts of the *Epodes*.[117] The traditional Roman punishment for parricides was to be sewn into a sack full of serpents and other animals;[118] Horace's own stomach is full of snakes' blood (6), just like the malevolent ibis, whose name, it has been spotted, lies embedded in the first word of the *Epodes*. Many examples of the destruction of cities or enmity between cities in the *Epodes* are described in the language Horace uses of his own body, or contagious attacks against his body. His description

[114] The metaphor of stinking applied to words as though they were organically emitted can be seen in Var. *Men.* 63, *avi et atavi nostri cum verba alium ac cepe olerent* ('ancestors whose words stank of garlic and onion'); or Plaut. *Cas.* 727, *foetet tuo' mihi sermo* ('Your language stinks'). Cf. metaphorical uses of *olere* (e.g. Quint. 8. 1. 3), *redolere* (e.g. Cic. *Brut.* 285), and *sapere* (e.g. Quint. 11. 3. 182). See Assfahl (1932: 17); Bramble (1974: 50).

[115] Like the planetary exhalations lowering over Apulia in lines 15–16: *nec tantus umquam siderum insedit vapor / siticulosae Apuliae* (an etymological pun on *sidus*, planet, and *insidere*, to settle over, weigh down on; cf. Var. *LL* 7. 14, *sidera, quae quasi insidunt*).

[116] With the archaic subjunctive *edit*, Horace parodies Roman *leges* by decreeing that garlic is a suitable punishment for a parricide. This is either because the antisocial herb would make an outcast repellent; or so that Horace can jokingly make it more noxious than the traditional Greek punishment, hemlock (*cicutis alium nocentius*, 3).

[117] See Fitzgerald (1988) on the intestine as a civil metaphor.

[118] *Codex Justinianus* 9. 17: *insutus culleo et inter eius ferales angustias comprehensus serpentium contuberniis misceatur.*

of his burning *ilia* (4), for example, looks ahead to the burning of Ilium at 10. 13 (*Pallas usto vertit iram ab Ilio*) and (metaphorically) at 14. 13–14 (*quodsi non pulchrior ignis / accendit obsessam Ilion*).[119] The wrath of Achilles, the archetypal epic subject, is called *gravem / Pelidae stomachum* (literally: the heavy stomach) at *Ode* 1. 6. 5–6.[120] At 16. 2, Rome collapses under its own strength: *suis et ipsa Roma viribus ruit*, emasculated, like Horace (compare 8. 2: *viris quid enervet meas*). At 7. 5, Carthage becomes a malevolent witch or basilisk, *invidae Carthaginis* (literally, 'staring Carthage'), which duplicates the evil eye of 5. 9 (*quid . . . me intueris?*).

Epode 3 turns poetic creation into a physiological reaction, a process of ingestion, dyspepsia, anger, and breathing out. Garlic is transformed into a poison, causes and receives Horace's anger, and flavours his parting afflatus. The herb that shares the qualities of iambic *vis*[121]—heat, pungency, malodorousness, crudeness, virility—is translated into its pharmacological opposites (and familiars): snake-poison, witch's breath, animal bites. It is both the victim of the virulence it inspires and the agent of further revenge.[122] The smell of garlic, never mentioned but strongly implied, is one of many crudely physical suggestions in the *Epodes*, bad smells, sea-sickness, goats, pigs, cows' rumps,[123] and these are all self-conscious references to the offensive and unpleasing nature of iambic poetry. If wine-drinking in the *Odes* is

[119] Cf. *Od.* 1. 15. 35–6: *uret Achaicus / ignis Iliacas domos.*

[120] Where it is the contrast to Horace's chosen subject-matter, feasts and sexual battles: *nos convivia, nos proelia virginum* (17). For *stomachus* as another physiological metaphor for anger, see Onians (1954: 88).

[121] Kumaniecki (1935) uses adjectives of bitter flavour to describe Horace's tone and mood: 140, *acerbissima allii exsecratione*; *acribus cruciatibus*; 142, *acerrimis aculeis; quo poetae ioco quamvis acri et acerbo*; 144, *allium acerbissime exsecratus . . . acribus ironiae aculeis.*

[122] Perhaps we are meant to see a theory of contagious poisoning at work here. Cf. Plin. *NH* 28. 31–2: people who have once been bitten by a snake or dog can themselves addle hen's eggs and make cows miscarry: *tantum remanet virus ex accepto semel malo ut venefici fiant venena passi* ('the poison is so persistent that only one dose turns the victims into poisoners themselves').

[123] *Malus odor*, 12. 7–8; *odoror*, 12. 4; *olentem Maevium*, 10. 2; *nauseam*, 9. 35; *hircus*, 12. 5; *sus*, 12. 6; *podex velut crudae bovis*, 8. 6. Cf. *alium* mixed with other filthy smells at Plaut. *Most.* 39.

both the occasion and the inspiration for lyric writing,[124] garlic and other forms of filth can be said to be the 'food' or 'odour' of iambics.[125] The fetid breath of a garlic-eater is poles apart from the traditionally honeyed and nectared blandishments of poets or flatterers;[126] the act of criticizing and cursing a patron is a festive reversal of the tradition of flattery. Horace's garlic poem is the antithesis of his more respectful poems to his patron, and a reversal of the polite expression of thanks owed for a gift or a meal.[127] It is the antidote to Maecenas' *dona* and the anaphrodisiac to his *venenum*.

In spite of all this, the poem is only a parody of real wrath. The fact that Horace is bent double by the herb which fuels his rage is a hint that he may not have the stomach even for iambics, let alone Iliads: the iambic poet is hoist with his own petard. Despite its references to the wider themes of the *Epodes*—conflict, anger, love, war—*Epode* 3 is essentially an epitome. While pretending to be *nocens*, harmful, the poem is really *iocosum*, a joke.[128] Indirectly, it demonstrates the positive aspects of friendship, which can absorb such displays of feigned malice; it gives a glimpse of Maecenas' intimate life as friend, dining-companion, and lover. The context of the poem is an in-built remedy for its apparent virulence.[129] *Epode* 3 pretends to be the sort of venomous

[124] See Commager (1957).

[125] Cf. Ov. *Met.* 2. 769: Envy is shown eating snakes' flesh, the proper food of her viciousness: *vipereas carnes, vitiorum alimenta suorum*.

[126] See Bramble (1974: 52): e.g. Theoc. 7. 82 (sweet nectar from the Muses); Ar. *Av.* 908 (sweet-tongued poetry). Cf. Martial 7. 25, 10. 45: food images used of bland honorific poems and astringent epigrams.

[127] For more polite poems from Horace to Maecenas on the subject of meals see Pavlock (1982).

[128] For the principle that the ephithet applied by Horace to his addressee (here, *iocose*, 3. 20) usually has a bearing on the meaning of the poem, see Fraenkel (1932–3: 13–15).

[129] Gorgias, *Encomium* 10 speaks of the power of incantations (ἡ δύναμις τῆς ἐπῳδῆς) to sway the mind (after a passage in which he discusses poetry). At 14 he claims that the power of speech bears the same relation to the ordering of the mind as the ordering of drugs (φάρμακα) bears to the constitution of bodies; see Derrida (1981: 115). Plin. *NH* 28. 10 discusses whether words have any real power: *polleant ne aliquid verba et incantamenta carminum*; at 28. 12 in some cases he feels forced to accept the *vim carminum*. In the *Epodes* Horace exploits this uncertain power of words, leaving it open whether his ἐπῳδαί have the force of potent spells or whether they are powerless imitations. Plato, *Rep.* 10. 595 accuses the tragic poets of

carmen that Canidia might chant, but it is also the joke (*παίγνιον*) of Horace the *puer* at the table of Maecenas.[130] Even so, Horace's lashings of garlic have kept their potency. When one of the guests at Peregrine Pickle's Roman banquet shouts through streaming tears, 'Z——nds! this is the essence of a whole bed of garlic!', it is Horace, rather than Apicius, who has grated that flavour into the dish.[131]

corrupting their listeners, unless they possess an antidote (595b *φάρμακον*): 'knowledge of the real nature of things' (*τὸ εἰδέναι αὐτὰ οἷα τυγχάνει ὄντα* (quoted by Derrida 1981: 137). Maecenas' knowledge of the real nature of his relationship with Horace is the antidote to Horace's virulence in *Epode* 3.

[130] For the *puer* as an ambiguously impotent figure in the *Epodes*, see 5. 12 ff.: *puer / impube corpus, quale posset impia / mollire Thracum pectra* (so weak that he has the power to soften Thracian hearts) and 5. 86 ff., where, defeated, he lets out a Thyestean curse. At 6. 15–16 the blubbing *puer* is once again the unavenged victim of malice: *an si quis atro dente me petiverit, / inultus ut flebo puer?* With his malevolent *afflatus*, Horace is also the antitype of the boy Cupid (for *afflare* used of divine figures inspiring love, see Sil. 11. 420, *afflatus fallente Cupidine*; Tib. 2. 1. 80, *felix, cui placidus leniter afflat Amor*).

[131] Smollett, *Peregrine Pickle*, ch. 48.

References

ADAMIETZ, J. (1972), *Untersuchungen zu Juvenal* (*Hermes* Einz. 26; Wiesbaden 1972: 78–116).

ADAMS, J. (1982), *The Latin Sexual Vocabulary* (London).

AHL, F. (1985), *Metaformations* (Ithaca, NY).

ANDERSON, W. S. (1957), 'Studies in Book 1 of Juvenal', *YCS* 15: 33–90.

——(1958), 'Persius 1. 107–110', *CQ* 52: 195–7.

——(1963), 'The Roman Socrates: Horace and his Satires', in J. P. Sullivan (ed.), *Critical Essays on Roman Literature: Satire* (London).

——(1982), 'Persius and the Rejection of Society', in *Essays on Roman Satire* (Princeton, NJ), 169–93.

ANDRÉ, J. (1981), *L'Alimentation et la cuisine à Rome* (Paris).

ANGELI, E. S. de (1969), 'The Unity of Catullus 29', *CJ* 65: 81–4.

ARON, J.-P. (1979), 'The Art of Using Leftovers', in Forster and Ranum (1979), 98–108.

ARROWSMITH, W. S. (1966), 'Luxury and Death in the *Satyricon*', *Arion*, 5: 304–31.

ASSFAHL, G. (1932), *Vergleich und Metaphor bei Quintilian* (Stuttgart).

ASTIN, A. E. (1978), *Cato the Censor* (Oxford).

BABCOCK, B. (1978), *The Reversible World* (Ithaca, NY).

BAKHTIN, M. (1968), *Rabelais and his World*, tr. H. Iswolsky (Cambridge, Mass.).

BALSDON, J. P. V. D. (1969), *Life and Leisure in Ancient Rome* (London).

BARBER, C. (1959), *Shakespeare's Festive Comedy* (Princeton, NJ).

BARCHIESI, M. (1970), 'Plauto e il "metateatro" antico', *Il Verri*, 31: 113–30.

BARTHES, R. (1964), 'Le Monde-objet', in *Essais critiques* (Paris), 19–28.

——(1976), *The Pleasure of the Text*, tr. R. Miller (London).

——(1979), 'Toward a Psychology of Contemporary Food Consumption', in Forster and Ranum (1979), 166–73.

Bauer, G. (1988), 'Eating out: with Barthes', in Bevan (1988), 39–48.

Beard, M. (1985), *Writing and Ritual: A Study of Diversity and Expansion in the Arval Acta* (Papers of Brit. School at Rome 53; Rome), 114–62.

Beare, W. (1955), *The Roman Stage* (London).

Bek, L. (1983), 'Questiones Convivales: The Idea of the Triclinium and the Staging of Convivial Ceremony from Rome to Byzantium', *ARID* 12: 81–107.

Bernstein, W. H. (1985), 'A Sense of Taste: Catullus 13', *CJ* 80: 127–30.

Bertman, S. (1978), 'Oral Imagery in Catullus 7', *CQ* NS 28: 447–8.

Bevan, D. (1988) (ed.), *Literary Gastronomy* (Amsterdam).

Bianchi Bandinelli, R. (1970), *Rome the Centre of Power* (London).

Blanchard, M. E. (1981), 'On Still Life', *Yale French Studies*, 61: 276–98.

Bonnet, J. C. (1979), 'The Culinary System in the *Encyclopédie*', in Forster and Ranum (1979), 139–65.

Booth, A. C. (1980), 'Sur les sens obscènes de *sedere* dans Martial 11. 99', *Glotta* 58: 278–9.

Booth, W. C. (1961), *The Rhetoric of Fiction* (Chicago).

Bramble, J. (1974), *Persius and the Programmatic Satire* (Cambridge).

——(1982), 'Martial and Juvenal', in E. J. Kenney (ed.), *The Cambridge History of Classical Literature. II: Latin Literature* (Cambridge), 597–623.

Braund, S. (1986), 'Juvenal on how to (Tr)eat People', *Omnibus*, Mar.

Brink, C. O. (1963), *Horace on Poetry*, i. *Prolegomena to the Literary Epistles* (Cambridge).

——(1971), *Horace on Poetry*, ii. *The Ars Poetica* (Cambridge).

——(1982), *Horace on Poetry*, iii. *Epistles Book II* (Cambridge).

Brotherton, B. (1978), *The Vocabulary of Intrigue in Roman Comedy* (New York/London).

Brothwell, D., and Brothwell, P. (1969), *Food in Antiquity* (London).

Brown, F. E. (1964), 'Hadrianic Architecture', in L. F. Sandler (ed.), *Essays in Memory of Karl Lehmann* (New York), 55–8.

Brown, J. (1984), *Fictional Meals and their Function in the French Novel 1789–1848* (Toronto).

——(1987) (ed.), 'Littérature et nourriture', *Dalhousie French Studies*, 11.

BROWN, L. (1985), *Alexander Pope* (Oxford).
BROWN, P. (1989), *The Body and Society* (London).
BRYSON, N. (1981), *Word and Image* (Cambridge).
—— (1990), *Looking at the Overlooked* (London).
BUCHHEIT, V. (1975), 'Catullus Literarkritik und Kallimachos', *Grazer Beiträge*, 4: 21–50.
BÜCHNER, K. (1970), *Studien zur römischen Literatur, viii. Werkanalysen 50–96: Die* Epoden *des Horaz* (Wiesbaden).
BURKE, P. (1978), *Popular Culture in Early Modern Europe* (London).
BURNETT, A. P. (1983), *Three Archaic Poets* (London).
CAIRNS, F. (1972), *Generic Composition in Greek and Latin Poetry* (Edinburgh).
CAMERON, A. (1969), 'Petronius and Plato', *CQ* NS 19: 367–70.
CAMPBELL, A. Y. (1945), 'Pike and Eel: Juvenal 5, 103–6', *CQ* 39: 46–8.
—— (1970), *A New Interpretation of Horace* (London).
CARRUBBA, R. W. (1969), *The Epodes of Horace: A Study in Poetic Arrangement* (The Hague).
CAVE, T. (1979), *The Cornucopian Text* (Oxford).
CÈBE, J.-P. (1966), *La caricature et la parodie dans le monde romain antique* (Paris).
CHIARINI, G. (1979), *La recita: Plauto, la farsa, la festa* (Bologna).
CLARK, P. P. (1975), 'Thoughts for Food 1: French Cuisine and French Culture', *French Review*, 49 (Oct.): 32–41.
CLASSEN, C. J. (1978), 'Horace—a Cook?', *CQ*, NS 28, 2: 333–48.
CLAUSEN, W. V. (1964), 'Callimachus and Roman Poetry', *GRBS* 5: 194–5.
CLAYMAN, D. L. (1975), 'Horace's Epodes VIII and XII: More than Clever Obscenity?', *CW* 69: 55–61.
—— (1980), *Callimachus' Iambi* (Leiden).
CLEMENTE, G. (1981), 'Le leggi sul lusso', in A. Giardina and A. Schiavone (eds.), *Società romana e produzione schiavistica* (3rd edn.; Bari), 1–14.
CODY, J. V. (1976), *Horace and Callimachean Aesthetics* (Brussels).
COFFEY, M. (1963), 'Juvenal Report for the Years 1941–1961', *Lustrum*, 8: 161–215.
—— (1989), *Roman Satire* (London; 1st edn. 1976).
COLTON, R. E. (1965), 'A Dinner Invitation, Juvenal XI, 56–208', *CB* 41: 39–45.
COMMAGER, S. (1957), 'The Function of Wine in Horace's Odes', *TAΓA* 88: 68 80.
—— (1962), *The Odes of Horace: A Critical Study* (New Haven, Conn./London).

COPLEY, F. O. (1970), 'Plautus, *Poenulus* 53–55', *AJP* 91: 77–8.

CORNFORD, F. (1914), *Origins of Attic Comedy* (London).

COURTNEY, E. (1980), *A Commentary on the Satires of Juvenal* (London).

D'ARMS, J. H. (1984), 'Control, Companionship and Clientela: Some Social Functions of the Roman Communal Meal', *Echos du monde classique*, 28: 327–48.

——(1990), 'The Roman *Convivium* and the Idea of Equality', in Murray (1990), 308–20.

DEONNA, W., and RENARD, M. (1961), *Croyances et superstitions de table dans la Rome antique* (Collection Latomus; Brussels).

DERRIDA, J. (1981), 'Plato's Pharmacy', in *Dissemination* (Chicago), 63–171.

DEROUX, C. (1983), 'Domitian, the King Fish and the Prodigies: A Reading of Juvenal's Fourth Satire', in C. Deroux (ed.) *Studies in Latin Literature and Roman History* (Collection Latomus; Brussels), 283–98.

DETIENNE M. (1977), *The Gardens of Adonis* (Hassocks, Sussex).

——(1981), 'Between Beasts and Gods', in R. Gordon (ed.), *Myth, Religion and Society* (Cambridge), 215–28.

—— and VERNANT, J. P. (1979), *La cuisine du sacrifice en pays grec* (Paris).

DIACONESCU, T. (1980), 'Structuri metaforice şi univers Saturnalic în Comedia lui Plautus', *Studii classice* 19: 61–70.

DIEHL, E. (1910), *Pompeianische Wandinschriften und Verwandtes* (Bonn).

DOHM, H. (1964), *Mageiros: Die Rolle des Kochs in der Griechisch-Römischen Komödie* (Zetemata 32).

DOSI, A., and SCHNELL, F. (1984), *A tavola con i Romani antichi* (Rome).

DOUGLAS, M. (1970), *Purity and Danger* (Harmondsworth).

——(1975), 'Deciphering a Meal', in *Implicit Meanings* (London), 249–75.

DUPONT, F. (1977), *Le plaisir et la loi* (Paris).

ECKSTEIN, F. (1957), *Untersuchungen über die Stilleben aus Pompeji und Herculaneum* (Berlin).

EDEN, P. T. (1984), *Seneca Apocolocyntosis* (Cambridge).

EDMUNDS, L. (1980), 'Ancient Roman and Modern American Food: A Comparative Sketch of Two Semiological Systems' *Comp. Civilizations Review*, 5 (Fall): 52–68.

ELIAS, N. (1978), *The Civilizing Process*, i. *The History of Manners* (New York).

ELLIOTT, R. C. (1960), *The Power of Satire: Magic, Ritual, Art* (Princeton, NJ).

ELLIS, R. (1867), *Catulli Veronensis Liber* (Oxford).
FANTHAM, E. (1972), *Comparative Studies in Republican Imagery* (Toronto).
FEDELI, P. (1979), 'Il V Epodo e i Giambi d'Orazio come espressione d'arte alessandrina', *MPhL* 3: 67–138.
FEHR, B. (1990), 'Entertainers at the *Symposium*: The *Akletoi* in the Archaic Period', in Murray (1990), 185–95.
FERGUSON, J. (1979), *Juvenal: The Satires* (London).
FITZGERALD, W. (1988), 'Power and Impotence in Horace's Epodes', *Ramus*, 17, 2: 176–91.
FLINTOFF, E. (1982), 'Food for Thought: Some Imagery in Persius Satire 2', *Hermes*, 110: 341–54.
FORDYCE, C. J. (1961), *Catullus* (Oxford).
FOREHAND, W. E. (1972), '*Pseudolus* 868–872: *Ut Medea Peliam concoxit*', *CJ* 67: 293–8.
FORSTER, R., and RANUM, O. (1979), *Food and Drink in History* (Baltimore).
FOUCAULT, M. (1990), *The History of Sexuality*, iii. *Care of the Self*, tr. R. Hurley (London).
FRAENKEL, E. (1932–3), 'Das Pindargedicht des Horaz', *SHAW* 23, 2: 13–15.
——(1957), *Horace* (Oxford).
——(1960), *Elementi Plautini in Plauto* (Florence).
FRIEDLÄNDER, L. (1886), *M. Valerii Martialis Epigrammaton Liber* (Leipzig).
——(1909), *Life and Manners under the Early Empire*, ii, tr. L. A. Magnus and J. H. Freese (London).
FULLER, J. (1976), 'Carving Trifles: William King's Imitation of Horace', *Proc. Brit. Acad.* 62: 269–91.
GAMBERINI, F. (1983), *Stylistic Theory and Practice in the Younger Pliny* (Hildesheim).
GARNSEY, P. (1988), *Famine and Food-Supply in the Graeco-Roman World* (Cambridge).
GIANGRANDE, G. (1968), 'Sympotic Literature and Epigram', in *L'Épigramme grecque* (Entretiens Hardt; Geneva), 93–174.
GIANNINI, A. (1960), 'La figura del cuoco nella commedia greca', *Acme*, 13: 135–216.
GILL, D. (1974), 'Trapezomata: A Neglected Aspect of Greek Sacrifice', *Harvard Theol. Rev.* 67: 117–37.
GLAZEWSKI, J. (1971), '*Plenus Vitae Conviva*: A Lucretian Concept in Horace's Satires', *CB* 40, 7: 85–8.
GLEASON, M. W. (1986), 'Festive Satire: Julian's *Misopogon* and the New Year at Antioch', *JRS* 76: 106–19.

GOLDHILL, S. (1986), '*Framing and Polyphony: Readings in Hellenistic Poetry*', *PCPS* 212, NS 32: 25–52.

GOMBRICH E. (1963), 'Tradition and Expression in Western Still Life', in *Meditations on a Hobby-Horse and Other Essays on the Theory of Art* (London), 94–105.

GOMME, A. W; and SANDBACH, F. H. (1973), *Menander: A Commentary* (Oxford).

GOODY, J. (1977), *The Domestication of the Savage Mind* (Cambridge).

——(1982), *Cooking, Cuisine and Class* (Cambridge).

GOSLING, J. C. B., and TAYLOR, C. C. W. (1981), *The Greeks and Pleasure* (Oxford).

GOURÉVITCH, D. (1974), 'Le menu de l'homme libre: recherches sur l'alimentation et la digestion dans les œuvres en prose de Sénèque le philosophe', in *Mélanges P. Boyancé* (Rome), 311–44.

GRATWICK, A. S. (1973), 'Titus Maccius Plautus', *CQ*, NS 23: 78–84.

——(1982), 'Drama' in E. J. Kenney (ed.), *Cambridge History of Classical Literature. II: Latin Literature* (Cambridge), 77–137.

GRAY, P. (1987), *Honey from a Weed* (London).

GREENOUGH, J. B. (1890), *HCSP* 1: 191–2.

GRIFFIN, J. (1985), *Latin Poets and Roman Life* (London).

GUILLEMIN A.-M. (1934), 'L'Inspiration du repas ridicule d'Horace', *Humanités*, 10: 377–80.

HALL, E. (1989), *Inventing the Barbarian* (Oxford).

HALLETT, J. P. (1978), 'Divine Unction: Some Further Thoughts on Catullus 13', *Latomus*, 37: 747–8.

HALLIDAY, M. A. K. (1961), 'Categories of the Theory of Grammar', *Word*, 17: 241–91.

HANDLEY, E. W. (1968), *Menander and Plautus: A Study in Comparison* (London).

HEILMAN, W. (1967), 'Zur Komposition der vierten Satire und des ersten Satirenbuchs Juvenals', *RhMus* 110: 358–70.

HELLER, J. L. (1939), 'Festus on *Nenia*', *TAPA* 70: 357–67.

——(1943), '*Nenia* "παίγνιον"', *TAPA* 74: 215–68.

HELM, J. T. (1980–1), 'Poetic Structure and Humor: Catullus 13', *CW* 74: 213–17.

HELMBOLD, W. C., and O'NEIL, E. N. (1956), 'The Structure of Juvenal IV', *AJP* 77: 68–73.

HÉMARDINQUER, J.-J. (1970), *Pour une histoire de l'alimentation* (Cahiers des Annales 28; Paris).

——(1979), 'The Family Pig of the Ancien Regime: Myth or Fact?', in Forster and Ranum (1979), 50–72.

HENDERSON, J. (1975), *The Maculate Muse* (New Haven, Conn.).

HENDERSON, J. G. W. (1987), 'Suck it and See (Horace *Epode* 8)' in M. Whitby, P. Hardie, and M. Whitby (eds.), *Homo Viator* (Bristol/Oak Park, Ill.), 105–18.

HENDRICKSON, G. L. (1897), 'Are the Letters of Horace Satires?', *AJP* 18: 313–24.

HENISCH, B. A. (1976), *Fast and Feast* (University Park, Penn.)

——(1984), *Cakes and Characters* (London).

HENSE, O. (1906), 'Eine Menippea des Varro', *RhMus* 61: 1–18.

HERESCU, N. I. (1960), 'Sur le sens érotique de *sedere*', *Glotta* 38: 125–34.

HERRMANN, L. (1953) (ed. and tr.), *Horace, Epodes, Collection* Latomus 14 (Brussels).

HERZOG, R. (1989), 'Fest, Terror und Tod in Petrons *Satyrica*', *Poetik und Hermeneutik*, 14 (*Das Fest*): 120–50.

HICKS, A. H. (1986), 'The Mystique of Garlic: History, Uses Superstitions and Revelations', in T. Jaine (ed.), *Oxford Symposium on Food and Cookery 1984–1985: Cookery: Science, Lore, and Books* (London), 140–61.

HIGHET, G. (1954), *Juvenal the Satirist* (London).

HODGART, M. J. C., and WORTHINGTON, M. P. (1959), *Song in the Works of James Joyce* (New York).

HOUSMAN, A. E. (1913), 'Notes on Persius', *CQ* 7: 12–32.

HUBBARD, T. K. (1981), 'The Structure and Programmatic Intent of Horace's First Satire', *Latomus*, 40: 305–21.

HUDSON, N. (1989), 'Food in Roman Satire' in S. Braund (ed.), *Satire and Society in Ancient Rome* (Bristol), 69–87.

HUNTER, R. L. (1983), *Eubulus: The Fragments* (Cambridge).

IHM, M. (1890), 'Römische Spieltafeln', in R. Kekule (ed.), *Aufsätze aus der Altertumswissenschaft*. (Bonner Studien; Bonn).

JACKSON, H. (1883), *PCPS*, 18 Oct.: 25–6.

JAMES, A. (1982), 'Confections, Concoctions and Conceptions', in B. Waites, T. Bennet, and G. Martin (eds.), *Popular Culture: Past and Present* (London), 294–307.

JEANNERET, M. (1988), *Des mets et des mots: Banquets et propos de table à la Renaissance* (Paris); tr. J. Whiteley and E. Hughes as *A Feast of Words* (Cambridge, 1991).

JOHNSON, L. W. (1985), 'La Salade tourangelle de Pierre Ronsard', in Tobin (1985), 149–73.

KAY, N. M. (1985), *Martial Book XI: A Commentary* (London).

KENNEY, E. J. (1958), 'Nequitiae Poeta', in Herescu (ed.) *Ovidiana* (Paris), 201–9.

——(1962), 'The First Satire of Juvenal', *PCPhS* NS 8: 29–40.

——(1970), 'Doctus Lucretius', *Mnemosyne*, 23: 366–92.

——(1984), *The Ploughman's Lunch: Moretum* (Bristol).

KHAN, H. A. (1969), 'Image and Symbol in Catullus 17', *CPh* 64: 88–97.

KILPATRICK, R. S. (1973), 'Juvenal's Patchwork Satires', *YCS* 23: 229–41.

——(1986), *The Poetry of Friendship: Horace* Epistles *1* (Edmonton, Alta.).

KNAPP, C. (1979), 'References in Plautus and Terence to Plays, Players, and Playwrights', *CPhil* 14: 35–55.

KOENEN, L. (1977), 'Horaz, Catull und Hipponax', *ZPE* 26: 73–93.

KONSTAN, D. (1979), 'An Interpretation of Catullus 21', in C. Deroux (ed.), *Studies in Latin Literature and Roman History* (Brussels), 214–16.

KUMANIECKI, C. F. (1935), *De Epodis quibusdam Horationis: Commentationes Horatianae* (Cracow), 140–4.

LAFLEUR, R. (1981), 'Horace and *onomasti komodein*: The Law of Satire', in *Aufstieg und Niedergang der Römischen Welt* (Berlin/New York), ii. 1790–1876.

LEACH, E. (1961), 'Time and False Noses', in *Rethinking Anthropology*, London School of Economics Monographs on Social Anthropology, 22 (London), 132–6.

LE BONNIEC, H. (1958), *Le Culte de Cérès à Rome*, Études et commentaires 27 (Paris)

LEE, G., and BARR, W. (1987), *The Satires of Persius* (Liverpool).

LEHRER, A. (1972), 'Cooking Vocabularies and the Culinary Triangle of Lévi-Strauss', *Anthropological Linguistics*, 14: 155–71.

LEO, F. (1895–6), *T. Maccius Plautus: Comoediae* (Berlin).

LÉVI-STRAUSS, C. (1965), 'Le Triangle culinaire', *L'Arc*, 26: 19–29.

——(1970), *The Raw and the Cooked* (London).

LINCOLN, B. (1985), 'Of Meat and Society, Sacrifice and Creation, Butchers and Philosophy', *L'Uomo*, 9–19.

LING, R. (1990), 'Street Plaques at Pompeii', in M. Henig (ed.), *Architecture and Architectural Sculpture in the Roman Empire* (Oxford), 51–66.

LITTMAN, R. J. (1977), 'The Unguent of Venus: Catullus 13', *Latomus*, 36: 123–6.

LOVEJOY, A. O., and BOAS, G. (1965), *Primitivism and Related Ideas in Antiquity* (New York).
LOWE, J. C. B. (1985*a*), 'Cooks in Plautus', *Class. Ant.* 4: 72–102.
——(1985*b*), 'The Cook Scene of Plautus' *Pseudolus*', *CQ* 35: 411–16.
LUCAS, D. W. (1968), *Aristotle: Poetics* (Oxford).
LUKINOVICH, A. (1990) 'The Play of Reflections between Literary Form and the Sympotic Theme in the *Deipnosophistae* of Athenaeus,' in Murray (1990), 263–71.
MACCARY, T., and WILLOCK, M. M. (1978), *Plautus Casina* (Cambridge).
MCGANN, M. J. (1969), *Studies in Horace's First Book of Epistles*, (Collection *Latomus* 100; Brussels).
MARCOVICH, M. (1982), 'Catullus 13 and Philodemus 23', *QUCC*, NS 11: 131–8.
MARIN, L. (1989), *Food for Thought*, tr. M. Hiort (Baltimore).
MARTYN, J. (1970), 'A New Approach to Juvenal's First Satire', *Antichthon* 4: 53–61.
MAURACH, G. (1975), *Plauti Poenulus* (Heidelberg).
MAUSS, M. (1954), *The Gift* (London).
MAYOR, J. E. B. (1881–6), *Juvenal: Thirteen Satires* (London).
MEIGGS, R. (1973), *Roman Ostia* (Oxford).
MENNELL, S. (1985), *All Manners of Food* (Oxford).
MERRILL, E. (1935), *Selected Letters of the Younger Pliny* (London).
MERWIN, W. S. (1961) (tr.), *Persius: The Satires*, introd. and notes by W. S. Anderson (Bloomington, Ind.).
METTE, J. (1961), '"Genus tenue" und "mensa tenuis" bei Horaz', *MH* 18: 136–9.
MILAZZO, V. (1982), 'Polisemia e parodia nel Iudicium coci et pistoris di Vespa', *Orpheus*, 3: 250–74.
MINYARD, J. D. (1971), 'Critical Notes on Catullus 29', *CPh* 66: 174–81.
MONTANARI, M. (1988), *Alimentazione e cultura nel medioevo* (Rome/Bari).
MORFORD, M. (1977), 'Juvenal's Fifth Satire', *AJPh* 98, 219–45.
MUECKE, F. (1985), 'Names and Players: The Sycophant Scene of the "Trinummus" (*Trin.* 4. 2)', *TAPA* 115: 167–86.
MURRAY, O. (1982), 'Symposium und Männerbund', *Concilium Eirene* (Prague), 16.
——(1985), 'Symposium and Genre in the Poetry of Horace', *JRS* 75: 39–50.
——(1990) (ed.), *Sympotica* (Oxford).

MURRAY, P. (1963), '"A Taint of Elegance" (Horace *Satire* II, viii)', *Arion*, 2, 4: 63–5.

NAUTA, R. R. (1987), 'Seneca's *Apocolocyntosis* as Saturnalian Literature', *Mnemosyne*, 40: 69–96.

NEWMAN, W. L. (1887) *Politics of Aristotle*, 3 vols. (Oxford).

NISBET, R. G. M., and HUBBARD, M. (1975), *A Commentary on Horace: Odes Book I* (Oxford).

NORDEN, E. (1898), *Die Antike Kunstprosa*, 2 vols. (Leipzig).

NORMANN, R., and HAARBERG J. (1980), *Nature and Language: A Semiotic Study of Gourds and Cucurbits in Literature* (London).

ONIANS, R. B. (1954), *Origins of European Thought* (Cambridge).

OWEN, J. H. (1977), 'Philosophy in the Kitchen; or Problems in Eighteenth-Century Culinary Aesthetics', *Eighteenth-Century Life*, 3, 3 (Mar.): 77–9.

PALMER, A. (1925), *The Satires of Horace* (London).

PARKER, P. (1987), *Literary Fat Ladies* (London).

PASOLI, E. (1970–2), 'Cuochi, convitati, carta nella critica letteraria di Marziale', *MusCrit* 5–7: 188–93.

PAVLOCK, B. (1982), 'Horace's Invitation Poems to Maecenas: Gifts to a Patron', *Ramus*, 11: 72–98.

PERINI, B. (1975), 'Aceto italiano e poesia luciliana', in *Scritti in onore di Carlo Diano* (Bologna), 1–24.

PETER, C. (1873), 'Zur Chronologie der Briefen des jüngeren Plinius', *Philologus*, 32: 689–710.

PETRONE, G. (1983), *Teatro antico e inganno: finzioni plautine* (Palermo).

POLHEMUS, T. (1978) (ed.), *Social Aspects of the Human Body* (London).

PRESTON, K. (1920), 'Martial and Formal Literary Criticism', *CPh* 15: 340–52.

PUELMA, M. (1988), 'Plautus und der Titel der Casina', *MH* 45: 13–27.

QUINN, K. (1970), *Catullus: The Poems* (London).

RACE, W. H. (1978), '*Odes* 1. 20: An Horatian Recusatio', *CSCA* 11: 179–96.

RAMAGE, E. S. (1973), *Urbanitas: Ancient Sophistication and Refinement* (Norman, Okla.).

RANKIN, H. D. (1962), 'Apophoreta and Saturnalian Word-Play', *C&M* 23: 134–42.

REVEL, J. F. (1979), *Un festin en paroles: Histoire littéraire de la sensibilité gastronomique de l'antiquité à nos jours* (Paris).

RICHARDSON, N. J. (1979), *The Homeric Hymn to Demeter* (Oxford).

RICHLIN, A. (1981) *The Garden of Priapus: Sexuality and Aggression in Roman Humor* (New Haven, Conn.).

——(1988), 'Systems of Food Imagery in Catullus', *CW* 81, 5: 355–67.

RICKS, C. (1974), *Keats and Embarrassment* (Oxford).

RICOTTI, E. SALZA PRINA (1983), *L'arte del convito nella Roma antica* (Rome).

RIOS, A. (1986), 'Garlic: A Kitchen Amulet', in T. Jaine (ed.), *Oxford Symposium on Food and Cookery 1984–1985: Cookery: Science, Lore and Books* (London), 164–74.

ROCCA, S. (1979), '*Mellitus* tra lingua familiare e lingua letteraria', *Maia* 31, 37–43.

ROOS, E. (1959), *Die Person des Nasidienus bei Horatius* (Lund).

ROSATI, G. (1983), 'Trimalchione in Scena', *Maia* 35.3, 213–227.

ROSS, D. O. (1969), *Style and Tradition in Catullus* (Cambridge).

ROUSSELLE, A. (1983), *Porneia: de la maitrise du corps à la privation sensorielle* (Paris).

RUDD, N. (1960), 'The Names in Horace's Satires', *CQ* NS 10: 161–80.

——(1966), *The Satires of Horace* (Cambridge).

——(1970), 'Persiana', *CR*, NS 20: 282–8.

——(1986), *Themes in Roman Satire* (London).

SALEMME, C. (1976), *Marziale e la 'Poetica' degli oggetti* (Naples).

SALLER, R. (1983), 'Martial on Patronage and Literature', *CQ* NS 33: 246–57.

SANTINI, C. (1985), 'Il lessico della spartizione nel sacrificio romano', *L'Uomo* 9, 1–2: 63–73.

SCHAMA, S. (1979), 'The Unruly Realm: Appetite and Restraint in Seventeenth-Century Holland', *Daedalus*, 108, 3: 103–23.

——(1987), *The Embarrassment of Riches* (London).

SCHMIDT, E. A. (1977), '*Amica vis pastoribus*: Der Iambiker Horaz in seiner Epodenbuch', *Gymnasium*, 84: 401–23.

SCHMIDT, M. (1937), 'Mitteilungen zu Hor. *Sat.* 2. 8', *Philologisches Wochenschrift*, 37/8: 1071–2.

SCHUSTER, M. (1925), 'Zur Auffassung von Catull. 13 Gedicht', *WS* 44: 227–34.

SCOTT, W. C. (1971), 'Catullus and Caesar: Catullus 29', *CPh* 66: 17–25.

SCULLARD, H. H. (1981), *Festivals and Ceremonies of the Roman Republic* (London).

SEAGER, R. (1983), 'Aristophanes *Thesm.* 493–496 and the Comic Possibilities of Garlic', *Philologus* 127, 139–42.

SEGAL, E. (1987), *Roman Laughter* (Oxford).

SETAOLI, A. (1981), 'Gli "Epodi" di Orazio nella critica dal 1937 al

1972', in *Aufstieg und Niedergang der römischen Welt* (Berlin/New York), ii. 1674–788 (*Epode* 3, 1701–3).
SHACKLETON BAILEY, D. R. (1982), *Profile of Horace* (London).
SHERO, L. R. (1923), 'The *Cena* in Roman Satire', *CPh* 18: 126–45.
—— (1929), 'Lucilius's *Cena Rustica*', *AJP* 50: 64–70.
SHERWIN-WHITE, A. N. (1985), *The Letters of Pliny* (Oxford).
SHIPP, G. P. (1960), *Terentius: Andria* (Melbourne).
SIDER, S. (1978), 'On Stuffing Quilts: Plautus *Epid.* 455', *AJP* 99: 41–4.
SIEBERS, T. (1983), *The Mirror of Medusa* (Berkeley, Calif.).
SKEAT, T. E. (1975), 'Brief Communications', *JEA* 61: 246–54.
SKINNER, M. B. (1979), 'Parasites and Strange Bedfellows: A Study in Catullus' Political Imagery', *Ramus*, 8: 139–52.
SLATER, N. W. (1985), *Plautus in Performance* (Princeton, NJ).
SMITH, M. S. (1982), *Petronii Arbitri Cena Trimalchionis* (Oxford).
SOLOMON, J. and J. (1977), *Ancient Roman Feasts and Recipes* (Miami, Fla.).
SPERBER, D. (1975), *Rethinking Symbolism* (Cambridge).
STALLYBRASS, P., and WHITE, A. (1986), *The Politics and Poetics of Transgression* (London).
STERLING, C. (1981), *Still Life Painting from Antiquity to the Twentieth Century* (New York).
STOLLER, P. (1989), *The Taste of Ethnographic Things* (Philadelphia).
SWEET, D. (1979), 'Juvenal's Satire IV: Poetic Use of Indirection', *CSCA* 12: 283–303.
TAILLARDAT, J. (1965), *Les Images d'Aristophane* (Paris).
TANNER, T. (1979), *Adultery in the Novel* (Baltimore).
THOMPSON, D'A. W. (1936), *A Glossary of Greek Birds* (Oxford).
—— (1957), *A Glossary of Greek Fishes* (Oxford).
THOMSON, D. F. S. (1964), 'Catullus 95. 8: "Et laxas scombris saepe dabunt tunicas" ', *Phoenix*, 18: 30–6.
TOBIN, R. W. (1985) (ed.), *Littérature et gastronomie*, *Biblio* 17, 23.
TRIMPI, W. (1962), *Ben Jonson's Poems: A Study of the Plain Style* (Stanford, Calif.).
TUCKER, L. (1984), *Stephen and Bloom at Life's Feast: Alimentary Symbolism and the Creative Process in James Joyce's 'Ulysses'* (Columbus, Ohio).
TURNER B. (1984), *The Body and Society* (Oxford).
TUROLLA, E. (1957) (ed. and tr.), *I Giambi* (Turin).
ULLMAN, B. L. (1913), 'Satura and Satire', *CPh* 8: 172–94.
—— (1914), 'Dramatic "Satura" ', *CPh* 9: 1–23.
—— (1920), 'The Present Status of the *Satura* Question', *Studies in Philology*, 17: 379–401.

VALÉRI, R. (1977), *Le confit et son rôle dans l'alimentation traditionelle du Sud-Ouest de la France* (Lund).

VAN HOOK, L. (1905), *The Metaphorical Terminology of Greek Rhetoric and Literary Criticism* (Chicago).

VAN ROOY, C. A. (1965), *Studies in Classical Satire and Related Literary Theory* (Leiden).

VEGETTI, M. (1981), 'Lo spettacolo della natura: circo, teatro e potere in Plinio', *Aut Aut*, 184–5: 111–25.

VEHLING, J. D. (1977), *Apicius: Cookery and Dining in Imperial Rome* (New York/London).

VERSNEL, H. S. (1970), *Triumphus* (Leiden).

VESSEY, D. T. W. C. (1971), 'Thoughts on Two Poems of Catullus: 13 and 30', *Latomus*, 30: 45–55.

VEYNE, P. (1961), 'Vie de Trimalcion', *Annales*, 2.

——(1990), *Bread and Circuses* (tr. B. Pearce from *Le Pain et le cirque* (Paris, 1976); London).

VIDAL-NAQUET, P. (1981a), 'Land and Sacrifice in the Odyssey: A Study of Religious and Mythical Meanings', in R. Gordon (ed.), *Myth, Religion and Society* (Cambridge), 80–94.

——(1981b), 'Recipes for Greek Adolescence', in R. Gordon (ed.), *Myth, Religion and Society* (Cambridge), 163–85 (first published as 'Les Jeunes, le cru, l'enfant grec, et le cuit' in J. Le Goff and P. Nora (eds.), *Faire de l'histoire* (Paris, 1971), 137–68).

WATSON, L. (1983), 'Two Problems in Horace *Epode* 3', *Philologus*, 127: 80–6.

WATT, I. (1976), *The Rise of the Novel* (London).

WEISINGER, K. (1972), 'Irony and Moderation in Juvenal XI', *CSCA* 5, 227–40.

WEST, D. (1974), '"Of Mice and Men": Horace *Satires* 2. 6. 77–117', in T. Woodman and D. West (eds.), *Quality and Pleasure in Latin Poetry* (Cambridge), 67–80.

——(1977), *Reading Horace* (Edinburgh).

WHEATON, B. K. (1983), *Savoring the Past: The French Kitchen and Table from 1300 to 1789* (Philadelphia).

WHITE, K. D. (1976), 'Food Requirements and Food Supplies in Classical Times in Relation to the Diet of the Various Classes', *Progress in Food and Nutrition Science*, 2, 4: 143–91.

WILLCOCK, M. (1987), *Plautus: Pseudolus* (Bristol).

WILLIAMS, G. (1968), *Tradition and Originality in Roman Poetry* (Oxford).

WILTSHIRE, S. F. (1977), 'Catullus Venustus', *CW*, 319–26.

WIMMEL, W. (1960), *Kallimachos in Rom* (*Hermes* Einz.; Wiesbaden).

WINNICZUCK, L. (1966), '"Urbanitas" nelle lettere di Plinio il Giovane', *Eos* 56: 198–205.

WISEMAN, T. P. (1988), 'Satyrs in Rome? The Background to Horace's *Ars Poetica*', *JRS* 78: 1–13.

WITKE, C. (1980), 'Catullus 13: A Reexamination', *CPh* 75: 325–31.

WOYTEK, E. (1982), *Plautus: Persa* (Vienna).

WRIGHT, J. (1974), *Dancing in Chains: The Stylistic Unity of the Comoedia Palliata*', Am. Acad. in Rome Papers and Monographs, 25.

——(1975), 'The Transformations of *Pseudolus*', *TAPA* 105: 403–16.

WYKE, M. (1984), 'The Elegiac Woman and her Male Creators: Propertius and the Written Cynthia', unpub. Ph.D. thesis, Cambridge.

Index